Particles, Sources, and Fields

Volume II

ADVANCED BOOK CLASSICS
David Pines, Series Editor

Particles, Sources, and Fields

Volume II

Julian Schwinger
late, University of California at Los Angeles

CRC Press
Taylor & Francis Group
Boca Raton London New York

CRC Press is an imprint of the
Taylor & Francis Group, an **informa** business

ADVANCED BOOK PROGRAM

First published 1973 by Perseus Books Publishing

Published 2018 by CRC Press
Taylor & Francis Group
6000 Broken Sound Parkway NW, Suite 300
Boca Raton, FL 33487-2742

CRC Press is an imprint of the Taylor & Francis Group, an informa business

Visit the Taylor & Francis Web site at
http://www.taylorandfrancis.com

and the CRC Press Web site at
http://www.crcpress.com

Library of Congress Catalog Card Number: 98-87896

ISBN 13: 978-0-7382-0054-5 (pbk)

Cover design by Suzanne Heiser

Editor's Foreword

Perseus Books's *Frontiers in Physics* series has, since 1961, made it possible for leading physicists to communicate in coherent fashion their views of recent developments in the most exciting and active fields of physics—without having to devote the time and energy required to prepare a formal review or monograph. Indeed, throughout its nearly forty-year existence, the series has emphasized informality in both style and content, as well as pedagogical clarity. Over time, it was expected that these informal accounts would be replaced by more formal counterparts—textbooks or monographs—as the cutting-edge topics they treated gradually became integrated into the body of physics knowledge and reader interest dwindled. However, this has not proven to be the case for a number of the volumes in the series: Many works have remained in print on an on-demand basis, while others have such intrinsic value that the physics community has urged us to extend their life span.

The *Advanced Book Classics* series has been designed to meet this demand. It will keep in print those volumes in *Frontiers in Physics* or its sister series, *Lecture Notes and Supplements in Physics*, that continue to provide a unique account of a topic of lasting interest. And through a sizable printing, these classics will be made available at a comparatively modest cost to the reader.

These lecture notes by Julian Schwinger, one of the most distinguished theoretical physicists of this century, provide both beginning graduate students and experienced researchers with an invaluable introduction to the author's perspective on quantum electrodynamics and high-energy particle physics. Based on lectures delivered during the period 1966 to 1973, in which Schwinger developed a point of view (the physical source concept) and a technique that emphasized the unity of particle physics, electrodynamics, gravitational theory, and many-body theory, the notes serve as both a textbook on source theory and an informal historical record of the author's approach to many of the central problems in physics. I am most pleased that *Advanced Book Classics* will make these volumes readily accessible to a new generation of readers.

David Pines
Aspen, Colorado
July 1998

Vita

Julian Schwinger

University Professor, University of California, and Professor of Physics at the
University of California, Los Angeles since 1972, was born in New York City on
February 12, 1918. Professor Schwinger obtained his Ph.D. in physics from
Columbia University in 1939. He has also received honorary doctorates in science
from four institutions: Purdue University (1961), Harvard University (1962),
Brandeis University (1973), and Gustavus Adolphus College (1975). In addition
to teaching at the University of California, Professor Schwinger has taught at
Purdue University (1941–43), and at Harvard University (1945–72). Dr.
Schwinger was a Research Associate at the University of California, Berkeley,
and a Staff Member of the Massachusetts Institute of Technology Radiation
Laboratory. In 1965 Professor Schwinger became a co-recipient (with Richard
Feynman and Sin Itiro Tomonaga) of the Nobel Prize in Physics for work in
quantum electrodynamics. A National Research Foundation Fellow (1939–40)
and a Guggenheim Fellow (1970), Professor Schwinger was also the recipient of
the C. L. Mayer Nature of Light Award (1949); the First Einstein Prize Award
(1951); a J. W. Gibbs Honorary Lecturer of the American Mathematical Society
(1960); the National Medal of Science Award for Physics (1964); a Humboldt
Award (1981); the Premio Citta di Castiglione de Sicilia (1986); the Monie A.
Ferst Sigma Xi Award (1986); and the American Academy of Achievement
Award (1987).

Special Preface

Isaac Newton used his newly invented method of fluxious (the calculus) to compare the implications of the inverse square law of gravitation with Kepler's empirical laws of planetary motion. Yet, when the time came to write the *Principia*, he resorted entirely to geometrical demonstrations. Should we conclude that calculus is superfluous?

Source theory—to which the concept of renormalization is foreign—and *renormalized operator field theory* have both been found to yield the same answers to electrodynamic problems (which disappoints some people who would prefer that source theory produce new—and wrong—answers). Should we conclude that source theory is thus superfluous?

Both questions merit the same response: the simpler, more intuitive formation, is preferable.

This edition of *Particles, Sources, and Fields* is more extensive than the original two volumes of 1970 and 1973. It now contains four additional sections that finish the chapter entitled, "Electrodynamics II." These sections were written in 1973, but remained in partially typed form for fifteen years. I am again indebted to Mr. Ronald Bohm, who managed to decipher my fading scribbles and completed the typescript. Particular attention should be directed to Section 5-9, where, in a context somewhat larger than electrodynamics, a disagreement between source theory and operator field theory finally does appear.

Readers making their first acquaintance with source theory should consult the Appendix in Volume I. This Appendix contains suggestions for threading one's way through the sometimes cluttered pages.

Los Angeles, California   J. S.
April 1988

Preface

The writing of this second volume has spread over two years and two continents. A substantial portion was penned in Tokyo where, thanks to a sabbatical leave from Harvard University and additional support from the Guggenheim Foundation, I spent the first eight months of 1970. Some day, when not preoccupied with the writing of a book, I shall return to Japan and fully savor its delights. The book was completed during the 1971–1972 period when I was Visiting Professor at the University of California in Los Angeles.

This volume is almost exclusively concerned with quantum electrodynamics. As such it is retrospective in its subject matter. But the concentration on this relatively simple dynamical situation is directed toward the exploration and elaboration of viewpoints and techniques that should be viable in the domains of strong and weak interactions. And it is intended that the self-contained source theory development, with its significant conceptual and computational simplifications, shall be the face of quantum electrodynamics for future generations. No longer need surveyors of the subject couple ecstatic remarks about the accuracy of the theory with rumblings about its unsatisfactory conceptual basis.

Perhaps a word of explanation is needed for the historically oriented vignettes that occasionally appear in these pages. They are not priority claims. After all these years? But I do wish to place on record some aspects of my personal recollection of various events, which are not likely to be forthcoming from any other source.

One element of the organization of the book, or the lack of it, requires comment. A topic that has apparently been concluded is sometimes taken up again to explore some point in greater detail. This represents the historical evolution of the subject, in which various subtleties became clarified only after some time. The alternative to the plan actually followed was to rewrite various sections as greater understanding became available. But, since much experience had taught that a program of constant rewriting led to no book at all, this alternative was rejected.

Finally, I must again thank devoted and talented typists, Mrs. Susan Wagenseil at Harvard and Mr. Ronald Bohn at UCLA.

Los Angeles, California
March 1972

J. S.

Contents

4. Electrodynamics I

If you can't join 'em,
beat 'em.

Particles, Sources,
and Fields

4

ELECTRODYNAMICS I

Causality and space-time uniformity are the creative principles of source theory. Uniformity in space-time also has a complementary momentum-energy implication. It is illustrated by the extended source concept. Not only for that special balance of energy and momentum involved in the emission or absorption of a single particle is the source defined and meaningful. Given a sufficient excess of energy over momentum—an excess of mass—several particles can be emitted or absorbed. The description of the coupling between sources must include an account of such multi-particle exchange acts.

The additional single photon emission, or absorption, that can accompany the working of a charged particle source has already been considered. It is described by the primitive interaction. That same primitive interaction also permits an extended photon source to emit, or absorb, a neutral pair of charged particles. These two-particle processes are the simplest examples of the multi-particle exchanges that supplement single-particle exchange, with corresponding additions to the appropriate propagation function. The replacement of the initial single-particle propagation functions by the modified ones is of universal applicability but does not exhaust the implications of multi-particle exchange. The propagation function modification gives an account of what is common to all realistic sources of a specified type, but does not represent their individual characteristics. To illustrate this additional aspect, let us consider the emission of a pair of charged particles by an extended photon source. This process can be represented as the conversion of a virtual photon into a pair of real particles. The description provided by the primitive interaction refers to circumstances in which the two particles do not interact. The subsequent interaction of the particles must be invoked to supply the mechanism that introduces the modified propagation function for the virtual photon. As we have seen in Sections 3–12 and 3–13, where particles of spin 0 and $\frac{1}{2}$ were considered, particle-antiparticle scattering involves two distinct mechanisms. In one, an ordinary scattering act that also occurs in particle-particle scattering, the particles persist while exchanging a space-like virtual photon (a photon that carries space-like momentum). The other is an annihilation of the particle-antiparticle pair, producing a time-like virtual photon, which quickly decays back into particles. It is the latter process that supplies the modification of the photon propagation function, for, connecting the final pair of charged particles with the extended photon source is a chain of events in which a virtual photon produces a pair of particles that eventually recombine to produce a virtual photon. Here is just the additional coupling that operates between extended photon sources. But the introduction of the modified photon propagation function does not describe the effect of particle interactions completely since the simple scattering process of space-like photon exchange is not included.

The primitive interaction can be characterized as a local coupling of the fields. Two charged particle fields and the electromagnetic vector potential, all referring to the same space-time point, are multiplied together. The introduction of modified propagation functions alters the quantitative meaning of the fields, but does not change the local character of the field coupling. This complete locality disappears when we consider the ordinary scattering of the oppositely charged particles, which can take place at some distance (in time and space) from the point at which the virtual photon decays into the particle pair. The final particles are described by fields referring to the region in which these particles come into being, as distinguished from the coordinate values in the electromagnetic potential vector, which represents the initial particle creation act. This non-local modification of the primitive interaction alters the electromagnetic properties assigned to the charged particles. New couplings can appear, and the non-locality of all couplings is the introduction of form factors, which convey the effective space-time distribution of the particle's electromagnetic properties. It is the quantitative discussion of such refinements, appearing at the dynamical level that incorporates two-particle exchange, to which the developing techniques of Source Theory will now be applied.

4-1 CHARGED PARTICLE PROPAGATION FUNCTIONS

The description of the emission or absorption of a charged particle and a photon by an extended particle source involves the choice of an electromagnetic model for the source. As discussed in Sections 3-10 and 3-11, the most natural, covariant, source model suppresses charge acceleration radiation. With this choice, photons do not accompany spin 0 charged particles and there is no corresponding modification in the propagation function. Radiation does occur for spin $\frac{1}{2}$ particles, however, and we shall examine that situation.

The primitive interaction incorporated in (3-10.63) contains an arbitrary gyromagnetic ratio g. For the particular value $g = 2$, which is very nearly that exhibited by the electron and the muon, the effective photon-particle emission source is [Eq. (3-11.79)]

$$i J_2^\mu(k)\eta_2(p)|_{\text{eff.}} = (1/2kp)\sigma^{\mu\nu}ik_\nu eq\eta_2(P), \qquad (4\text{-}1.1)$$

where we have omitted the charge acceleration term

$$\frac{p^\mu}{kp} - if^\mu(k, P) = \frac{p^\mu}{kp} - \frac{P^\mu}{kP} = -\frac{k^\mu}{kp}, \qquad (4\text{-}1.2)$$

since it does not contribute to the radiation:

$$e^{\mu}_{k\lambda}{}^{*} k_{\mu} = 0. \tag{4-1.3}$$

The probability amplitude for the two-particle emission process, obtained as

$$\langle 1_{k\lambda} 1_{p\sigma q} | 0_- \rangle^{n_1} = i(d\omega_k)^{1/2} e^{\mu}_{k\lambda}{}^{*} J_{2\mu}(k) i(2m\, d\omega_p)^{1/2} u^{*}_{p\sigma q} \gamma^0 \eta_2(p)\big|_{\text{eff.}}, \tag{4-1.4}$$

is

$$\langle 1_{k\lambda} 1_{p\sigma q} | 0_- \rangle^{n_1} = ieq(d\omega_k\, 2m\, d\omega_p)^{1/2} u^{*}_{p\sigma q} \gamma^0 \gamma k \gamma e^{*}_{k\lambda} \eta_2(P) \frac{1}{P^2 + m^2}, \tag{4-1.5}$$

which utilizes the equivalent forms

$$P^2 + m^2 = 2kp, \qquad \gamma k \gamma e^{*}_{k\lambda} = i e^{\mu}_{k\lambda}{}^{*} \sigma_{\mu\nu} k^{\nu}. \tag{4-1.6}$$

The derivation could be repeated for the two-particle absorption process. The required probability amplitude can be obtained directly, however, as the negative complex conjugate of the emission amplitude:

$$\langle 0_+ | 1_{k\lambda} 1_{p\sigma q} \rangle^{n_1} = ieq(d\omega_k\, 2m\, d\omega_p)^{1/2} \frac{1}{P^2 + m^2} \eta_1(P)^{*} \gamma^0 \gamma e_{k\lambda} \gamma k u_{p\sigma q}. \tag{4-1.7}$$

The basis for this is the same as for single-particle emission — the maintenance of orthogonality between the vacuum state and the class of states into which the weak sources emit and from which they absorb.

We now consider a causal arrangement of extended particle sources, and evaluate the contribution to the vacuum amplitude of two-particle exchange:

$$\sum_{k\lambda,p\sigma q} \langle 0_+ | 1_{k\lambda} 1_{p\sigma q} \rangle^{n_1} \langle 1_{k\lambda} 1_{p\sigma q} | 0_- \rangle^{n_1}$$

$$= -e^2 \int d\omega_p\, d\omega_k \sum_{\lambda} \eta_1(P)^{*} \gamma^0 \gamma e_{k\lambda} \gamma k (m - \gamma p) \gamma k \gamma e^{*}_{k\lambda} \eta_2(P) \frac{1}{(P^2 + m^2)^2}, \tag{4-1.8}$$

in which we have introduced the expression of completeness for the various particle states of specified momentum,

$$\sum_{\sigma q} u_{p\sigma q} u^{*}_{p\sigma q} \gamma^0 = (1/2m)(m - \gamma p). \tag{4-1.9}$$

The algebraic property

$$(\gamma k)^2 = -k^2 = 0 \tag{4-1.10}$$

implies that

$$\gamma k (m - \gamma p) \gamma k = 2kp \gamma k, \tag{4-1.11}$$

while

$$\{\gamma e_{k\lambda}, \gamma k\} = -2e_{k\lambda}k = 0. \tag{4-1.12}$$

Then, if we exercise the option of using real polarization vectors, for which

$$(\gamma e_{k\lambda})^2 = -1, \tag{4-1.13}$$

we get

$$\sum_{\lambda} \gamma e_{k\lambda} \gamma k(m - \gamma p)\gamma k \gamma e_{k\lambda} = \sum_{\lambda} 2kp\gamma k$$

$$= 2(P^2 + m^2)\gamma k. \tag{4-1.14}$$

Since the sources refer only to the total momentum P, one must integrate over all partitions of this momentum between the particle and the photon. That is facilitated by introducing the total momentum as an integration variable, the kinematical relation with the momenta p and k being conveyed by the appropriate delta function. Thus, we insert the unit factor

$$(2\pi)^3 \int \frac{(dP)}{(2\pi)^3} \delta(P - p - k) = (2\pi)^3 \int d\omega_P \, dM^2 \, \delta(P - p - k), \tag{4-1.15}$$

where

$$M^2 = -P^2 \tag{4-1.16}$$

replaces the energy as an integration variable in accordance with

$$2P^0 \, dP^0 = dM^2. \tag{4-1.17}$$

The vacuum amplitude contribution now appears as

$$2e^2 \int dM^2 \, d\omega_P \, \frac{1}{M^2 - m^2} \eta_1(P)^* \gamma^0 \gamma^\mu \eta_2(P) I_\mu(P), \tag{4-1.18}$$

with

$$I_\mu(P) = (2\pi)^3 \int d\omega_p \, d\omega_k \, \delta(P - p - k) k_\mu. \tag{4-1.19}$$

Only the vector P_μ is available to supply a preferred direction for k_μ in this integral, and k_μ can be replaced by its projection onto P_μ:

$$k_\mu \to \frac{kP}{P^2} P_\mu = \frac{M^2 - m^2}{2M^2} P_\mu. \tag{4-1.20}$$

Accordingly,

$$I_\mu(P) = \frac{M^2 - m^2}{2M^2} P_\mu I(M), \tag{4-1.21}$$

where

$$I(M) = (2\pi)^3 \int d\omega_p \, d\omega_k \, \delta(P - p - k) \qquad (4\text{--}1.22)$$

defines a scalar, a function of $-P^2 = M^2 \geqslant m^2$.

A related integral has been encountered, for arbitrary masses m_a and m_b, in the discussion of scattering cross sections. On completing the angular integration in (3–12.75) and dividing by 2π, we get

$$I(M, m_a, m_b) = (2\pi)^3 \int d\omega_a \, d\omega_b \, \delta(P - p_a - p_b)$$

$$= \frac{1}{(4\pi)^2} \left[1 - \left(\frac{m_a + m_b}{M}\right)^2 \right]^{1/2} \left[1 - \left(\frac{m_a - m_b}{M}\right)^2 \right]^{1/2} . \qquad (4\text{--}1.23)$$

Some special examples are

$$I(M, m, m) = \frac{1}{(4\pi)^2} \left(1 - \frac{4m^2}{M^2} \right)^{1/2} \qquad (4\text{--}1.24)$$

and

$$I(M, m, 0) = \frac{1}{(4\pi)^2} \left(1 - \frac{m^2}{M^2} \right) . \qquad (4\text{--}1.25)$$

This evaluation was performed in the rest frame of the time-like vector P^μ. In the interests of exploring various computation techniques, we shall repeat the calculation, using a quite different coordinate system, with which we also have had some experience. It is one in which the speed associated with the vector P^μ approaches that of light. Such coordinate systems are often known as infinite momentum frames. All momenta are decomposed into longitudinal (L) and transverse (T) components relative to the direction provided by the vector $\mathbf{P}$. Denoting the magnitude of this vector by P_L, we state the associated energy as

$$P^0 = (P_L{}^2 + M^2)^{1/2} \cong P_L + (M^2/2P_L), \qquad (4\text{--}1.26)$$

where $\cong$ indicates statements that become exact asymptotically, as $P_L \to \infty$. For the individual particles, a and b, we shall write, for example,

$$p_{aL} = P_L u_a,$$

$$p_a^0 = (p_{aL}{}^2 + p_{aT}{}^2 + m_a{}^2)^{1/2} \cong P_L |u_a| + \frac{p_{aT}{}^2 + m_a{}^2}{2P_L |u_a|} . \qquad (4\text{--}1.27)$$

The invariant momentum space measures appear as

$$d\omega_a = \frac{(d\mathbf{p}_{aT})\, du_a\, P_L}{(2\pi)^3 2p_a^0} \cong \frac{1}{(2\pi)^3} \frac{(d\mathbf{p}_{aT})\, du_a}{2|u_a|}, \tag{4-1.28}$$

while the four-dimensional delta function becomes:

$$\delta(P - p_a - p_b) \cong \delta(\mathbf{p}_{aT} + \mathbf{p}_{bT})\, \delta(P_L(1 - u_a - u_b))$$

$$\times \delta\left(P_L(1 - |u_a| - |u_b|) + \frac{M^2}{2P_L} - \frac{\mathbf{p}_{aT}^2 + m_a^2}{2P_L|u_a|} - \frac{\mathbf{p}_{bT}^2 + m_b^2}{2P_L|u_b|}\right), \tag{4-1.29}$$

or

$$\delta(P - p_a - p_b) \cong \delta(\mathbf{p}_{aT} + \mathbf{p}_{bT})\, \delta(1 - u_a - u_b)\, \eta(u_a)\eta(u_b)$$

$$\times 2\delta\left(\frac{\mathbf{p}_{aT}^2 + m_a^2}{u_a} + \frac{\mathbf{p}_{bT}^2 + m_b^2}{u_b} - M^2\right). \tag{4-1.30}$$

The second version involves the recognition that the energy and longitudinal momentum factors are incompatible, unless u_a and u_b are positive. Also used is the delta function property

$$|\lambda|\delta(\lambda x) = \delta(x). \tag{4-1.31}$$

Since the integration over the variables of one particle is immediate, we now get

$$I(M, m_a, m_b) = \frac{1}{(4\pi)^2} \int_0^1 du \int_0^\infty d\left(\frac{\mathbf{p}_T^2}{u(1-u)}\right) \delta\left(\frac{\mathbf{p}_T^2}{u(1-u)} + \frac{m_a^2}{u} + \frac{m_b^2}{1-u} - M^2\right)$$

$$= \frac{1}{(4\pi)^2} \int_0^1 du\, \eta\left(M^2 - \frac{m_a^2}{u} - \frac{m_b^2}{1-u}\right), \tag{4-1.32}$$

where the step function contains the conditions for the nonvanishing of the integral. In general ($m_a, m_b \neq 0$), the argument of the step function is negative at both integration limits, and becomes positive in the interior of the interval only if $M > m_a + m_b$. The value of the u-integral is the range of positiveness, which is found by subtracting the two roots of the quadratic equation

$$M^2 u(1 - u) - m_a^2(1 - u) - m_b^2 u = 0. \tag{4-1.33}$$

The result is (4-1.23), of course.

We return to the vacuum amplitude contribution (4-1.18) and use the evaluation (4-1.25). On introducing

$$\alpha = e^2/4\pi, \tag{4-1.34}$$

this contribution becomes

$$\frac{\alpha}{4\pi} \int \frac{dM^2}{M^2}\, d\omega_P \left(1 - \frac{m^2}{M^2}\right) \eta_1(P)^* \gamma^0 \gamma P \eta_2(P). \tag{4–1.35}$$

The causal space-time arrangement is made explicit by writing

$$\eta_2(P) = \int (dx')\, \exp(-iPx')\, \eta_2(x'),$$

$$\eta_1(P)^* = \int (dx)\eta_1(x)\, \exp(iPx), \tag{4–1.36}$$

which gives the space-time coupling:

$$-i\frac{\alpha}{4\pi} \int \frac{dM^2}{M^2}\left(1 - \frac{m^2}{M^2}\right) \int (dx)(dx')\eta_1(x)\gamma^0$$

$$\times \left(\gamma \frac{1}{i}\partial\right)\left[i\int d\omega_P \exp[iP(x-x')]\right]\eta_2(x'). \tag{4–1.37}$$

We recognize, in the form appropriate to the causal order $x^0 > x^{0\prime}$, the invariant propagation function

$$x^0 > x^{0\prime}: \quad \Delta_+(x-x', M^2) = i\int d\omega_P \exp[iP(x-x')], \tag{4–1.38}$$

where the variable mass M has been made explicit. The addition of this coupling to the one associated with single-particle exchange is represented by the modified propagation function

$$\bar{G}_+(x-x') = G_+(x-x') + \frac{\alpha}{4\pi} \int \frac{dM^2}{M^2}\left(1 - \frac{m^2}{M^2}\right)\left(-\gamma\frac{1}{i}\partial\right)\Delta_+(x-x', M^2),$$

$$\tag{4–1.39}$$

which possesses the required F.D. antisymmetry:

$$(\gamma^0 \bar{G}_+(x'-x))^T = -\gamma^0 \bar{G}_+(x-x'). \tag{4–1.40}$$

The momentum space transcription is

$$\bar{G}_+(p) = \frac{1}{\gamma p + m - i\varepsilon} + \frac{\alpha}{4\pi} \int \frac{dM^2}{M^2}\left(1 - \frac{m^2}{M^2}\right)\frac{-\gamma p}{p^2 + M^2 - i\varepsilon}$$

$$= \frac{1}{\gamma p + m - i\varepsilon} + \frac{\alpha}{4\pi} \int_m^\infty \frac{dM}{M}\left(1 - \frac{m^2}{M^2}\right)\left[\frac{1}{\gamma p + M - i\varepsilon} + \frac{1}{\gamma p - M + i\varepsilon}\right].$$

$$\tag{4–1.41}$$

The latter form makes explicit that the spectral mass integral begins at $M = m$, and can be extended tentatively to large mass values since, as is more evident in the first form, the integration is quite convergent.

An alternative presentation of this, and related calculations, begins with the term in the vacuum amplitude for noninteracting particles that represents the exchange of one photon and one charged particle:

$$\langle 0_+|0_-\rangle^{Jn} = 1 + \cdots + i \int (d\xi)(d\xi') J_1^\mu(\xi) D_+(\xi - \xi') J_{2\mu}(\xi')$$

$$\times i \int (dx)(dx') \eta_1(x) \gamma^0 G_+(x - x') \eta_2(x') + \cdots. \qquad (4\text{--}1.42)$$

Into that term,

$$\int (d\xi) \cdots (dx') i J_1^\mu(\xi) \eta_1(x) \gamma^0 D_+(\xi - \xi') G_+(x - x') i J_{2\mu}(\xi') \eta_2(x'), \quad (4\text{--}1.43)$$

one introduces the effective source [Eq. (3–11.66), with $g = 2$]

$$i J^\mu(\xi) \eta(x)|_{\text{eff.}} = eq[\delta(\xi - x)\gamma^\mu \psi(x) - i f^\mu(\xi - x)\eta(x)], \qquad (4\text{--}1.44)$$

which is equally applicable to emission and absorption. The preferred form for the latter process, however, is

$$i J^\mu(\xi) \eta(x) \gamma^0|_{\text{eff.}} = [\psi(x)\gamma^0 \gamma^\mu \delta(x - \xi) - i\eta(x)\gamma^0 f^\mu(x - \xi)]eq. \qquad (4\text{--}1.45)$$

The momentum space versions are:

$$i J_2^\mu(k) \eta_2(p)|_{\text{eff.}} = eq[\gamma^\mu \psi_2(P) - i f^\mu(k)\eta_2(P)],$$

$$i J_1^\mu(-k) \eta_1(-p) \gamma^0|_{\text{eff.}} = [\psi_1(-P)\gamma^0 \gamma^\mu - i\eta_1(-P)\gamma^0 f^\mu(k)]eq. \qquad (4\text{--}1.46)$$

First, let us pick out the contribution that does not involve photons emitted or absorbed by the sources, and therefore has no reference to the electromagnetic source model. Inserting in (4–1.43) the propagation function forms appropriate to the causal situation, we find

$$- e^2 \int d\omega_p\, d\omega_k\, \psi_1(-P)\gamma^0 \gamma^\mu (m - \gamma p)\gamma_\mu \psi_2(P)$$

$$= - e^2 \int dM^2\, d\omega_P\, \psi_1(-P)\gamma^0 \gamma^\mu F(P)\gamma_\mu \psi_2(P), \qquad (4\text{--}1.47)$$

with

$$F(P) = (2\pi)^3 \int d\omega_p\, d\omega_k\, \delta(P - p - k)(m - \gamma p)$$

$$= \frac{1}{(4\pi)^2}\left(1 - \frac{m^2}{M^2}\right)\left(m - \frac{M^2 + m^2}{2M^2}\gamma P\right). \qquad (4\text{--}1.48)$$

The last evaluation differs from that of $I_\mu(P)$ only in using the effective projection

$$p_\mu \rightarrow \frac{pP}{P^2} P_\mu = \frac{M^2 + m^2}{2M^2} P_\mu, \qquad (4\text{-}1.49)$$

which also follows from (4–1.20), since $p = P - k$. The matrix $F(P)$ appears in the combination

$$\gamma^\mu F(P)\gamma_\mu = \frac{1}{(4\pi)^2}\left(1 - \frac{m^2}{M^2}\right)\left(-4m - \frac{M^2 + m^2}{M^2}\gamma P\right), \qquad (4\text{-}1.50)$$

which exhibits the evaluations

$$\gamma^\mu \gamma_\mu = -4,$$

$$\gamma^\mu \gamma^\nu \gamma_\mu = (-2g^{\mu\nu} - \gamma^\nu \gamma^\mu)\gamma_\mu = 2\gamma^\nu. \qquad (4\text{-}1.51)$$

This portion of the vacuum amplitude thus becomes:

$$\frac{\alpha}{4\pi}\int dM^2\, d\omega_P \left(1 - \frac{m^2}{M^2}\right)\psi_1(-P)\gamma^0\left(4m + \frac{M^2 + m^2}{M^2}\gamma P\right)\psi_2(P)$$

$$= \frac{\alpha}{4\pi}\int dM^2\, d\omega_P \left(1 - \frac{m^2}{M^2}\right)\eta_1(-P)\gamma^0 \frac{1}{\gamma P + m}\left(4m + \frac{M^2 + m^2}{M^2}\gamma P\right)\frac{1}{\gamma P + m}\eta_2(P).$$
$$(4\text{-}1.52)$$

The matrix that occurs in the second line is simplified algebraically by introducing the projection matrices for γP relative to its eigenvalues $\pm M$,

$$\frac{1}{(\gamma P + m)^2}\left(4m + \frac{M^2 + m^2}{M^2}\gamma P\right)$$

$$= \frac{\gamma P + M}{2M^2}\left[1 + \frac{2mM}{(M + m)^2}\right] + \frac{\gamma P - M}{2M^2}\left[1 - \frac{2mM}{(M - m)^2}\right]. \qquad (4\text{-}1.53)$$

Note that the resulting source coupling reduces to (4–1.35) for $M \gg m$. The space-time extrapolation proceeds as before, and since the factors are fixed by the asymptotic ($M \gg m$) agreement with the earlier calculation, we can immediately write the momentum space form of the incomplete modified propagation function:

$$G_+'(p) = \frac{1}{\gamma p + m - i\varepsilon} + \frac{\alpha}{4\pi}\int_{\rightarrow m}^\infty \frac{dM}{M}\left(1 - \frac{m^2}{M^2}\right)\left[\frac{1 - \dfrac{2mM}{(M - m)^2}}{\gamma p + M - i\varepsilon} + \frac{1 + \dfrac{2mM}{(M + m)^2}}{\gamma p - M + i\varepsilon}\right].$$
$$(4\text{-}1.54)$$

In contrast with (4–1.41), there is now an infrared divergence, a logarithmic singularity at the lower limit of the spectral integral. It would be premature, however, to accept the quantitative details of this phenomenon, since the above propagation function is incomplete. Note also that, again in contrast with (4–1.41), the weight factor of $(\gamma p + M)^{-1}$ is not always positive.

This calculation enables us to emphasize physical requirements that have remained implicit thus far. Suppose the space-time extrapolation had been based on the first, field coupling, version of (4–1.52) rather than the second, source coupling. The additional action term obtained in that way would be

$$-\tfrac{1}{2}\int (dx)(dx')\psi(x)\gamma^0 M(x-x')\psi(x'), \tag{4–1.55}$$

with

$$M(p) = M(\gamma p)$$

$$= -\frac{\alpha}{4\pi}\int_m^\infty \frac{dM}{M}\left(1 - \frac{m^2}{M^2}\right)\left[\frac{(M-m)^2 - 2mM}{\gamma p + M - i\varepsilon} + \frac{(M+m)^2 + 2mM}{\gamma p - M + i\varepsilon}\right]. \tag{4–1.56}$$

Two things are wrong here. First, the modified propagation function that now emerges, on expressing the fields in terms of sources, is

$$\frac{1}{\gamma p + m - i\varepsilon} - \frac{1}{\gamma p + m - i\varepsilon} M(\gamma p) \frac{1}{\gamma p + m - i\varepsilon}. \tag{4–1.57}$$

Since $M(\gamma p)$ does not vanish for $\gamma p + m = 0$, the behavior of the propagation function is drastically modified in the neighborhood: $\gamma p + m \sim 0$. This contradicts the phenomenological basis of the theory, as we discussed extensively in Section 3–9, under the heading of mass normalization. Second, the space-time extrapolation that was performed in reaching (4–1.55) is invalid, since the integral of (4–1.56) does not exist when extended up to indefinitely large M values.

Source couplings that are inferred through space-time extrapolations of causal arrangements can always be supplemented by contact interactions. Unless additional physical considerations can be adduced, the contact terms remain arbitrary and may be omitted. But, when fields replace sources such local interaction terms do have physical content; their existence must be recognized and related to the accompanying physical requirements. Since contact couplings in coordinate space appear as polynomial functions of momenta in momentum space, the correct form of $M(\gamma p)$ supplements (4–1.56) by a polynomial in $\gamma p + m$. Quadratic and higher powers of this convenient combination modify the propagation function (4–1.57) by constant or polynomial functions of momenta. We are not interested in contact additions to the propagation function, and it suffices to use constant and linear terms in $\gamma p + m$. These are fixed by the physical requirement that the second term of (4–1.57) does not have singularities in the neighborhood of $\gamma p + m = 0$, as stated by:

$$M(\gamma p = -m) = 0, \qquad (d/d\gamma p)M(\gamma p = -m) = 0. \qquad (4\text{–}1.58)$$

This is achieved, for each value of M, by the contact modifications

$$\frac{1}{\gamma p + M - i\varepsilon} \rightarrow \frac{1}{\gamma p + M - i\varepsilon} - \frac{1}{M - m} + \frac{\gamma p + m}{(M - m)^2}$$

$$= \frac{(\gamma p + m)^2}{(M - m)^2} \frac{1}{\gamma p + M - i\varepsilon}, \qquad (4\text{–}1.59)$$

and

$$\frac{1}{\gamma p - M + i\varepsilon} \rightarrow \frac{1}{\gamma p - M + i\varepsilon} + \frac{1}{M + m} + \frac{\gamma p + m}{(M + m)^2}$$

$$= \frac{(\gamma p + m)^2}{(M + m)^2} \frac{1}{\gamma p - M + i\varepsilon}. \qquad (4\text{–}1.60)$$

With this corrected interpretation,

$$M(\gamma p) = -(\gamma p + m)^2 \frac{\alpha}{4\pi} \int_{\rightarrow m}^{\infty} \frac{dM}{M}\left(1 - \frac{m^2}{M^2}\right)\left[\frac{1 - \dfrac{2mM}{(M - m)^2}}{\gamma p + M - i\varepsilon} + \frac{1 + \dfrac{2mM}{(M + m)^2}}{\gamma p - M + i\varepsilon}\right],$$

$$(4\text{–}1.61)$$

and the propagation function that now emerges from (4–1.57) is just (4–1.54).

The partial propagation function we have been discussing has its counterpart for spin 0. The momentum space presentation of the effective sources is [Eq. (3–11.16)]

$$iJ_2^\mu(k)K_2(p)\big|_{\text{eff.}} = eq[(p^\mu + P^\mu)\phi_2(P) - if^\mu(k)K_2(P)],$$

$$iJ_1^\mu(-k)K_1(-p)\big|_{\text{eff.}} = [\phi_1(-P)(p^\mu + P^\mu) - K_1(-P)if^\mu(k)]eq. \qquad (4\text{–}1.62)$$

When the f^μ terms are omitted in the two-particle exchange coupling,

$$\int (d\xi) \cdots (dx') iJ_1^\mu(\xi)K_1(x)\big|_{\text{eff.}} D_+(\xi - \xi')\Delta_+(x - x') iJ_{2\mu}(\xi')K_2(x')\big|_{\text{eff.}}, \qquad (4\text{–}1.63)$$

the resulting vacuum amplitude contribution is

$$-e^2 \int d\omega_p\, d\omega_k\, \phi_1(-P)(p + P)^2\phi_2(P) = e^2 \int d\omega_p\, d\omega_k\, 2\frac{M^2 + m^2}{(M^2 - m^2)^2} K_1(-P)K_2(P)$$

$$= \frac{\alpha}{2\pi} \int \frac{dM^2}{M^2} \frac{M^2 + m^2}{M^2 - m^2} d\omega_P \, K_1(-P)K_2(P). \qquad (4\text{-}1.64)$$

The incomplete modified propagation function implied by space-time extrapolation of this coupling is expressed by

$$\Delta_+'(p) = \frac{1}{p^2 + m^2 - i\varepsilon} - \frac{\alpha}{2\pi} \int \frac{dM^2}{M^2} \frac{M^2 + m^2}{M^2 - m^2} \frac{1}{p^2 + M^2 - i\varepsilon}, \qquad (4\text{-}1.65)$$

where the M^2 integration can be extended up to infinity, but has an infrared singularity as $M^2 \to m^2$.

Let us use the simpler situation of spin 0 first, in order to indicate the structure of a modified propagation function that refers to an electromagnetic source model for which charge acceleration radiation does occur. When inserted in (4-1.62), the choice [Eq. (3-10.44)]

$$if^\mu(k) = n^\mu/nk \qquad (4\text{-}1.66)$$

modifies the vacuum amplitude calculation of (4-1.64) in the manner given by

$$-2 \frac{M^2 + m^2}{(M^2 - m^2)^2} \to -2 \frac{M^2 + m^2}{(M^2 - m^2)^2} + \frac{2}{M^2 - m^2}\left(2\frac{nP}{nk} - 1\right) - \frac{1}{(nk)^2}. \qquad (4\text{-}1.67)$$

The inverse powers of nk must be averaged over all partitions of the total momentum P, as indicated by the notation

$$(2\pi)^3 \int d\omega_p \, d\omega_k \, \delta(P - p - k) \, (nk)^{-1,2} = \frac{1}{(4\pi)^2}\left(1 - \frac{m^2}{M^2}\right)\langle(nk)^{-1,2}\rangle. \qquad (4\text{-}1.68)$$

In the rest frame of the vector P^μ the only significant variable is z, the cosine of the angle between the relative momentum vector of the particles and $\mathbf{n}$. Thus,

$$\left\langle\frac{1}{(nk)^2}\right\rangle = \left(\frac{2M}{M^2 - m^2}\right)^2 \frac{1}{2} \int_{-1}^{1} dz \, \frac{1}{(n^0 - |\mathbf{n}|z)^2}, \qquad (4\text{-}1.69)$$

which also exhibits the photon energy in the P^μ rest frame, as inferred from

$$-kP/M = (M^2 - m^2)/2M. \qquad (4\text{-}1.70)$$

This integral depends only upon the combination

$$(n^0)^2 - |\mathbf{n}|^2 = 1, \qquad (4\text{-}1.71)$$

and

$$\left\langle\frac{1}{(nk)^2}\right\rangle = \left(\frac{2M}{M^2 - m^2}\right)^2. \qquad (4\text{-}1.72)$$

The integral that evaluates $\langle (nk)^{-1} \rangle$ in the rest frame of P^μ contains $|\mathbf{n}|^2$ explicitly, however. This quantity is presented covariantly as

$$|\mathbf{n}|^2 = (n^0)^2 - 1 = (nP/M)^2 - 1 = Q^2/M^2, \qquad (4\text{–}1.73)$$

where

$$Q^2 = (nP)^2 - n^2 P^2 = -\tfrac{1}{2}(n^\mu P^\nu - n^\nu P^\mu)(n_\mu P_\nu - n_\nu P_\mu). \qquad (4\text{–}1.74)$$

Although this second integral is also an elementary one, we prefer to leave it in integral form, as given by

$$\frac{M^2 - m^2}{2M^2}\left\langle \frac{nP}{nk} \right\rangle = \frac{1}{2}\int_{-1}^{1} dz\, \frac{n^0}{n^0 - |\mathbf{n}|z}$$

$$= \frac{1}{2}\int_{-1}^{1} dz\, \frac{(n^0)^2}{(n^0)^2 - |\mathbf{n}|^2 z^2}$$

$$= 1 + \frac{Q^2}{M^2}\int_0^1 dz\, \frac{z^2}{1 + (Q^2/M^2)(1 - z^2)}. \qquad (4\text{–}1.75)$$

The substitution of (4–1.67) now appears effectively as

$$-2\frac{M^2 + m^2}{(M^2 - m^2)^2} \rightarrow 8\frac{M^2}{(M^2 - m^2)^2}\frac{Q^2}{M^2}\int_0^1 dz\, \frac{z^2}{1 + (Q^2/M^2)(1 - z^2)}. \qquad (4\text{–}1.76)$$

The result thus deduced from the incomplete propagation function (4–1.65) is

$$\bar{\Delta}_+(p, q) = \frac{1}{p^2 + m^2 - i\varepsilon} + \frac{2\alpha}{\pi}\int_{\to m^2}^{\infty} \frac{dM^2}{M^2 - m^2}\int_0^1 dz\, \frac{1}{p^2 + M^2 - i\varepsilon}\frac{(q^2/M^2)z^2}{1 + (q^2/M^2)(1 - z^2)}, \qquad (4\text{–}1.77)$$

where q^2 is formed from the arbitrary vector p^μ in the manner of (4–1.74). Note that $q^2 > 0$ continues to be valid since n^μ is a time-like vector. In contrast with (4–1.65), the weight factor of each $(p^2 + M^2 - i\varepsilon)^{-1}$ is now a positive number. The M^2 integral in (4–1.77) converges rapidly with increasing M^2, but has an infrared singularity as $M^2 \to m^2$. There is another presentation of (4–1.77) that uses the variable

$$M'^2 = M^2/(1 - z^2), \qquad (4\text{–}1.78)$$

namely,

$$\bar{\Delta}_+(p, q) = \frac{1}{p^2 + m^2 - i\varepsilon} + \frac{\alpha}{\pi}\int_{\to m^2}^{\infty} \frac{dM^2}{M^2 - m^2}\int_{M^2}^{\infty} \frac{dM'^2}{M'^2}\left(1 - \frac{M^2}{M'^2}\right)^{1/2}$$

$$\times \frac{1}{p^2 + M^2 - i\varepsilon}\frac{q^2}{q^2 + M'^2}. \qquad (4\text{–}1.79)$$

Structures such as (4–1.41) and (4–1.65), describing the mass spectrum of one- and two-particle excitations, are single spectral forms. In contrast, (4–1.79) contains a double spectral form. But, to what kind of excitation the mass M' refers cannot be clear from the purely mathematical origin of this structure.

Since spectral forms represent the possibility of excitations with variable masses, the weight factor of a standard propagation function, $(p^2 + M^2 - i\varepsilon)^{-1}$ for example, must be a measure of the effectiveness of the source, a relative probability, for that kind of excitation. The unphysical nature of partial propagation functions, illustrated by (4–1.65), is thus manifest in the appearance of negative weight factors. As a check of this intuitive probability interpretation, we shall extract a known result by examining the propagation function (4–1.77) for $M^2 \sim m^2$, corresponding to a particle accompanied by a soft photon:

$$M^2 = -(p + k)^2 = m^2 - 2pk$$

$$= m^2 + 2mk^0 \sim m^2. \tag{4–1.80}$$

The quantity $k^0(\ll m)$ is the photon energy in the rest frame of the particle. This discussion is restricted, for simplicity, to the situation

$$q^2/m^2 \ll 1, \tag{4–1.81}$$

which, interpreted in the rest frame of the unit vector n^μ,

$$q^2/m^2 = \mathbf{p}^2/m^2 = \mathbf{v}^2 \ll 1, \tag{4–1.82}$$

corresponds to considering slowly moving charged particles. Then, the portion of the propagation function that is selected by (4–1.80) and (4–1.81) reads

$$\bar{\Delta}_+(p, q) \cong \frac{1}{p^2 + m^2 - i\varepsilon} + \frac{2\alpha}{3\pi} \mathbf{v}^2 \int \frac{dk^0}{k^0} \frac{1}{p^2 + (m^2 + 2mk^0) - i\varepsilon}. \tag{4–1.83}$$

Since multiple photon emission has been neglected in arriving at this result, the differential probability exhibited here, $(2\alpha/3\pi)\mathbf{v}^2(dk^0/k^0)$, states the average number of photons emitted. As such, it is in complete agreement with the calculation of (3–11.60). The discussion given in connection with the latter equation emphasizes that the mathematical singularity at $k^0 = 0$ is spurious. The integration in (4–1.83) must be stopped at such a value of k^0 that m^2 and $m^2 + 2mk^0$ are no longer experimentally distinguishable. The contribution of all smaller values of k^0 is included already in the description of the emission and absorption of the charged particle without a detectable accompanying photon. The infrared situation is of general occurrence and will be commented on from time to time in the course of various applications.

To perform the analogous calculation referring to spin $\frac{1}{2}$ we return to the effective sources (4–1.46) and insert the electromagnetic model function (4–1.66).

The change produced in the vacuum amplitude (4–1.52) is indicated by

$$\frac{1}{\gamma P + m}\left(4m + \frac{M^2 + m^2}{M^2}\gamma P\right)\frac{1}{\gamma P + m}$$

$$\rightarrow -\left\langle\left(\frac{1}{\gamma P + m}\gamma^\mu - \frac{n^\mu}{nk}\right)(m - \gamma p)\left(\gamma_\mu \frac{1}{\gamma P + m} - \frac{n_\mu}{nk}\right)\right\rangle, \quad (4\text{–}1.84)$$

where the notation $\langle\ \rangle$ is that introduced in (4–1.68). When the right-hand side is written out and some known integrals inserted, it reads:

$$\frac{1}{\gamma P + m}\left(4m + \frac{M^2 + m^2}{M^2}\gamma P\right)\frac{1}{\gamma P + m} + \frac{4M^2}{(M^2 - m^2)^2}(m - \gamma P) + \left\langle\frac{\gamma k}{(nk)^2}\right\rangle$$

$$+ \frac{4}{\gamma P + m}\left(\left\langle\frac{nP}{nk}\right\rangle - 1\right) + 2\gamma n\left\langle\frac{1}{nk}\right\rangle - \frac{1}{\gamma P + m}\left\langle\frac{\gamma k}{nk}\right\rangle\gamma n$$

$$- \gamma n\left\langle\frac{\gamma k}{nk}\right\rangle\frac{1}{\gamma P + m}. \quad (4\text{–}1.85)$$

We have already evaluated [cf. Eq. (4–1.75)]

$$\frac{M^2 - m^2}{2M^2}\left\langle\frac{nP}{nk}\right\rangle = 1 + \frac{Q^2}{M^2}Z\left(\frac{Q^2}{M^2}\right), \quad (4\text{–}1.86)$$

which uses the abbreviation

$$Z(Q^2/M^2) = \int_0^1 dz\,\frac{z^2}{1 + (Q^2/M^2)(1 - z^2)} = \frac{1}{2}\int_{M^2}^\infty \frac{dM'^2}{M'^2}\left(1 - \frac{M^2}{M'^2}\right)^{1/2}\frac{M^2}{Q^2 + M'^2}. \quad (4\text{–}1.87)$$

Also required here are $\langle\gamma k/(nk)^{1,2}\rangle$ where γ enters just as an arbitrary constant vector. After integration only the combinations γn and γP appear, and these structures are fixed by the results obtained when γ is replaced by n and by P. The latter are all known and no further integrations are needed. This leads to the evaluations

$$\left\langle\frac{\gamma k}{nk}\right\rangle = \frac{\gamma P}{nP} - \left(\gamma n + \frac{\gamma P}{nP}\right)Z\left(\frac{Q^2}{M^2}\right),$$

$$\frac{M^2 - m^2}{2M^2}\left\langle\frac{\gamma k}{(nk)^2}\right\rangle = -\frac{\gamma n}{nP} + \left(\frac{\gamma P}{M^2} + \frac{\gamma n}{nP}\right)Z\left(\frac{Q^2}{M^2}\right). \quad (4\text{–}1.88)$$

The algebraic reduction of the various terms in (4–1.85) brings about the introduction of the space-like vector

$$Q^\mu = P^\mu + n^\mu nP, \qquad nQ = 0. \quad (4\text{–}1.89)$$

It is appropriately designated since

$$Q^\mu Q_\mu = (nP)^2 - n^2 P^2 = Q^2.$$ (4-1.90)

Alternative presentations of the result are given by

$$\frac{\gamma P}{M^2} + \frac{8M^2}{(M^2 - m^2)^2}(\gamma P - m)\frac{Q^2}{M^2}Z\left(\frac{Q^2}{M^2}\right) + \frac{4}{M^2 - m^2}\gamma Q Z\left(\frac{Q^2}{M^2}\right)$$

$$= \frac{\gamma P + M}{2M}\frac{1}{M}\left[1 + \frac{8}{M^2 - m^2}\frac{M}{M + m}Q^2 Z\right]$$

$$+ \frac{\gamma P - M}{2M}\frac{1}{M}\left[1 + \frac{8}{M^2 - m^2}\frac{M}{M - m}Q^2 Z\right] + \frac{4}{M^2 - m^2}\gamma Q Z.$$ (4-1.91)

On setting $Q = 0$ this reduces to $\gamma P/M^2$, as exhibited in (4-1.35). The modified propagation function that now appears is a mixture of single and double spectral forms:

$$\bar{G}_+(p, q) = \frac{1}{\gamma p + m - i\varepsilon} + \frac{\alpha}{4\pi}\int_{\to m}^{\infty}\frac{dM}{M}\left[\frac{1 - \dfrac{m^2}{M^2} + \dfrac{8}{M(M - m)}q^2 Z\left(\dfrac{q^2}{M^2}\right)}{\gamma p + M - i\varepsilon}\right.$$

$$\left. + \frac{1 - \dfrac{m^2}{M^2} + \dfrac{8}{M(M + m)}q^2 Z\left(\dfrac{q^2}{m^2}\right)}{\gamma p - M + i\varepsilon}\right] - \frac{\alpha}{\pi}\gamma q\int_{m^2}^{\infty}\frac{dM^2}{M^2}\frac{Z(q^2/M^2)}{p^2 + M^2 - i\varepsilon}.$$

(4-1.92)

The infrared singularity is concentrated in one term. When we retain only the portion of the spectral integral for which

$$k^0 = M - m \ll m$$ (4-1.93)

and also introduce the simplifying restriction of (4-1.81, 82), where

$$q^2/m^2 \ll 1: \qquad Z(q^2/m^2) = \tfrac{1}{3},$$ (4-1.94)

the propagation function becomes [the soft photon regime is $k^0 \ll (q^2)^{1/2} \ll m$]

$$\bar{G}_+(p, q) \cong \frac{1}{\gamma p + m - i\varepsilon} + \frac{2\alpha}{3\pi}\mathbf{v}^2\int\frac{dk^0}{k^0}\frac{1}{\gamma p + (m + k^0) - i\varepsilon}.$$ (4-1.95)

As expected, one finds the same differential probability for soft photon emission as in the spin 0 result of Eq. (4-1.83).

The spectral presentation of Eq. (4-1.92) obscures one aspect of the propagation characteristics exhibited by the two-particle excitations, that of parity. The basic spin $\tfrac{1}{2}$ propagation function $(\gamma p + M - i\varepsilon)^{-1}$ differs from $(\gamma p + m - i\varepsilon)^{-1}$

in mass only. Both represent excitations which, in the appropriate rest frame, are characterized by $\gamma^{0\prime} = +1$ and thus have the same parity. The propagation function $(\gamma p - M + i\varepsilon)^{-1}$ describes spin $\frac{1}{2}$ excitations with the opposite parity. This remark explains, incidentally, why terms of the latter type are missing in the restricted propagation function of (4–1.95). There must be flexibility in distinguishing between the description of an electron accompanied by detectable and by undetectable photons, which is possible only when both situations have a common parity. But, the last term of (4–1.92) does not have the parity classification of the preceding terms. We therefore seek another presentation in which the two excitation types are more in evidence.

For this purpose let us return to (4–1.91) and introduce the space-like vector $Q^{\prime \mu}$,

$$Q' = Q + P(Q^2/M^2), \qquad Q'P = 0,$$

$$Q'^2 = Q^2[1 + (Q^2/M^2)]. \tag{4–1.96}$$

After multiplying (4–1.91) by M, to make it dimensionless, it reads

$$-8\frac{mM}{(M^2-m^2)^2}Q^2Z + \left[1 + 4\frac{M^2+m^2}{(M^2-m^2)^2}Q^2Z\right]\frac{\gamma P}{M} + 4\frac{M^2}{M^2-m^2}Z\frac{\gamma Q'}{M}. \tag{4–1.97}$$

Now, γP and $\gamma Q'$ anticommute, since Q' and P are orthogonal, which means that the square of the linear combination of γP and $\gamma Q'$ is a multiple of the unit matrix. We shall call that number N^2, and

$$N^2 = \left[1 + 4\frac{M^2+m^2}{(M^2-m^2)^2}Q^2Z\right]^2 - \left[4\frac{M^2}{M^2-m^2}Z\right]^2\frac{Q^2}{M^2}\left(1 + \frac{Q^2}{M^2}\right)$$

$$= \left(1 + 8\frac{M^2}{(M^2-m^2)^2}Q^2Z\right)\left(1 + 8\frac{m^2}{(M^2-m^2)^2}Q^2Z\right) - 16\frac{M^2}{(M^2-m^2)^2}Q^2Z^2.$$

$$\tag{4–1.98}$$

The weak inequality

$$Z(Q^2/M^2) < \tfrac{1}{2} \tag{4–1.99}$$

suffices to show that

$$N^2 > 1 + \left[8\frac{Mm}{(M^2-m^2)^2}Q^2Z\right]^2. \tag{4–1.100}$$

The $\gamma Q'$ term can be removed by a matrix Lorentz transformation. This is expressed by

$$\left[1 + 4\frac{M^2+m^2}{(M^2-m^2)^2}Q^2Z\right]\frac{\gamma P}{M} + 4\frac{M^2}{M^2-m^2}Z\frac{\gamma Q'}{M} = L^{-1}\left[N\frac{\gamma P}{M}\right]L, \tag{4–1.101}$$

where

$$L^T \gamma^0 L = \gamma^0. \tag{4-1.102}$$

The transformation matrix has the following realization,

$$L = a + (b/M^2)\gamma P \gamma Q', \tag{4-1.103}$$

in which the parameters given by

$$2a^2 = \frac{1}{N}\left[1 + 4\frac{M^2 + m^2}{(M^2 - m^2)^2}Q^2 Z\right] + 1,$$

$$2b^2\frac{Q^2}{M^2}\left(1 + \frac{Q^2}{M^2}\right) = \frac{1}{N}\left[1 + 4\frac{M^2 + m^2}{(M^2 - m^2)^2}Q^2 Z\right] - 1 \tag{4-1.104}$$

obey

$$1 = a^2 - b^2(Q^2/M^2)[1 + (Q^2/M^2)]. \tag{4-1.105}$$

The positiveness of N is implicit in this realization and the product ab must also be positive, which permits us to give positive values to both a and b.

These observations enable us to write (4–1.97) in the form

$$L^{-1}\left[N_+\frac{\gamma P - M}{2M} + N_-\frac{\gamma P + M}{2M}\right]L, \tag{4-1.106}$$

where

$$\tfrac{1}{2}(N_+ + N_-) = N, \qquad \tfrac{1}{2}(N_+ - N_-) = \frac{8mM}{(M^2 - m^2)^2}Q^2 Z, \tag{4-1.107}$$

and the inequality (4–1.100) provides the assurance that both N_+ and N_- are positive quantities. Since $(M \pm \gamma P)/2M$ are projection matrices for γP, one can give (4–1.106) another appearance, based on the structures

$$L = a + b(Q^2/M^2) + (b/M^2)\gamma P \gamma Q,$$

$$L^{-1} = a + b(Q^2/M^2) + (b/M^2)\gamma Q \gamma P. \tag{4-1.108}$$

It is

$$N_+ L_+\frac{\gamma P - M}{2M}L_+ + N_- L_-\frac{\gamma P + M}{2M}L_-, \tag{4-1.109}$$

where

$$L_\pm = a + b[(Q^2/M^2) \mp \gamma Q/M]. \tag{4-1.110}$$

These matrices are such that

$$(\gamma^0 L_\pm)^T = -\gamma^0 L_\mp. \tag{4-1.111}$$

They also obey

$$L_+ \frac{M - \gamma P}{2M} L_+ + L_- \frac{M + \gamma P}{2M} L_- = 1. \tag{4-1.112}$$

The propagation function now emerges in the desired form:

$$\bar{G}_+(p, q) = \frac{1}{\gamma p + m - i\varepsilon} + \int_{\to m}^\infty dM \left[A_+ L_+ \frac{1}{\gamma p + M - i\varepsilon} L_+ \right.$$

$$\left. + A_- L_- \frac{1}{\gamma p - M + i\varepsilon} L_- \right], \tag{4-1.113}$$

where

$$A_\pm = \frac{\alpha}{4\pi} \frac{1}{M} \left(1 - \frac{m^2}{M^2} \right) N_\pm. \tag{4-1.114}$$

Both the positive numbers $A_\pm$ and the matrices $L_\pm$ are functions of M and of the vector

$$q^\mu = p^\mu + n^\mu n p, \qquad nq = 0, \tag{4-1.115}$$

which are produced by substituting q for Q. The normalization relation (4-1.112) continues to apply with P replaced by the arbitrary p, since it involves only the property

$$pq = q^2. \tag{4-1.116}$$

The F.D. antisymmetry property here reads

$$[\gamma^0 \bar{G}_+(-p, -q)]^T = -\gamma^0 \bar{G}_+(p, q), \tag{4-1.117}$$

where the sign reversal of q is implied by that of p; it is explicitly needed to restore L_+ and L_- which have been interconverted by transposition [Eq. (4-1.111)]. The presence of the additional transformation matrices $L_\pm$ does not alter the interpretation of (4-1.113) as a superposition of excitations having variable mass and parity, with $dM A_\pm$ supplying relative probability measures. This is also consistent with the soft photon limit where, simplified by $q^2/m^2 \ll 1$,

$$M - m = k^0 \ll (q^2)^{1/2}: \quad N_- \ll N_+ \simeq \frac{4}{3} \frac{q^2}{(k^0)^2}, \tag{4-1.118}$$

and

$$dMA_+ \cong \frac{2\alpha}{3\pi} \frac{q^2}{m^2} \frac{dk^0}{k^0} \tag{4-1.119}$$

restates the soft photon emission probability.

4-2 A MAGNETIC MOMENT CALCULATION

In the course of the preceding section we have introduced a method of calculation in which fields play the role of sources. It forced us to use explicitly the physical requirements that accompany the consideration of multi-particle exchange processes—that the initial phenomenological description of single particle processes must not be altered. It is not too soon to recognize that this requirement refers to the behavior of free particles. When motion in applied electromagnetic fields is considered, deviations from the primitive electromagnetic interaction can appear. The reason is that the primitive electromagnetic interaction refers specifically to an elementary process such that no subsequent interaction takes place. When, in different experimental arrangements, interactions do occur, they imply modifications in the effective electromagnetic coupling. This will be made explicit in later considerations. But even now we can draw an experimentally significant consequence of this fact by a simple modification in the calculation of Section 4–1.

First let us consider some features of the spin $\frac{1}{2}$ propagation function $G_+^A(x, x')$ that refer to a weak, homogeneous electromagnetic field. When we use the kind of matrix notation that was introduced in Section 3–12, the defining differential equation [Eq. (3–12.2)] appears as

$$(\gamma \Pi + m)G_+^A = 1, \tag{4-2.1}$$

where

$$\Pi = p - eqA. \tag{4-2.2}$$

Now we shall write

$$G_+^A = (m - \gamma \Pi)\Delta_+^A \tag{4-2.3}$$

and deduce that

$$[-(\gamma \Pi)^2 + m^2]\Delta_+^A = [\Pi^2 - eq\sigma F + m^2]\Delta_+^A = 1. \tag{4-2.4}$$

This involves the algebraic properties

$$-(\gamma \Pi)^2 = -\tfrac{1}{2}\{\gamma^\mu, \gamma^\nu\}\Pi_\mu \Pi_\nu - \tfrac{1}{2}[\gamma^\mu, \gamma^\nu]\tfrac{1}{2}[\Pi_\mu, \Pi_\nu]$$

$$= g^{\mu\nu}\Pi_\mu \Pi_\nu + i\sigma^{\mu\nu}\tfrac{1}{2}ieqF_{\mu\nu}, \tag{4-2.5}$$

where

$$[\Pi_\mu, \Pi_\nu] = [p_\mu, - eqA_\nu] - [p_\nu, - eqA_\mu] = ieqF_{\mu\nu}, \qquad (4\text{-}2.6)$$

and we have introduced the simplified notation (note the factor of $\frac{1}{2}$)

$$\sigma F = \tfrac{1}{2}\sigma^{\mu\nu}F_{\mu\nu}. \qquad (4\text{-}2.7)$$

The matrix viewpoint also makes clear that the construction of (4–2.3) can be presented as

$$G_+^A = \Delta_+^A(m - \gamma\Pi), \qquad (4\text{-}2.8)$$

for Δ_+^A is a function of the matrix $\gamma\Pi$ only.

The defining equation for Δ_+^A is a familiar one. It differs from the spin 0 Green's function equation of (3–11.40) merely in the appearance of the $- eq\sigma F$ term. And, if $F_{\mu\nu}$ is a homogeneous field, the additional term is a constant matrix that can be grouped with m^2. We also know that after the phase transformation of (3–11.37) is performed, the restriction to a homogeneous field [Eq. (3–11.46)] that is weak, permitting the neglect of quadratic field terms in (3–11.52), reduces the transformed propagation function to the free particle form. Thus, the construction of Δ_+^A for a weak, homogeneous electromagnetic field is given by

$$\Delta_+^A(x, x') = \exp[ieq\varphi(x, x')]\Delta_+(x - x', m^2 - eq\sigma F) \qquad (4\text{-}2.9)$$

with

$$\varphi(x, x') = \int_{x'}^x d\xi^\mu A_\mu(\xi) \qquad (4\text{-}2.10)$$

denoting a straight line path integral.

Turning to G_+^A, we encounter

$$(- i\partial_\mu - eqA_\mu(x)) \exp[ieq\varphi(x, x')] = \exp[ieq\varphi(x, x')](- i\partial_\mu + eq\tfrac{1}{2}F_{\mu\nu}(x - x')^\nu),$$
$$(4\text{-}2.11)$$

which expresses the gauge transformation that produces the vector potential (3–11.46). Another such relation is obtained by interchanging x and x' while reversing the sign of $F_{\mu\nu}$:

$$\exp[ieq\varphi(x, x')](- i\partial_\mu'^T - eqA_\mu(x'))$$
$$= (- i\partial_\mu'^T + eq\tfrac{1}{2}F_{\mu\nu}(x' - x)^\nu) \exp[ieq\varphi(x, x')], \qquad (4\text{-}2.12)$$

where we write ∂^T as an indicator of differentiation to the left with an associated minus sign. If we use the average of the two constructions given in (4–2.3) and (4–2.8), the $\gamma^\mu F_{\mu\nu}(x - x')^\nu$ term, appearing with opposite signs in the applications of (4–2.11) and (4–2.12), will fail to cancel only because of noncommutativity with σF. But this contribution is quadratic in the field strength and will be omitted. Accordingly, we get

$$G_+^A(x, x') = \exp[ieq\varphi(x, x')] \int \frac{(dp)}{(2\pi)^4} \exp[ip(x - x')] (m - \gamma p) \frac{1}{p^2 + m^2 - eq\sigma F - i\varepsilon},$$

(4–2.13)

where symmetrization in multiplication of the two matrix factors is understood.

We now return to the two-particle exchange coupling of (4–1.43) and ask how it is modified by the presence of a weak, homogeneous electromagnetic field. The change in physical circumstances is conveyed by the introduction of the charged particle propagation function that we have just constructed. As in (4–1.47), we shall omit reference to photons emitted or absorbed directly by the sources, which gives the vacuum amplitude coupling term

$$e^2 \int (dx)(dx')\psi_1(x)\gamma^0\gamma^\mu D_+(x - x')G_+^A(x, x')\gamma_\mu\psi_2(x').$$

(4–2.14)

Using an arrangement that is more convenient for our present purposes, we summarize some significant aspects of the earlier calculation in

$$x^0 > x^{0'}: \quad D_+(x - x')G_+(x - x') = -\int dM^2\, d\omega_P \exp[iP(x - x')]F(P),$$

(4–2.15)

with

$$F(P) = \frac{1}{(4\pi)^2}\left(1 - \frac{m^2}{M^2}\right)\left(m - \frac{M^2 + m^2}{2M^2}\gamma P\right),$$

(4–2.16)

and its space-time extrapolation

$$D_+(x - x')G_+(x - x')$$

$$= \frac{i}{(4\pi)^2} \int_{m^2}^{\to\infty} dM^2 \left(1 - \frac{m^2}{M^2}\right)\left(m - \frac{M^2 + m^2}{2M^2}\gamma\left(\frac{1}{i}\right)\partial\right)\Delta_+(x - x', M^2)$$

$$+ \text{ contact terms}.$$

(4–2.17)

The structure of (4–2.13) shows that the introduction of G_+^A is produced by supplying the phase factor

$$\Phi = \exp[ieq\varphi(x, x')],$$

(4–2.18)

replacing m^2 by $m^2 - eq\sigma F$, and thereby M^2 by $M^2 - eq\sigma F$, while positioning γ^μ at the extremities of all products in a symmetrical way:

$$D_+(x - x')G_+^A(x, x') = \Phi \frac{i}{(4\pi)^2} \int_{m^2}^{\to \infty} dM^2 \frac{M^2 - m^2}{M^2 - eq\sigma F}$$

$$\times \left[m - \left(1 - \frac{1}{2} \frac{M^2 - m^2}{M^2 - eq\sigma F}\right) \gamma\left(\frac{1}{i}\right) \partial \right] \Delta_+(x - x', M^2 - eq\sigma F)$$

$$+ \text{ contact terms.} \tag{4–2.19}$$

Note that we have not troubled to isolate the γ matrices; the sense of their multiplication should be kept in mind.

Since attention is restricted to weak fields, one can replace (4–2.19) by the equivalent version

$$D_+(x - x')G_+^A(x, x')$$

$$= \Phi \frac{i}{(4\pi)^2} \int_{m^2}^{\to \infty} dM^2 \left(1 - \frac{m^2}{M^2}\right) \left(m - \frac{M^2 + m^2}{2M^2} \gamma\left(\frac{1}{i}\right) \partial\right) \Delta_+(x - x', M^2 - eq\sigma F)$$

$$+ \Phi \frac{i}{(4\pi)^2} \int_{m^2}^{\to \infty} \frac{dM^2}{M^2} \left(1 - \frac{m^2}{M^2}\right) eq\sigma F \left(m - \frac{m^2}{M^2} \gamma\left(\frac{1}{i}\right) \partial\right) \Delta_+(x - x', M^2)$$

$$+ \text{ contact terms.} \tag{4–2.20}$$

Before we can apply the physical considerations that fix the contact terms, the calculation must be completed by supplying the γ^μ factors that appear in (4–2.14). A basic property of the Dirac matrices is the following,

$$\gamma^\mu \sigma_{\lambda\nu} \gamma_\mu = 0, \tag{4–2.21}$$

for two of the γ matrices commute and two anticommute with any σ matrix. This also implies that

$$\gamma^\mu \Delta_+(x - x', M^2 - eq\sigma F) \gamma_\mu = -4\Delta_+(x - x', M^2), \tag{4–2.22}$$

in view of the limitation to linear field effects. The other needed combinations are

$$\gamma^\mu \gamma \partial \sigma F \gamma_\mu = -2\gamma \partial \sigma F \tag{4–2.23}$$

and

$$\gamma^\mu \gamma \partial \Delta_+(x - x', M^2 - eq\sigma F) \gamma_\mu = -2\gamma \partial \Delta_+(x - x', M^2 - eq\sigma F)$$

$$+ 4\gamma \partial \Delta_+(x - x', M^2). \tag{4–2.24}$$

We shall also want to restore the $M^2 - eq\sigma F$ combination, in accordance with

$$\Delta_+(x - x', M^2) = \Delta_+(x - x', M^2 - eq\sigma F) + eq\sigma F \frac{d}{dM^2} \Delta_+(x - x', M^2), \tag{4–2.25}$$

which can be followed by partial integration on the variable M^2; there are no contributions at the integration boundaries, m^2, and infinity. When this has been done we reinstate the propagation function (4–2.9), now referring to the variable mass M^2:

$$\gamma^\mu D_+(x - x') G_+^A(x, x') \gamma_\mu$$

$$= -\frac{i}{(4\pi)^2} \int_{m^2}^{\to\infty} dM^2 \left(1 - \frac{m^2}{M^2}\right)\left(4m + \frac{M^2 + m^2}{M^2}\gamma\Pi\right) \Delta_+^A(x, x', M^2)$$

$$+ \frac{i}{(4\pi)^2} \int_{m^2}^\infty \frac{dM^2}{M^2} \frac{2m^2}{M^2}\left(2m + \frac{M^2 + m^2}{M^2}\gamma\Pi\right) eq\sigma F \Delta_+^A(x, x', M^2)$$

$$+ \text{ contact terms.} \tag{4–2.26}$$

The formal matrix construction

$$\Delta_+^A(M^2) = \frac{1}{-(\gamma\Pi)^2 + M^2 - i\varepsilon} \tag{4–2.27}$$

enables us to present (4–2.26) as

$$\frac{i}{(4\pi)^2} \int_m^{\to\infty} \frac{dM}{M}\left(1 - \frac{m^2}{M^2}\right)\left[\frac{(M - m)^2 - 2mM}{\gamma\Pi + M - i\varepsilon} + \frac{(M + m)^2 + 2mM}{\gamma\Pi - M + i\varepsilon}\right]$$

$$- \frac{i}{(4\pi)^2} \int_m^\infty \frac{dM}{M} \frac{2m^2}{M^2} eq\sigma F \left[\frac{(1 - m/M)^2}{\gamma\Pi + M - i\varepsilon} + \frac{(1 + m/M)^2}{\gamma\Pi - M + i\varepsilon}\right]$$

$$+ \text{ contact terms,} \tag{4–2.28}$$

with symmetrized matrix multiplication still understood. The physical requirements that determine the contact terms apply in the absence of the electromagnetic field. They are expressed by the substitutions of (4–1.59, 60) although we retain the gauge covariant symbol Π, which can be identified as p in any fixed gauge. The result is to correct the first term of (4–2.28):

$$\frac{i}{(4\pi)^2}(\gamma\Pi + m)^2 \int_{\to m}^\infty \frac{dM}{M}\left(1 - \frac{m^2}{M^2}\right)\left[\frac{1 - \dfrac{2mM}{(M - m)^2}}{\gamma\Pi + M - i\varepsilon} + \frac{1 + \dfrac{2mM}{(M + m)^2}}{\gamma\Pi - M + i\varepsilon}\right]. \tag{4–2.29}$$

So much, and no more, is required by the phenomenological constraints. The additional action term that is now obtained, replacing (4–1.55), is

$$-\tfrac{1}{2} \int (dx)(dx')\psi(x)\gamma^0 M(x, x', F)\psi(x') \tag{4–2.30}$$

with

$$M(F) = -(\gamma\Pi + m)^2 \frac{\alpha}{4\pi} \int_{\to m}^{\infty} \frac{dM}{M}\left(1 - \frac{m^2}{M^2}\right)\left[\frac{1 - \dfrac{2mM}{(M-m)^2}}{\gamma\Pi + M - i\varepsilon} + \frac{1 + \dfrac{2mM}{(M+m)^2}}{\gamma\Pi - M + i\varepsilon}\right]$$

$$+ \frac{\alpha}{2\pi}\int_m^{\infty}\frac{dM}{M}\frac{m^2}{M^2}eq\sigma F\left[\frac{(1 - m/M)^2}{\gamma\Pi + M - i\varepsilon} + \frac{(1 + m/M)^2}{\gamma\Pi - M + i\varepsilon}\right]. \tag{4-2.31}$$

This supplements the initial action expression

$$-\tfrac{1}{2}\int (dx)\psi(x)\gamma^0(\gamma\Pi + m)\psi(x). \tag{4-2.32}$$

Let us confine our attention to the motion of a particle far from its source, where, in the absence of the weak electromagnetic field, ψ obeys

$$(\gamma\Pi + m)\psi = 0. \tag{4-2.33}$$

The weak field limitation justifies the use of (4-2.33) in simplifying (4-2.30, 31), which leads to

$$M(F) \to -\frac{\alpha}{2\pi}\frac{1}{2m}eq\sigma F\int_m^{\infty}\frac{dM}{M}\frac{m^2}{M^2}\left(\frac{2m}{M}\right)^2 = -\frac{\alpha}{2\pi}\frac{1}{2m}eq\sigma F. \tag{4-2.34}$$

The effective action under these circumstances, combining (4-2.32) and (4-2.30, 34):

$$-\frac{1}{2}\int (dx)\psi(x)\gamma^0\left(\gamma\Pi + m - \frac{\alpha}{2\pi}\frac{eq}{2m}\sigma F\right)\psi(x), \tag{4-2.35}$$

identifies the additional spin magnetic moment of $\alpha/2\pi$ magnetons. In terms of the g factor introduced in (3-10.63), we have

$$g = 2\left(1 + \frac{\alpha}{2\pi}\right). \tag{4-2.36}$$

The fine structure constant value

$$\alpha = 1/137.036 \tag{4-2.37}$$

gives

$$\alpha/2\pi = 0.00116141 \tag{4-2.38}$$

and values of $\tfrac{1}{2}g$ that are in remarkable accord with those measured for the electron [1.0011596] and the muon [1.001166]. The tiny discrepancies, which are real, are reasonably assigned to the as yet unconsidered exchange processes involving more than two particles. Our conclusion concerning the electron and the muon is that the g factor of the primitive interaction does not equal the measured value, but rather, $g_{\text{prim.}} = 2$, exactly.

The inference that the small deviation of $\frac{1}{2}g$ from unity is dynamical in origin, and should not be assigned to the local primitive interaction, is reinforced by the following additional consideration. The action expression (4–2.30, 32) relates the field ψ to its source by means of a modified propagation function,

$$\psi = \bar{G}_+(F)\eta, \tag{4–2.39}$$

where

$$[\gamma\Pi + m + M(F)]\bar{G}_+(F) = 1 \tag{4–2.40}$$

continues the use of space-time coordinate matrix notation. In this more general form, the electromagnetic field dependent part of $M(F)$, the second term of (4–2.31), is nonlocal since it is constructed from the propagation functions $[\gamma\Pi \pm$ $\pm (M - i\varepsilon)]^{-1}$. When this distributed magnetic moment interaction is examined over very short time intervals, as represented by the complementary limit $\gamma\Pi \to$ $\to \infty$, it becomes asymptotically

$$\frac{3\alpha}{4\pi} eq\sigma F \frac{1}{\gamma\Pi}, \tag{4–2.41}$$

and thus vanishes without a residual local interaction.

There is another way of presenting the magnetic moment calculation which is still more phenomenological; the initial electromagnetic interaction now contains a deviation of $\frac{1}{2}g$ from unity as observed. The additional hypothesis that the extra spin magnetic moment is a dynamical effect, which is suppressed over very short time intervals, then determines it. The modified Green's function equation (4–2.40) now contains, added to m, the magnetic moment interaction

$$- (\tfrac{1}{2}g - 1) \frac{eq}{2m} \sigma F. \tag{4–2.42}$$

The dynamical term $M(F)$ is certainly altered by the changed electromagnetic properties, but if one only retains effects of order $\alpha/2\pi$, such changes can be neglected as of higher order. This time, however, we must impose a phenomenological normalization requirement on $M(F)$, according to the assumption that g is the observed g factor. One must supplement $M(F)$ by a contact term designed to remove the electromagnetic field dependence under the conditions symbolized by $\gamma\Pi + m = 0$. The necessary modifications are:

$$\frac{1}{\gamma\Pi \pm (M - i\varepsilon)} \to \frac{1}{\gamma\Pi \pm (M - i\varepsilon)} \mp \frac{1}{M \mp m} = \mp \frac{\gamma\Pi + m}{M \mp m} \frac{1}{\gamma\Pi \pm (M - i\varepsilon)}.$$

$$\tag{4–2.43}$$

Thus the complete magnetic moment term becomes

$$-\frac{eq}{2m}\sigma F \left\{ \tfrac{1}{2}g - 1 + (\gamma\Pi + m)\frac{\alpha}{2\pi}\int_m^\infty \frac{dM}{M}\left(\frac{m}{M}\right)^2 \frac{2m}{M} \right.$$
$$\left. \times \left[\frac{1-(m/M)}{\gamma\Pi + M - i\varepsilon} - \frac{1+(m/M)}{\gamma\Pi - M + i\varepsilon} \right] \right\}. \tag{4-2.44}$$

The asymptotic limit of the factor in braces, for very short time intervals $(\gamma\Pi \to \infty)$, is

$$\tfrac{1}{2}g - 1 - \frac{\alpha}{2\pi}. \tag{4-2.45}$$

If this is required to vanish, as the sign of a dynamical origin for $\tfrac{1}{2}g - 1$, the successful result (4-2.36) is recovered. The viewpoint just discussed has the advantage of suggesting that some of the small corrections of order $(\alpha/2\pi)^2$ are aspects of self-consistency; the g value is computed in terms of dynamical processes that are characterized in part by the value of g. But we shall not attempt such an improved calculation now.

One point demands a comment, however. No mention has been made of photon emission or absorption directly from the sources, although we know such processes to be important for the physical consistency of the description. The reason is, of course, that the magnetic moment calculation is unaffected by these processes, with their elements of arbitrariness. In the latter, one or both fields ψ is replaced by the source η, or effectively $(\gamma\Pi + m)\psi$. Thus, such effects do not contribute in the physical circumstances of a magnetic moment measurement, as expressed by the condition $\gamma\Pi + m = 0$. They will alter the detailed structure of $G_+(F)$, however, although consistency demands that the short time behavior of the magnetic interaction remain qualitatively as before.

4-3 PHOTON PROPAGATION FUNCTION

The primitive electromagnetic interaction for spin $\tfrac{1}{2}$ charged particles identifies the electromagnetic vector potential as an effective extended source for the two-particle emission or absorption of oppositely charged particles, provided the mass threshold at $2m$ is exceeded,

$$- k^2 = M^2 > (2m)^2. \tag{4-3.1}$$

Since the physical context is that of noninteracting particles, we compare the vacuum amplitude for the elementary process:

$$i \int (dx)\tfrac{1}{2}\psi(x)\gamma^0 eq\gamma^\mu \psi(x) A_\mu(x), \tag{4-3.2}$$

with the vacuum amplitude that refers to the noninteracting propagation of two particles:

$$\tfrac{1}{2}\left[i\int(dx)\psi(x)\gamma^0\eta(x)\right]^2 = -\tfrac{1}{2}\int(dx)(dx')\psi(x)\gamma^0\eta(x)\eta(x')\gamma^0\psi(x'). \quad (4\text{-}3.3)$$

The latter is the quadratic term in the expansion of

$$\exp\left[i\int(dx)(dx')\eta_1(x)\gamma^0 G_+(x-x')\eta_2(x')\right], \quad (4\text{-}3.4)$$

with $\eta(x)$ designated as the source of interest in emission or absorption while $\psi(x)$ is the associated particle field. The comparison of (4-3.2) and (4-3.3) supplies the matrix

$$i\eta(x)\eta(x')\big|_{\text{eff.}} = \delta(x-x')eq\gamma^\mu\gamma^0 A_\mu(x). \quad (4\text{-}3.5)$$

Note that the antisymmetry of the left side for an interchange of all indices, expressing F.D. statistics, is matched on the right-hand side, specifically by the antisymmetry of the charge matrix.

The coupling between two-particle emission and absorption sources is described by the quadratic term of (4-3.4), first written as

$$\tfrac{1}{2}\left[i\int(dx)(dx')\eta_1(x)\gamma^0 G_+(x-x')\eta_2(x')\right]^2$$

$$= \tfrac{1}{2}\int(dx)\cdots(dx''')i\eta_2(x''')\gamma^0 G_+(x'''-x'')i\eta_1(x'')\eta_1(x)\gamma^0 G_+(x-x')\eta_2(x'),$$

$$(4\text{-}3.6)$$

and then rearranged entirely in matrix form:

$$-\tfrac{1}{2}\int(dx)\cdots(dx''')\text{tr}[i\eta_1(x'')\eta_1(x)\gamma^0 G_+(x-x')i\eta_2(x')\eta_2(x''')\gamma^0 G_+(x'''-x'')].$$

$$(4\text{-}3.7)$$

This uses the possibility of writing

$$\eta(x)M\eta(x'') = \eta_a(x)M_{ab}\eta_b(x'')$$

$$= -M_{ab}\eta_b(x'')\eta_a(x) = -\text{tr}[M\eta(x'')\eta(x)] \quad (4\text{-}3.8)$$

where, as in the situation of interest, M commutes with the anticommuting sources. Further rearrangements are made in accordance with the cyclic property of the trace. The introduction of the two effective sources, related through the vector potentials $A^\mu_{1,2}$ to the extended photon sources $J^\mu_{1,2}$, then gives the vacuum amplitude representing two-particle exchange between extended photon sources:

$$- \tfrac{1}{2} \int (dx)(dx') \, \mathrm{tr}[eq\gamma A_1(x)G_+(x - x')eq\gamma A_2(x')G_+(x' - x)]$$

$$= - \tfrac{1}{2} \int (dx)(dx') A_1^\mu(x) \, \mathrm{tr}[eq\gamma_\mu G_+(x - x')eq\gamma_\nu G_+(x' - x)]A_2^\nu(x'). \quad (4\text{--}3.9)$$

Supplying the appropriate causal forms of the propagation functions converts this into

$$- \tfrac{1}{2} \int d\omega_p \, d\omega_{p'} A_1^\mu(- k) \, \mathrm{tr}[eq\gamma_\mu(m - \gamma p)eq\gamma_\nu(- m - \gamma p')]A_2^\nu(k), \quad (4\text{--}3.10)$$

where

$$k = p + p' \quad (4\text{--}3.11)$$

is the momentum exchanged between the extended photon sources. With the insertion of the unit factor

$$1 = \int d\omega_k \, dM^2 (2\pi)^3 \delta \, (p + p' - k), \quad (4\text{--}3.12)$$

the vacuum amplitude coupling term becomes

$$- e^2 \int dM^2 \, d\omega_k \, A_1^\mu(- k) I_{\mu\nu}(k) A_2^\nu(k), \quad (4\text{--}3.13)$$

in which

$$I_{\mu\nu}(k) = \int d\omega_p \, d\omega_{p'} (2\pi)^3 \delta(p + p' - k) \, \mathrm{tr}[\gamma_\mu(m - \gamma p)\gamma_\nu(- m - \gamma p')]. \quad (4\text{--}3.14)$$

Note carefully that the trace of (4–3.14) refers only to the four-dimensional space of the Dirac matrices. The additional multiplicity of 2 associated with the charge space has been made explicit.

The tensor $I_{\mu\nu}(k)$ has two important properties. It is symmetrical in μ and ν, and it obeys

$$k^\mu I_{\mu\nu}(k) = 0. \quad (4\text{--}3.15)$$

Concerning the symmetry of the tensor, we observe that

$$\mathrm{tr}[\gamma_\mu(m - \gamma p)\gamma_\nu(- m - \gamma p')] = - \, \mathrm{tr}[\gamma_5\gamma_\nu(- m - \gamma p')\gamma_\mu(m - \gamma p)\gamma_5]$$

$$= \mathrm{tr}[\gamma_\nu(m - \gamma p')\gamma_\mu(- m - \gamma p)]; \quad (4\text{--}3.16)$$

the interchange of p and p' in the integral completes the verification:

$$I_{\mu\nu}(k) = I_{\nu\mu}(k). \quad (4\text{--}3.17)$$

To confirm (4–3.15), we note the relation

$$\gamma k = (\gamma p + m) + (\gamma p' - m), \tag{4-3.18}$$

and that each term is annulled in the trace, by the factors $m - \gamma p$ and $- m - \gamma p'$, respectively. These properties are necessary on physical grounds. In particular, (4–3.15) expresses the gauge invariance of the coupling. The explicit tensor form that satisfies these requirements,

$$I_{\mu\nu}(k) = \left(g_{\mu\nu} + \frac{k_\mu k_\nu}{M^2}\right) I(M^2), \tag{4-3.19}$$

is a reminder that a massive excitation emitted by a vector source acts like a unit spin particle. The scalar function $I(M^2)$ can be computed from the trace of $I_{\mu\nu}(k)$,

$$3I(M^2) = g^{\mu\nu} I_{\mu\nu}(k)$$

$$= \int d\omega_p \, d\omega_{p'} \cdot (2\pi)^3 \delta(p + p' - k) \, \mathrm{tr}[\gamma^\mu(m - \gamma p)\gamma_\mu(- m - \gamma p')], \tag{4-3.20}$$

where the Dirac matrix trace is simply

$$\mathrm{tr}[(- 4m - 2\gamma p)(- m - \gamma p')] = 4(4m^2 - 2pp')$$

$$= 4(M^2 + 2m^2). \tag{4-3.21}$$

The remaining integration in (4–3.20) is just (4–1.24), and

$$I(M^2) = \frac{4}{3} \frac{1}{(4\pi)^2} (M^2 + 2m^2)\left(1 - \frac{4m^2}{M^2}\right)^{1/2}. \tag{4-3.22}$$

In view of the gauge invariance of the coupling, the vector potential can be chosen as

$$A^\mu(k) = \frac{1}{k^2} J^\mu(k) = - \frac{1}{M^2} J^\mu(k), \tag{4-3.23}$$

and the continued requirement of vanishing source divergence,

$$k_\mu J^\mu(k) = 0, \tag{4-3.24}$$

enables the vacuum amplitude to be presented as follows:

$$i\frac{\alpha}{3\pi} \int \frac{dM^2}{M^2}\left(1 + \frac{2m^2}{M^2}\right)\left(1 - \frac{4m^2}{M^2}\right)^{1/2} J_1^\mu(-k) i \, d\omega_k J_{2\mu}(k). \tag{4-3.25}$$

We recognize, in

$$\int J_1^\mu(-k) i\, d\omega_k J_{2\mu}(k) = \int (dx)(dx') J_1^\mu(x) \left[i \int d\omega_k \exp[ik(x-x')] \right] J_{2\mu}(x'), \quad (4\text{–}3.26)$$

the description of the causal exchange of an excitation with mass M between the sources. The introduction of the propagation function $\Delta_+(x-x', M^2)$ produces the necessary space-time extrapolation. Adding this coupling to the one representing photon exchange, we get the modified photon propagation function

$$\bar{D}_+(x-x') = D_+(x-x') + \frac{\alpha}{3\pi} \int_{(2m)^2}^\infty \frac{dM^2}{M^2} \left(1 + \frac{2m^2}{M^2}\right)\left(1 - \frac{4m^2}{M^2}\right)^{1/2} \Delta_+(x-x', M^2).$$

$$(4\text{–}3.27)$$

The momentum space transcription is

$$\bar{D}_+(k) = \frac{1}{k^2 - i\varepsilon} + \frac{\alpha}{3\pi} \int_{(2m)^2}^\infty \frac{dM^2}{M^2} \left(1 + \frac{2m^2}{M^2}\right)\left(1 - \frac{4m^2}{M^2}\right)^{1/2} \frac{1}{k^2 + M^2 - i\varepsilon}.$$

$$(4\text{–}3.28)$$

Since the M^2 integral is quite convergent, it has been extended from the threshold at $(2m)^2$ up to infinity.

We have performed this calculation for spin $\frac{1}{2}$ particles since the important physical application refers to the charged particle of smallest mass, the electron. Nevertheless, it is interesting to see a parallel calculation for charged spin 0 particles. Note that the various kinds of charged particles give additive contributions to $\bar{D}_+(k)$ when just two-particle exchange is considered. The spin 0 primitive interaction supplies the vacuum amplitude

$$i \int (dx)\phi(x) eq(1/i)\partial^\mu \phi(x) A_\mu(x), \quad (4\text{–}3.29)$$

which is to be compared with

$$\tfrac{1}{2} \left[i \int (dx) K(x)\phi(x) \right]^2 = -\tfrac{1}{2} \int (dx)(dx')\phi(x) K(x) K(x')\phi(x'). \quad (4\text{–}3.30)$$

That gives

$$iK(x)K(x')|_{\text{eff.}} = eq(A^\mu(x) + A^\mu(x'))(1/i)\partial_\mu \delta(x-x'), \quad (4\text{–}3.31)$$

where the B.E. symmetry of the left side matches that of the right side. The description of two-particle exchange is contained in

$$\tfrac{1}{2} \left[i \int (dx)(dx') K_1(x)\Delta_+(x-x') K_2(x') \right]^2$$

$$= \tfrac{1}{2} \int (dx) \cdots (dx''') \; \mathrm{tr}[iK_1(x'')K_1(x)\Delta_+(x-x')iK_2(x')K_2(x''')\Delta_+(x'''-x'')],$$

$$(4\text{-}3.32)$$

where the trace refers to charge space, and the effective sources of (4–3.31) are to be inserted. For simplicity of writing, we present this coupling in the momentum space form appropriate to the causal situation, using the momentum equivalents of (4–3.31):

$$iK(p)K(p')\big|_{\mathrm{eff.}} = iK(-p')K(-p)\big|_{\mathrm{eff.}}$$

$$= eq(p-p')^\mu A_\mu(k), \qquad (4\text{-}3.33)$$

which make explicit the reference to oppositely charged particles. The vacuum amplitude is

$$-\tfrac{1}{2}\int d\omega_p \, d\omega_{p'} A_1^\mu(-k) \, \mathrm{tr}[eq(p-p')_\mu eq(p-p')_\nu] A_2^\nu(k)$$

$$= -e^2 \int dM^2 \, d\omega_k \, A_1^\mu(-k) I_{\mu\nu}(k) A_2^\nu(k), \qquad (4\text{-}3.34)$$

where, now,

$$I_{\mu\nu}(k) = \int d\omega_p \, d\omega_{p'} (2\pi)^3 \delta(p+p'-k)(p-p')_\mu (p-p')_\nu \qquad (4\text{-}3.35)$$

is evidently symmetrical in μ and ν, and has the necessary gauge invariance property (4–3.15) since

$$k(p-p') = p^2 - p'^2 = 0. \qquad (4\text{-}3.36)$$

The scalar function of (4–3.19) is evaluated as

$$I(M^2) = \frac{1}{3}\frac{1}{(4\pi)^2} M^2 \left(1 - \frac{4m^2}{M^2}\right)^{3/2}, \qquad (4\text{-}3.37)$$

since

$$(p-p')^2 = M^2 - 4m^2. \qquad (4\text{-}3.38)$$

The modified photon propagation function, in which only spin 0 charged particles are taken into account, is, therefore,

$$D_+(k) = \frac{1}{k^2 - i\varepsilon} + \frac{\alpha}{12\pi}\int_{(2m)^2}^{\infty} \frac{dM^2}{M^2}\left(1 - \frac{4m^2}{M^2}\right)^{3/2}\frac{1}{k^2 + M^2 - i\varepsilon}. \qquad (4\text{-}3.39)$$

Several direct uses of the modified photon propagation function can be made. The first is a dynamical application of the vacuum amplitude

$$\langle 0_+|0_-\rangle^J = \exp\left[i\tfrac{1}{2}\int (dx)(dx')J^\mu(x)D_+(x-x')J_\mu(x')\right], \qquad (4\text{–}3.40)$$

yielding the probability that the vacuum is disturbed, as the complement of the vacuum persistence probability. For a weak extended source this gives the probability of single pair emission by the source:

$$|\langle 0_+|0_-\rangle^J|^2 = 1 - \int \frac{(dk)}{(2\pi)^4}J^\mu(k)^* \,\mathrm{Im}\,D_+(k)J_\mu(k), \qquad (4\text{–}3.41)$$

where, using the spin $\tfrac{1}{2}$ structure,

$$-k^2 = M^2 > (2m)^2: \quad \mathrm{Im}\,D_+(k) = \frac{\alpha}{3}\frac{1}{M^2}\left(1+\frac{2m^2}{M^2}\right)\left(1-\frac{4m^2}{M^2}\right)^{1/2} \qquad (4\text{–}3.42)$$

The pair emission probability thus obtained,

$$\frac{\alpha}{3}\int \frac{(dk)}{(2\pi)^4}\frac{1}{M^2}\left(1+\frac{2m^2}{M^2}\right)\left(1-\frac{4m^2}{M^2}\right)^{1/2} J^\mu(k)^* J_\mu(k), \qquad (4\text{–}3.43)$$

is also the result produced by direct calculation from (4–3.2).

The quadratic source combination that appears here is positive, incidentally, as one sees from

$$J^\mu(k)^* g_{\mu\nu}J^\nu(k) = J^\mu(k)^*(g_{\mu\nu} + (1/M^2)k_\mu k_\nu)J^\nu(k)$$
$$= \sum_\lambda |e_{k\lambda}^{\mu*} J_\mu(k)|^2, \qquad (4\text{–}3.44)$$

which also indicates that the two particles have unit angular momentum in the k rest frame. That gives some insight into the difference between the spin $\tfrac{1}{2}$ and spin 0 results. For spin $\tfrac{1}{2}$ particles created nearly at rest, the unit angular momentum can be realized with zero orbital angular momentum, in a ^3S state. The transition probability near threshold will vary as the relative velocity of the particles, $\sim (M - 2m)^{1/2}$. With spinless particles, on the other hand, the unit angular momentum forces the particles into the P state of orbital angular momentum, and the threshold behavior contains two additional powers of relative momentum, $\sim (M - 2m)^{3/2}$.

The introduction of the modified photon propagation function implies a change in the interaction between static charges—the Coulomb potential. Replacing it is [Eq. (2–3.92)]

$$\mathcal{D}(\mathbf{x}) = \int_{-\infty}^{\infty} dx^0\, D_+(x)$$

$$= \frac{1}{4\pi|\mathbf{x}|} + \frac{\alpha}{3\pi}\int_{(2m)^2}^{\infty} \frac{dM^2}{M^2}\left(1+\frac{2m^2}{M^2}\right)\left(1-\frac{4m^2}{M^2}\right)^{1/2}\frac{\exp(-M|\mathbf{x}|)}{4\pi|\mathbf{x}|}$$

$$= \frac{1}{4\pi|\mathbf{x}|}\left[1 + \frac{\alpha}{\pi}\int_0^1 dv\, \frac{v^2(1 - \frac{1}{3}v^2)}{1 - v^2}\exp(-(1 - v^2)^{-1/2}2m|\mathbf{x}|)\right], \quad (4\text{-}3.45)$$

where the last version uses the variable

$$v = \left(1 - \frac{4m^2}{M^2}\right)^{1/2}. \quad (4\text{-}3.46)$$

In writing (4–3.45) we have recognized that time integration reduces the four-dimensional Green's function to a three-dimensional one, specifically of the differential operator $-\nabla^2 + M^2$. Its form can be obtained, for example, from the integral (3–14.35) by placing $p^0 = 0$. At distances such that

$$2m|\mathbf{x}| \gg 1, \quad (4\text{-}3.47)$$

the Coulomb potential is not significantly altered. In the other limit,

$$2m|\mathbf{x}| \ll 1: \quad \bar{\mathcal{D}}(\mathbf{x}) = \frac{1}{4\pi|\mathbf{x}|}\left[1 + \frac{2\alpha}{3\pi}\left(\log\frac{1}{m|\mathbf{x}|} - C - \frac{5}{6}\right)\right], \quad (4\text{-}3.48)$$

in which

$$C = 0.57721\ldots \quad (4\text{-}3.49)$$

is the Eulerian constant. The additional logarithmic behavior under these circumstances comes from the interval of M integration such that

$$|\mathbf{x}|^{-1} \gg M \gg 2m, \quad (4\text{-}3.50)$$

where the integral of (4–3.45) reduces to $2\int dM/M$. The evaluation of (4–3.48) is obtained by partitioning the integral at some value of M that satisfies the inequalities of (4–3.50).

The effect we are discussing, usually referred to as vacuum polarization, increases the strength of the Coulomb interaction with diminishing distance. The increase is quite small, however, at any realizable distance. Thus, with $2m|\mathbf{x}| \sim 10^{-3}$, which represents a distance of roughly 10^{-14} cm when the electron mass is used, it is approximately one percent. In view of the logarithmic dependence on distance, this order of magnitude cannot be changed significantly by any conceivable improvement in experimental prowess; a ten-percent increase in interaction strength requires dropping to a distance $\sim 10^{-37}$ cm! And long before such distances could be approached, the situation would change qualitatively through the growing importance of particles that are heavier than the electron. Despite their smallness, vacuum polarization effects are measurable at the present level of experimental technique. The most elementary situation is that of hydrogenic atoms where the strengthened attraction between electron and nucleus depresses the energy values of zero orbital angular momentum states, these being the ones in which the electron spends appreciable time near the nucleus.

Simple perturbation theory can be applied to the change in interaction energy,

$$\delta V(\mathbf{x}) = -Ze^2 \delta\mathscr{D}(\mathbf{x}), \tag{4-3.51}$$

where $\delta\mathscr{D}(\mathbf{x})$ represents the difference between $\bar{\mathscr{D}}(\mathbf{x})$ and

$$\mathscr{D}(\mathbf{x}) = \frac{1}{4\pi|\mathbf{x}|}. \tag{4-3.52}$$

In a state with nonrelativistic wave function $\psi(\mathbf{x})$, appropriate to the restriction $Z\alpha \ll 1$, we have

$$\delta E = \int (d\mathbf{x})\,\delta V(\mathbf{x})|\psi(\mathbf{x})|^2$$

$$\cong -4\pi Z\alpha|\psi(0)|^2 \int (d\mathbf{x})\,\delta\mathscr{D}(\mathbf{x}), \tag{4-3.53}$$

which uses the fact that the perturbation is significant only over distances that are small compared with atomic dimensions. The integration that appears here is equivalent to evaluating the zero momentum limit of $\delta D_+(k)$, and

$$\int (d\mathbf{x})\,\delta\mathscr{D}(\mathbf{x}) = \frac{\alpha}{3\pi}\int_{(2m)^2}^{\infty} \frac{dM^2}{(M^2)^2}\left(1 + \frac{2m^2}{M^2}\right)\left(1 - \frac{4m^2}{M^2}\right)^{1/2}$$

$$= \frac{\alpha}{\pi}\frac{1}{(2m)^2}\int_0^1 dv\,v^2(1 - \tfrac{1}{3}v^2) = \frac{\alpha}{15\pi}\frac{1}{m^2}. \tag{4-3.54}$$

Only s-states need be considered. For principal quantum number n,

$$|\psi_{ns}(0)|^2 = \frac{1}{\pi}\left(\frac{Z\alpha}{n}m\right)^3 \tag{4-3.55}$$

and

$$\delta E_{ns} = -\frac{4}{15\pi}\frac{Z^4\alpha^5}{n^3}m, \tag{4-3.56}$$

or

$$\frac{\delta E_{ns}}{\left(\dfrac{1}{2}\dfrac{Z^2\alpha^2}{n^2}m\right)} = -\frac{8}{15\pi}\frac{Z^2\alpha^3}{n}, \tag{4-3.57}$$

the latter giving a comparison with the Bohr energy values. More details will not be supplied now since, as will be seen later, this effect is rather minor compared to another that displaces the s-states in the opposite sense. The existence of the vacuum polarization effect must be inferred from the quantitative comparison with experiment; in its absence a small but significant discrepancy with experiment would remain.

It is particularly interesting, then, that observations exist which point directly to the existence of the vacuum polarization phenomena. They are the g-factor measurements of electron and muon, which disclose the small discrepancy

$$\tfrac{1}{2}g_\mu - \tfrac{1}{2}g_e = 0.66 \times 10^{-5}. \tag{4-3.58}$$

The displacement of $\tfrac{1}{2}g$ from unity is produced by the exchange of photons and charged particles, each described by the appropriate propagation function. The substitution of $\bar{D}_+$ for D_+ introduces one aspect of the more complicated exchange processes which, in this example, is the consideration of three-particle exchange. An asymmetry between electron and muon is created through the domination of the vacuum polarization processes in δD_+ by the light electron. The mass of each particle sets the scale for the significant exchange acts. The heavy muon is therefore more influenced by the vacuum polarization phenomena. To the extent that the massless photon and electron-positron pairs of mass $M \ll m_\mu$ are not significantly different kinematically, the modified photon propagation function in the muon application is, approximately,

$$\bar{D}_+(k) \sim \frac{1}{k^2 - i\varepsilon}\left[1 + \frac{2\alpha}{3\pi}\int_{m_e}^{m_\mu}\frac{dM}{M}\right]. \tag{4-3.59}$$

The integration limits indicate the range of M values for which the simple treatment, resembling the discussion of (4-3.48), is applicable. The factor that appears here,

$$1 + \frac{2\alpha}{3\pi}\log\frac{m_\mu}{m_e}, \tag{4-3.60}$$

measures the increase of $\tfrac{1}{2}g_\mu$ relative to $\tfrac{1}{2}g_e$,

$$\tfrac{1}{2}g_\mu - \tfrac{1}{2}g_e \cong \frac{\alpha}{2\pi}\frac{2\alpha}{3\pi}\log\frac{m_\mu}{m_e}$$
$$= (1.16 \times 10^{-3})(0.825 \times 10^{-2})$$
$$\sim 10^{-5}, \tag{4-3.61}$$

in qualitative agreement with (4-3.58). Later in this section we shall give a more precise treatment, which will further support the interpretation of the difference between g_μ and g_e as a vacuum polarization phenomenon.

But first let us return to the vacuum amplitude coupling term (4-3.13, 19),

$$- e^2 \int dM^2 \, I(M^2) \, d\omega_k \, A_1^{\mu}(- k) \left(g_{\mu\nu} + \frac{k_{\mu}k_{\nu}}{M^2} \right) A_2^{\nu}(k), \qquad (4\text{–}3.62)$$

and proceed to make its gauge invariance explicit by introducing field strengths

$$F_{\mu\nu}(k) = ik_{\mu}A_{\nu}(k) - ik_{\nu}A_{\mu}(k), \qquad (4\text{–}3.63)$$

according to the relation

$$- \tfrac{1}{2} F_1^{\mu\nu}(- k) F_{2\mu\nu}(k) = M^2 A_1^{\mu}(- k) \left(g_{\mu\nu} + \frac{k_{\mu}k_{\nu}}{M^2} \right) A_2^{\nu}(k). \qquad (4\text{–}3.64)$$

The resulting form of (4–3.62),

$$ie^2 \int \frac{dM^2}{M^2} \, I(M^2)(- \tfrac{1}{2}) F_1^{\mu\nu}(- k) i \, d\omega_k \, F_{2\mu\nu}(k), \qquad (4\text{–}3.65)$$

leads directly to a space-time extrapolation which we present as an action expression:

$$\int dM^2 \, M^2 a(M^2)(- \tfrac{1}{4}) \int (dx)(dx') F^{\mu\nu}(x) [\Delta_+(x - x', M^2) + \text{cont. term}] F_{\mu\nu}(x'), \qquad (4\text{–}3.66)$$

where

$$M^2 a(M^2) = \frac{4\pi\alpha}{M^2} \, I(M^2) = \frac{\alpha}{3\pi} \left(1 + \frac{2m^2}{M^2} \right) \left(1 - \frac{4m^2}{M^2} \right)^{1/2}; \qquad (4\text{–}3.67)$$

the last version refers to the spin $\tfrac{1}{2}$ example. The contact term indicated here is required to maintain a physical normalization condition—the action appropriate to photons ($k^2 = 0$) must not be altered. This is achieved by the combination

$$\Delta_+(x - x', M^2) - \frac{1}{M^2} \delta(x - x') = \frac{1}{M^2} \partial^2 \Delta_+(x - x', M^2) \qquad (4\text{–}3.68)$$

or, written in momentum space,

$$\frac{1}{k^2 + M^2 - i\varepsilon} - \frac{1}{M^2} = - \frac{k^2}{M^2} \frac{1}{k^2 + M^2 - i\varepsilon}. \qquad (4\text{–}3.69)$$

Thus, a more complete action for the electromagnetic field is given by

$$W = \int (dx)[J^{\mu}(x) A_{\mu}(x) - \tfrac{1}{4} F^{\mu\nu}(x) F_{\mu\nu}(x)]$$

$$- \int dM^2 \, a(M^2)(- \tfrac{1}{4}) \int (dx)(dx') \partial^{\lambda} F^{\mu\nu}(x) \Delta_+(x - x', M^2) \partial'_{\lambda} F_{\mu\nu}(x'). \qquad (4\text{–}3.70)$$

The locality necessary to define a Lagrange function no longer exists, in general. But, if we consider fields that vary slowly over the interval $1/M < 1/(2m)$, one can simplify (4–3.70) by substituting x for x' in the field structure. Then, using

$$\int (dx')\Delta_+(x - x', M^2) = \Delta_+(k = 0, M^2) = 1/M^2 \qquad (4\text{–}3.71)$$

together with [Eq. (4–3.54)]

$$\int_{(2m)^2}^{\infty} \frac{dM^2}{M^2}\, a(M^2) = \frac{\alpha}{15\pi}\frac{1}{m^2}, \qquad (4\text{–}3.72)$$

we can replace the last term of (4–3.70) with

$$-\frac{\alpha}{15\pi}\frac{1}{m^2}\int (dx)(-\tfrac{1}{4})\partial^\lambda F^{\mu\nu}(x)\, \partial_\lambda F_{\mu\nu}(x). \qquad (4\text{–}3.73)$$

In this limit there is a Lagrange function:

$$L = -\tfrac{1}{4}\left[F^{\mu\nu}F_{\mu\nu} - \frac{\alpha}{15\pi}\frac{1}{m^2}\partial^\lambda F^{\mu\nu}\, \partial_\lambda F_{\mu\nu}\right]. \qquad (4\text{–}3.74)$$

It implies the modified Maxwell field equation

$$\left(1 + \frac{\alpha}{15\pi}\frac{1}{m^2}\partial^2\right)\partial_\nu F^{\mu\nu}(x) = J^\mu(x). \qquad (4\text{–}3.75)$$

An exact solution of this equation would have no meaning, since it is restricted to the circumstances indicated by $\partial^2 \ll m^2$. Accordingly, we write the approximate solution as

$$A_\mu(x) = \left(1 - \frac{\alpha}{15\pi}\frac{1}{m^2}\partial^2\right)\int (dx')D_+(x - x')J_\mu(x')$$

$$= \int (dx')D_+(x - x')J_\mu(x') + \frac{\alpha}{15\pi}\frac{1}{m^2}J_\mu(x), \qquad (4\text{–}3.76)$$

apart from a gauge term. The explicit expression for W is now given by

$$W = \tfrac{1}{2}\int (dx)J^\mu(x)A_\mu(x)$$

$$= \tfrac{1}{2}\int (dx)(dx')J^\mu(x)D_+(x - x')J_\mu(x') + \frac{\alpha}{15\pi m^2}\frac{1}{2}\int (dx)J^\mu(x)J_\mu(x), \qquad (4\text{–}3.77)$$

which implies the modified interaction energy of two static charge-current distributions:

$$E_{\text{int.}} = -\int (dx)(dx')J_a^\mu(x)\mathscr{D}(x-x')J_{b\mu}(x') - \frac{\alpha}{15\pi}\frac{1}{m^2}\int (dx)J_a^\mu(x)J_{b\mu}(x). \quad (4\text{–}3.78)$$

The additional contact energy term restates what we have already seen in (4–3.53, 54), which is the example of the two charge distributions $Ze\,\delta(\mathbf{x})$ and $-e|\psi(\mathbf{x})|^2$. It is the slow variation of the latter charge density that validates the present approach.

In general, the field equations derived from (4–3.70) imply a modified propagation function,

$$A_\mu(x) = \int (dx')D_+(x-x')J_\mu(x') + \partial_\mu\lambda(x), \quad (4\text{–}3.79)$$

such that

$$k^2\left[1 - k^2\int dM^2\frac{a(M^2)}{k^2+M^2-i\varepsilon}\right]D_+(k) = 1. \quad (4\text{–}3.80)$$

Thus we now get

$$D_+(k) = \frac{1}{k^2-i\varepsilon}\frac{1}{1 - k^2\displaystyle\int dM^2\frac{a(M^2)}{k^2+M^2-i\varepsilon}}, \quad (4\text{–}3.81)$$

to be compared with the previous result [Eq. (4–3.28)]

$$D_+(k) = \frac{1}{k^2-i\varepsilon} + \int dM^2\frac{a(M^2)}{k^2+M^2-i\varepsilon}. \quad (4\text{–}3.82)$$

The latter would be reproduced by retaining only the first terms in an expansion of the inverse expression that appears in (4–3.81). The use of the action principle to derive modified field equations has given an improved account of the altered propagation characteristics that are implied by the vacuum polarization phenomena. It is in this context, incidentally, that such terminology acquires meaning; the modified field equations can be presented in Maxwellian form, with an additional current flowing in regions where the field is rapidly changing. According to the underlying physical picture it should be possible to exhibit the propagation function (4–3.81) in the form (4–3.82), but with a different positive weight factor:

$$D_+(k) = \frac{1}{k^2-i\varepsilon} + \int dM^2\frac{A(M^2)}{k^2+M^2-i\varepsilon}. \quad (4\text{–}3.83)$$

The function $A(M^2)$ can be determined by a comparison of imaginary parts for $-k^2 = M^2$, using the relation

$$\frac{1}{M'^2-M^2-i\varepsilon} = P\frac{1}{M'^2-M^2} + \pi i\,\delta(M'^2-M^2). \quad (4\text{–}3.84)$$

This gives

$$A(M^2) = \frac{a(M^2)}{\left[1 - M^2 P \displaystyle\int dM'^2 \frac{a(M'^2)}{M^2 - M'^2}\right]^2 + [\pi M^2 a(M^2)]^2}, \qquad (4\text{–}3.85)$$

which is a positive quantity.

As a simplification of this situation, which is not very drastic, let us treat $(2m)^2/M^2$ as negligible almost everywhere, in $a(M^2)$:

$$M^2 > (2m)^2: \quad a(M^2) \cong \frac{\alpha}{3\pi} \frac{1}{M^2}. \qquad (4\text{–}3.86)$$

That permits the elementary evaluations

$$k^2 \int dM^2 \frac{a(M^2)}{k^2 + M^2 - i\varepsilon} = \frac{\alpha}{3\pi} \log\left(\frac{k^2}{4m^2} + 1\right) \qquad (4\text{–}3.87)$$

and

$$M^2 P \int dM'^2 \frac{a(M'^2)}{M^2 - M'^2} = \frac{\alpha}{3\pi} \log\left(\frac{M^2}{4m^2} - 1\right). \qquad (4\text{–}3.88)$$

In applying (4–3.87) to negative values of $k^2 + 4m^2$, they should be reached from the lower half of the complex plane, according to the instruction contained in $k^2 - i\varepsilon$. That supplies the logarithm with an imaginary term of $-\pi i$, in agreement with the imaginary part of the left side, $\pi i(-M^2)a(M^2)$. This simplified version gives

$$\bar{D}_+(k) = \frac{1}{k^2 - i\varepsilon} \frac{1}{1 - \dfrac{\alpha}{3\pi} \log\left(\dfrac{k^2}{4m^2} + 1\right)}, \qquad (4\text{–}3.89)$$

and

$$\bar{D}_+(k) = \frac{1}{k^2 - i\varepsilon} + \frac{\alpha}{3\pi} \int_{(2m)^2}^{\infty} \frac{dM^2}{M^2} \frac{\left[\left(1 - \dfrac{\alpha}{3\pi} \log\left(\dfrac{M^2}{4m^2} - 1\right)\right)^2 + (\tfrac{1}{3}\alpha)^2\right]^{-1}}{k^2 + M^2 - i\varepsilon}. \qquad (4\text{–}3.90)$$

Are these two expressions really equivalent? This mathematical problem is best approached by regarding k^2 as a complex variable. The function (4–3.89) has a pole at $k^2 = 0$ and a branch point at $k^2 = -4m^2$, both of which are correctly represented in (4–3.90). But (4–3.89) also possesses a pole at

$$k^2 = 4m^2[\exp(3\pi/\alpha) - 1] \qquad (4\text{–}3.91)$$

which is not represented in (4–3.90), since it is an unphysical singularity at a space-like value of k. This is a mathematical failure—it is not a significant physical one. As in the discussion of (4–3.48), the values of momenta, or distances, that are required to have the logarithm overcome the smallness of $\alpha/3\pi$ are utterly beyond the level of any conceivable physical relevance. For all physical purposes, (4–3.89), or its more precise version (4–3.81), is correct, within the limitations of the physical processes considered. And there is no reason at this point to believe that the formal failure is other than physical incompleteness becoming apparent under the conditions of an outrageous extrapolation. Incidentally, the photon propagation function of (4–3.89) indicates that a more general statement of the g-factor asymmetry between electron and muon is given by the replacement

$$1 + \frac{2\alpha}{3\pi} \log \frac{m_\mu}{m_e} \rightarrow \frac{1}{1 - \frac{2\alpha}{3\pi} \log \frac{m_\mu}{m_e}}. \tag{4–3.92}$$

This would indeed represent the most important aspect of the higher-order effects if the logarithm were a large number; however, $\log(m_\mu/m_e) = 5.3$.

The last remark brings us back to the question of improving the estimate (4–3.61) by removing the oversimplification inherent in (4–3.59), where the masses of electron-positron pairs were ignored. We must repeat the calculation of Section 4–2, replacing the photon propagation function $D_+(x - x')$ by $\varDelta_+(x - x', M'^2)$, and then integrate over the spectral distribution of M'^2 that is contained in the modified propagation function $\bar{D}_+(x - x')$. The immediate kinematical changes in (4–2.17) are indicated by

$$\frac{M^2 + m^2}{2M^2} \rightarrow \frac{M^2 + m^2 -- M'^2}{2M^2}, \tag{4–3.93}$$

and

$$1 - \frac{m^2}{M^2} \rightarrow \left[1 - \left(\frac{m + M'}{M}\right)^2\right]^{1/2} \left[1 - \left(\frac{m - M'}{M}\right)^2\right]^{1/2}$$

$$= \int_0^1 du\, \eta\left(M^2 - \frac{m^2}{u} - \frac{M'^2}{1 - u}\right), \tag{4–3.94}$$

which uses (4–1.23, 32). The replacement of m^2 by $m^2 - eq\sigma F$ is again combined with the substitution $M^2 \rightarrow M^2 - eq\sigma F$. But instead of the explicit change of $1 - (m^2/M^2)$ that is evident in (4–2.19), we must now use (4–3.94) and the weak field expansion

$$\eta\left(M^2 - eq\sigma F - \frac{m^2 - eq\sigma F}{u} - \frac{M'^2}{1 - u}\right)$$

$$= \eta \left(M^2 - \frac{m^2}{u} - \frac{M'^2}{1-u} \right) + \frac{1-u}{u} eq\sigma F \, \delta \left(M^2 - \frac{m^2}{u} - \frac{M'^2}{1-u} \right). \quad (4\text{-}3.95)$$

This replaces (4–2.20) with

$$- i(4\pi)^2 \Phi^{-1} \Delta_+(x - x', M^2) G_+^A(x, x')$$

$$= \int dM^2 \, du \, \eta \left(m - \frac{M^2 + m^2 - M'^2}{2M^2} \gamma \left(\frac{1}{i} \right) \partial \right) \Delta_+(x - x', M^2 - eq\sigma F)$$

$$+ \int dM^2 \, du \, \delta \frac{1-u}{u} eq\sigma F \left(m - \frac{M^2 + m^2 - M'^2}{2M^2} \gamma \left(\frac{1}{i} \right) \partial \right) \Delta_+(x - x', M^2)$$

$$+ \int dM^2 \, du \, \eta eq\sigma F \frac{M^2 - m^2 + M'^2}{2(M^2)^2} \gamma \left(\frac{1}{i} \right) \partial \Delta_+(x - x', M^2), \quad (4\text{-}3.96)$$

where, for simplicity of presentation, some general factors have been transferred and the common argument of the step function η and the delta function is left unwritten. The pattern of calculation that gave (4–2.26) now produces

$$- i(4\pi)^2 \gamma^\mu \Delta_+(x - x', M^2) G_+^A(x, x') \gamma_\mu$$

$$= \int dM^2 \, du \, \delta \, eq\sigma F \left[4m + \frac{M^2 + m^2 - M'^2}{M^2} \frac{1+u}{u} \gamma\Pi \right] \Delta_+^A(x, x', M^2)$$

$$- \int dM^2 \, du \, \eta eq\sigma F \frac{M^2 + m^2 - M'^2}{(M^2)^2} \gamma\Pi \Delta_+^A(x, x', M^2), \quad (4\text{-}3.97)$$

where only the explicitly field dependent terms are exhibited. The step function that appears here can be converted into a delta function by performing a partial integration:

$$\int_0^1 du \, \eta = (u - 1)\eta \Big|_0^1 + \int_0^1 du(u - 1) \left(\frac{m^2}{u^2} - \frac{M'^2}{(1-u)^2} \right) \delta, \quad (4\text{-}3.98)$$

in which the integrated expression vanishes at both limits.

The effective value of the field dependent coupling that is appropriate to a ψ obeying (4–2.33) is attained through the substitutions

$$\gamma\Pi \to -m, \qquad \Delta_+^A = \frac{1}{M^2 - (\gamma\Pi)^2} \to \frac{1}{M^2 - m^2}. \quad (4\text{-}3.99)$$

The delta function will be used to determine M^2 as a function of the parameter u,

$$M^2 = \frac{m^2}{u} + \frac{M'^2}{1-u}. \quad (4\text{-}3.100)$$

We obtain in this way the following expression for the coefficient of $- (\alpha/2\pi)$ $\times (1/2m)eq\sigma F$ in the effective action:

$$2 \int_0^1 du \, \frac{u(1-u)^2}{(1-u)^2 + \lambda u} + \lambda \int_0^1 du \, u(1-u) \left[2u \, \frac{1}{1-u+\lambda u} \frac{1}{(1-u)^2 + \lambda u} \right.$$

$$\left. - \frac{1-u^2+\lambda u^2}{(1-u)^2 + \lambda u} \frac{1}{(1-u+\lambda u)^2} \right],$$ (4–3.101)

where

$$\lambda = M'^2/m^2.$$ (4–3.102)

Only algebraic rearrangements have been used to arrive at this form. But now we also recognize that the second integrand is a total differential, of the function

$$\frac{u(1-u)}{1+(\lambda-1)u} + \log \frac{(1-u)^2 + \lambda u}{1+(\lambda-1)u},$$ (4–3.103)

which vanishes at both endpoints. Here, then, is what replaces unity, for the photon, when an excitation of mass M' is considered:

$$2 \int_0^1 du \, \frac{u(1-u)^2}{(1-u)^2 + (M'/m)^2 u}.$$ (4–3.104)

In the approximate treatment that gave (4–3.60, 61) this factor was crudely represented by unity for $M' < m$, and zero for $M' > m$. Now we combine (4–3.104) with the weight factor of (4–3.28),

$$\frac{\alpha}{3\pi} \frac{dM'^2}{M'^2} \left(1 + 2 \frac{m'^2}{M'^2} \right) \left(1 - \frac{4m'^2}{M'^2} \right)^{1/2} = \frac{\alpha}{\pi} dv \, \frac{v^2(1 - \frac{1}{3}v^2)}{1-v^2},$$ (4–3.105)

where

$$v = \left(1 - \frac{4m'^2}{M'^2} \right)^{1/2}$$ (4–3.106)

and m' is the mass of the charged particle contributing to the vacuum polarization. The resulting double integral is

$$\frac{2\alpha}{\pi} \int_0^1 du \, u(1-u)^2 \int_0^1 dv \, v^2(1 - \tfrac{1}{3}v^2) \frac{1}{(1-u)^2(1-v^2) + (2m'/m)^2 u}.$$ (4–3.107)

We must note here that the electron vacuum polarization contribution to g_e and the muon vacuum polarization contribution to g_μ are equal ($m' = m$). The asymmetry is that between the electron vacuum polarization contribution to g_μ ($m' = m_e$, $m = m_\mu$) and the muon vacuum polarization contribution to g_e ($m' = m_\mu$, $m = m_e$). The large mass ratio permits the integral in the latter situation to be simplified:

$$\frac{\alpha}{2\pi}\left(\frac{m_e}{m_\mu}\right)^2 \int_0^1 du(1-u)^2 \int_0^1 dv\, v^2(1-\tfrac{1}{3}v^2) = \frac{2\alpha}{45\pi}\left(\frac{m_e}{m_\mu}\right)^2. \qquad (4\text{-}3.108)$$

This is a very small effect. For an approximate treatment of the situation $m'/m = m_e/m_\mu \ll 1$, the integral of (4-3.107) is rearranged as

$$\frac{2\alpha}{3\pi}\int_0^1 du\, u \int_0^1 dv\, \frac{2-(1-v^2)-(1-v^2)^2}{(1-v^2)+(2m_e/m_\mu)^2 u/(1-u)^2}, \qquad (4\text{-}3.109)$$

and $(2m_e/m_\mu)^2$ neglected in all but the first of the three terms. Performing the v integrations, we get

$$\frac{2\alpha}{3\pi}\int_0^1 du\, u \left[\log\left(\frac{m_\mu^2}{m_e^2}\frac{(1-u)^2}{u}\right)-\frac{5}{3}\right] = \frac{2\alpha}{3\pi}\left(\log\frac{m_\mu}{m_e}-\frac{25}{12}\right), \qquad (4\text{-}3.110)$$

and (4-3.61) is amended to

$$\tfrac{1}{2}g_\mu - \tfrac{1}{2}g_e = \frac{\alpha}{2\pi}\frac{2\alpha}{3\pi}\left(\log\frac{m_\mu}{m_e}-\frac{25}{12}\right). \qquad (4\text{-}3.111)$$

It is desirable and possible to improve this estimate by retaining terms of order m_e/m_μ. They appear since $(2m_e/m_\mu)^2 u/(1-u)^2$ ceases to be very small for u values near unity, within a range $\sim m_e/m_\mu$. One procedure for their evaluation begins by performing the v integration in (4-3.109) to produce

$$\frac{2\alpha}{3\pi}\int_0^1 du\, u\left[-\frac{5}{3}+x^2+(1-\tfrac{1}{2}x^2)(1+x^2)^{1/2}\log\frac{(1+x^2)^{1/2}+1}{(1+x^2)^{1/2}-1}\right], \qquad (4\text{-}3.112)$$

where

$$x = \frac{2m_e}{m_\mu}\frac{u^{1/2}}{1-u}. \qquad (4\text{-}3.113)$$

The u integral is then divided into two parts, in the first of which x is small everywhere compared to unity while, in the second one, x can assume large values but u is sufficiently near unity to validate the differential relation

$$u \sim 1: \quad du = \frac{2m_e}{m_\mu}\frac{dx}{x^2}. \qquad (4\text{-}3.114)$$

After rearranging the two terms to remove the arbitrary junction point, one obtains the following addition to (4-3.110):

$$\frac{4\alpha}{3\pi}\frac{m_e}{m_\mu}\int_0^\infty \frac{dx}{x^2}\left[x^2-\log\frac{4}{x^2}+(1-\tfrac{1}{2}x^2)(1+x^2)^{1/2}\log\frac{(1+x^2)^{1/2}+1}{(1+x^2)^{1/2}-1}\right]. \qquad (4\text{-}3.115)$$

Partial integration converts this into

$$\frac{4\alpha}{\pi}\frac{m_e}{m_\mu}\int_0^\infty dx\left[1-\frac{1}{2}\frac{x^2}{(1+x^2)^{1/2}}\log\frac{(1+x^2)^{1/2}+1}{(1+x^2)^{1/2}-1}\right],\qquad(4\text{-}3.116)$$

and, on restoring the v parametrization, we get as the factor of $(4\alpha/\pi)(m_e/m_\mu)$:

$$\int_0^\infty dx\int_0^1 dv\,\frac{1-v^2}{1-v^2+x^2}=\frac{\pi}{2}\int_0^1 dv(1-v^2)^{1/2}=\frac{\pi^2}{8}.\qquad(4\text{-}3.117)$$

This improved version of (4-3.111) is

$$\tfrac{1}{2}g_\mu-\tfrac{1}{2}g_e=\frac{\alpha}{2\pi}\frac{2\alpha}{3\pi}\left(\log\frac{m_\mu}{m_e}-\frac{25}{12}+\frac{3}{4}\pi^2\frac{m_e}{m_\mu}\right).\qquad(4\text{-}3.118)$$

The use of the numerical value

$$\log\frac{m_\mu}{m_e}-\frac{25}{12}+\frac{3}{4}\pi^2\frac{m_e}{m_\mu}=3.285\qquad(4\text{-}3.119)$$

changes the estimate (4-3.61) to

$$\tfrac{1}{2}g_\mu-\tfrac{1}{2}g_e=(1.161\times10^{-3})(5.085\times10^{-3})$$

$$=0.590\times10^{-5}.\qquad(4\text{-}3.120)$$

The general agreement with a recent experimental value, $(0.66\pm0.03)\times10^{-5}$, is striking confirmation that vacuum polarization gives the principal contribution to this effect. As to whether or not the residual difference is real, we remark that, in addition to various small corrections such as the one indicated in (4-3.92) [perhaps we shall return to this topic in later chapters], there are physical processes yet to be considered which involve particles other than electron and muon. The π-meson, for example, is not much heavier than the muon and might make a significant contribution to the asymmetry. This effect can be estimated by changing the weight function of (4-3.105) to accord with the spin 0 spectral form (4-3.39):

$$\frac{\alpha}{12\pi}\frac{dM'^2}{M'^2}\left(1-\frac{4m'^2}{M'^2}\right)^{3/2}=\frac{\alpha}{6\pi}dv\,\frac{v^4}{1-v^2}.\qquad(4\text{-}3.121)$$

The relevant integral, replacing (4-3.107), is then

$$\frac{\alpha}{3\pi}\int_0^1 du\,u(1-u)^2\int_0^1 dv\,v^4\frac{1}{(1-u)^2(1-v^2)+(2m_\pi/m_\mu)^2u}$$

$$<\frac{\alpha}{12}\left(\frac{m_\mu}{m_\pi}\right)^2\int_0^1 du(1-u)^2\int_0^1 dv\,v^4$$

$$=\frac{\alpha}{180\pi}\left(\frac{m_\mu}{m_\pi}\right)^2.\qquad(4\text{-}3.122)$$

Even this upper limit would barely change the last significant figure in (4–3.120). The calculation makes no reference to the strong interaction properties of the π-meson, however. A pair of oppositely charged π-mesons is strongly coupled to the neutral ρ-meson, which has the same quantum numbers as the photon. It is the direct coupling of the photon to ρ^0 (and ω, and ϕ) that may be most significant in vacuum polarization phenomena that are outside the electron-muon framework. Although we are not ready to give a quantitative discussion of this question, the stage can be set by noting the simplification of (4–3.104) for such massive particles:

$$\left(\frac{M'}{m}\right)^2 \gg 1: \quad 2\left(\frac{m}{M'}\right)^2 \int_0^1 du(1-u)^2 = \frac{2}{3}\left(\frac{m}{M'}\right)^2. \tag{4–3.123}$$

The strong interaction supplement to the photon effect is the weighted average of (4–3.123) over the heavy particle contribution in the modified photon propagation function. We shall make explicit an electromagnetic factor, α/π, in defining an average inverse squared mass. This gives the additional strong interaction contribution as

$$(\tfrac{1}{2}g_\mu - \tfrac{1}{2}g_e)_{\text{strong int.}} = \frac{\alpha}{2\pi}\frac{2\alpha}{3\pi}(m_\mu{}^2 - m_e{}^2)\left\langle\frac{1}{M'^2}\right\rangle. \tag{4–3.124}$$

The use of the ρ-mass in an estimate indicates a substantial contribution to the last significant figure in (4–3.120).

The final application of the modified photon propagation function that will be discussed in this section also lies somewhat outside the realm of photon-charged lepton phenomena. The neutral π-meson decays primarily into two photons. A small fraction of the decay processes involves a photon and an electron-positron pair; in a considerably smaller fraction of the events, two electron-positron pairs are emitted. A description of the relations among these processes is provided by the modified photon propagation function.

The effective coupling for the two-photon decay of the 0^- pion resembles (3–13.75), which is a description of the two-photon annihilation of an electron-positron pair; the pseudo-scalar pion field $\phi(x)$ replaces the quadratic field combination $\tfrac{1}{2}\psi(x)\gamma^0\gamma_5\psi(x)$:

$$W_{\pi^\bullet \to 2\gamma} = f\int (dx)\phi(x)(-\tfrac{1}{4})^*F^{\mu\nu}(x)F_{\mu\nu}(x), \tag{4–3.125}$$

where f is a suitable coupling constant. The role of the field $\phi(x)$ as an effective two-photon source is made explicit by comparing $iW_{\pi^\bullet \to 2\gamma}$ with

$$\tfrac{1}{2}\left[i\int (dx)J^\mu(x)A_\mu(x)\right]^2. \tag{4–3.126}$$

This gives

$$i J^\mu(x) J^\nu(x')|_{\text{eff.}} = f \varepsilon^{\mu\nu\kappa\lambda} \partial_\kappa \partial'_\lambda (\delta(x - x')\phi(x)). \tag{4–3.127}$$

Two such sources, causally arranged, have a two-particle exchange coupling that is given by

$$\frac{1}{2}\left[i \int (dx)(dx') J_1^\mu(x) D_+(x - x') J_{2\mu}(x') \right]^2 \Bigg|_{\text{eff.}}$$

$$= f^2 \int (dx)(dx') \phi_1(x) [\partial^\mu \partial^\nu D_+(x - x') \partial_\mu \partial_\nu D_+(x - x')$$

$$- \partial^2 D_+(x - x') \partial^2 D_+(x - x')] \phi_2(x'), \tag{4–3.128}$$

which uses the relation

$$- \frac{1}{2}\varepsilon^{\mu\nu\kappa\lambda} \varepsilon_{\mu\nu\alpha\beta} = \delta_\alpha^\kappa \delta_\beta^\lambda - \delta_\beta^\kappa \delta_\alpha^\lambda. \tag{4–3.129}$$

The combination in brackets is more usefully presented in the form

$$\frac{1}{4}(\partial^2)^2 (D_+(x - x'))^2 - \partial^2(D_+(x - x')\partial^2 D_+(x - x')) + \frac{1}{2}D_+(x - x')$$

$$\times (\partial^2)^2 D_+(x - x') - \frac{1}{2}\partial^2 D_+(x - x')\partial^2 D_+(x - x'). \tag{4–3.130}$$

The space-time extrapolation of this coupling, performed with due regard for mass normalization requirements, supplements the quadratic field structure of the action. When one considers a field far from its source, and essentially governed by

$$(- \partial^2 + m_\pi^2)\phi(x) = 0, \tag{4–3.131}$$

the effective Lagrange function becomes

$$- \frac{1}{2}[\partial^\mu \phi \partial_\mu \phi + (m_\pi^2 - im_\pi\gamma)\phi^2], \tag{4–3.132}$$

where $\frac{1}{2}im_\pi\gamma\phi^2$ is the reduced form of the additional term. (A change of m_π^2 is excluded by mass normalization.) The latter describes the instability of the particle. For weak instability

$$\gamma \ll m_\pi: \quad m_\pi^2 - im_\pi\gamma \cong (m_\pi - \tfrac{1}{2}i\gamma)^2, \tag{4–3.133}$$

and the time dependence of the field associated with a particle in its rest frame is such that

$$|\exp[- i(m_\pi - \tfrac{1}{2}i\gamma)x^0]|^2 = \exp(- \gamma x^0), \tag{4–3.134}$$

identifying γ, the reciprocal mean life time, with the total decay rate. The calculation of γ, with the aid of the modified propagation function D_+, makes explicit the contributions of the various processes involving two photons, one photon, or no photons.

Let us illustrate this procedure by evaluating the two-photon decay rate. Here, only D_+ is considered and, under the causal conditions assumed in (4–3.128, 130), $\partial^2 D_+ = 0$. The coupling thus reduces to

$$\tfrac{1}{4}f^2 \int (dx)(dx') \partial^2\phi_1(x)(D_+(x - x'))^2 \partial^2\phi_2(x')$$

$$\rightarrow \tfrac{1}{4}f^2 m_\pi^4 \int (dx)(dx')\phi_1(x)(D_+(x - x'))^2\phi_2(x'), \qquad (4\text{–}3.135)$$

where the second step anticipates the kind of field [Eq. (4–3.131)] to which it will be applied. Now observe that

$$x^0 > x^{0'}: \quad (D_+(x - x'))^2 = \left[i \int d\omega_k \exp[ik(x - x')] \right]^2$$

$$= i \int dM^2 \int d\omega_k\, d\omega_{k'} \cdot (2\pi)^3\, \delta(k + k' - P)|_{-P^2 = M^2}$$

$$\times i \int d\omega_P \exp[iP(x - x')]$$

$$= \frac{i}{(4\pi)^2} \int dM^2\, \Delta_+(x - x', M^2), \qquad (4\text{–}3.136)$$

according to the integral (4–1.23), with $m_a = m_b = 0$. When the space-time extrapolation is performed, the $\phi_1\phi_2$ structure becomes $\tfrac{1}{2}\phi\phi$, and, after removing the factor of i, we get the additional action expression. The restriction to a field obeying (4–3.131) implies the replacement

$$\int (dx')\, \Delta_+(x - x', M^2)\phi(x') = \frac{1}{M^2 - \partial^2 - i\varepsilon}\, \phi(x)$$

$$\rightarrow \frac{1}{M^2 - m_\pi^2 - i\varepsilon}\, \phi(x). \qquad (4\text{–}3.137)$$

Using the imaginary part of this factor, $\pi i\delta(M^2 - m_\pi^2)$, we pick out the addition to the action that is of interest,

$$i\, \frac{f^2}{64\pi}\, m_\pi^4\, \frac{1}{2} \int (dx)(\phi(x))^2, \qquad (4\text{–}3.138)$$

which gives

$$\gamma = (f^2/64\pi)m_\pi^3. \qquad (4\text{–}3.139)$$

A quick indication of the rates for the single and double pair processes can be inferred from the kind of simplification illustrated in (4–3.59), with m_π appearing as the upper limit to electron-positron pair masses. Since two propagation functions are multiplied together, the decay rate acquires the factor

$$\sim \left(1 + \frac{2\alpha}{3\pi} \log \frac{m_\pi}{m_e}\right)^2, \qquad (4\text{–}3.140)$$

which displays the relative rates of single and double pair processes,

$$\frac{\gamma'}{\gamma} \sim \frac{4\alpha}{3\pi} \log \frac{m_\pi}{m_e}, \qquad \frac{\gamma''}{\gamma} \sim \left(\frac{2\alpha}{3\pi} \log \frac{m_\pi}{m_e}\right)^2, \qquad (4\text{–}3.141)$$

respectively.

A more accurate evaluation of γ'/γ will now be given. After inserting the general propagation function form (4–3.83), we must extract contributions with one $D_+(x - x')$ function and one $\Delta_+(x - x', M'^2)$ function. Having in mind that m_π^2 will replace the ∂^2 operations that can be transferred to the fields, the effective form of (4–3.130) for this process becomes

$$\tfrac{1}{2} \int dM'^2 A(M'^2)(m_\pi^2 - M'^2)^2 D_+(x - x')\Delta_+(x - x', M'^2). \qquad (4\text{–}3.142)$$

These propagation function products are

$x^0 > x^{0\prime}$:

$$D_+(x - x')\,\Delta_+(x - x', M'^2) = i \int dM^2 \int d\omega_k\, d\omega_p (2\pi)^3\, \delta(k + p - P)|_{-P^2 = M^2}$$

$$\times i \int d\omega_P \exp[iP(x - x')]$$

$$= \frac{i}{(4\pi)^2} \int dM^2 \left(1 - \frac{M'^2}{M^2}\right)\Delta_+(x - x', M^2), \quad (4\text{–}3.143)$$

which uses the integral (4–1.25). Comparing this with the earlier calculation, we infer the ratio

$$\frac{\gamma'}{\gamma} = 2 \int_{(2m_e)^2}^{m_\pi^2} dM'^2 A(M'^2) \left(1 - \frac{M'^2}{m_\pi^2}\right)^3. \qquad (4\text{–}3.144)$$

The integral will be evaluated, with sufficient precision, for the weight function [Eq. (4–3.67)]

$$A(M'^2) = \frac{\alpha}{3\pi} \frac{1}{M'^2} \left(1 + \frac{2m_e^2}{M'^2}\right)\left(1 - \frac{4m_e^2}{M'^2}\right)^{1/2}. \qquad (4\text{–}3.145)$$

This is done by decomposing the integral at a value of $M'^2 = M_0^2$ such that

$$m_e^2 \ll M_0^2 \ll m_\pi^2, \tag{4-3.146}$$

which gives the simplifications exhibited in

$$\frac{\gamma'}{\gamma} = \frac{2\alpha}{3\pi} \int_{(2m_e)^2}^{M_0^2} \frac{dM'^2}{M'^2}\left(1 + \frac{2m_e^2}{M'^2}\right)\left(1 - \frac{4m_e^2}{M'^2}\right)^{1/2} + \frac{2\alpha}{3\pi}\int_{M_0^2}^{m_\pi^2} \frac{dM'^2}{M'^2}\left(1 - \frac{M'^2}{m_\pi^2}\right)^3. \tag{4-3.147}$$

The parametrization indicated in (4–3.105) can be used to evaluate the first integral, and

$$\frac{\gamma'}{\gamma} = \frac{2\alpha}{3\pi}\left(\log \frac{M_0^2}{m_e^2} - \frac{5}{3}\right) + \frac{2\alpha}{3\pi}\left(\log \frac{m_\pi^2}{M_0^2} - \frac{11}{6}\right)$$

$$= \frac{4\alpha}{3\pi}\left(\log \frac{m_\pi}{m_e} - \frac{7}{4}\right), \tag{4-3.148}$$

consistent with the estimate of (4–3.141). The numerical result obtained in this way,

$$\gamma'/\gamma = 1.18 \times 10^{-2}, \tag{4-3.149}$$

is in excellent agreement with the measured value, $(1.17 \pm 0.04) \times 10^{-2}$. Concerning γ'', we shall only remark that the improved calculation essentially maintains the simple relation indicated in (4–3.141),

$$\frac{\gamma''}{\gamma} \simeq \left(\frac{1}{2}\frac{\gamma'}{\gamma}\right)^2. \tag{4-3.150}$$

4-4 FORM FACTORS I. SCATTERING

At the beginning of this chapter, we observed that the introduction of modified propagation functions is an essential but incomplete characterization of multiparticle exchange processes. In discussing the primitive interaction and the interaction processes that produce the modified photon propagation function, it was indicated that these interaction processes have other consequences, which are represented by modified electromagnetic couplings. We now begin an extended consideration of these matters.

The spin 0 primitive interaction portrays the electromagnetic vector potential

$$A_\mu(k) = D_+(k)J_\mu(k) \tag{4-4.1}$$

as an effective two-particle emission source [Eq. (4–3.33)]

$$iK_2(p)K_2(p')|_{\text{eff.}} = eq(p - p')^\mu A_\mu(k), \tag{4-4.2}$$

provided the mass threshold is exceeded,

$$- k^2 > (2m)^2. \tag{4-4.3}$$

These oppositely charged particles are produced under noninteraction conditions. In the course of time, however, the particles can interact, in a way described by the scattering calculation of Section 3–12. We shall write the vacuum amplitude discussed there [Eq. (3–12.83)] in a more convenient form that restores the multicomponent sources in charge space:

$$ie^2 \tfrac{1}{2} \int d\omega_{p_1} d\omega_{p_1'} d\omega_{p_2} d\omega_{p_2'} K_1(-p_1)(K_2(p_2)K_2(p_2'))K_1(-p_1')$$

$$\times (2\pi)^4 \delta(p_1 + p_1' - p_2 - p_2') \left[-\frac{(p_1 + p_2)(p_1' + p_2')}{(p_1 - p_2)^2} + \frac{(p_1 - p_1')(p_2 - p_2')}{(p_1 + p_1')^2} \right].$$

$$\tag{4-4.4}$$

Its validity depends upon the selection of oppositely charged particles that is performed by (4–4.2). On inserting the latter this coupling becomes

$$- e^2 \tfrac{1}{2} \int d\omega_{p_1} d\omega_{p_1'} (2\pi)^4 \delta(p_1 + p_1' - k) \, dM^2 \, d\omega_k \, K_1(-p_1) eq K_1(-p_1') I^\mu A_\mu(k)$$

$$+ e^2 \tfrac{1}{2} \int d\omega_{p_1} d\omega_{p_1'} (2\pi)^4 \delta(p_1 + p_1' - k) \frac{dM^2}{(M^2)^2} d\omega_k$$

$$\times K_1(-p_1) eq(p_1 - p_1')^\mu K_1(-p_1') I_{\mu\nu} J^\nu(k), \tag{4-4.5}$$

where

$$I^\mu = \int d\omega_{p_2} d\omega_{p_2'} (2\pi)^3 \delta(p_2 + p_2' - k) \frac{(p_1 + p_2)(p_1' + p_2')}{(p_1 - p_2)^2} (p_2 - p_2')^\mu \tag{4-4.6}$$

and

$$I_{\mu\nu} = \int d\omega_{p_2} d\omega_{p_2'} (2\pi)^3 \delta(p_2 + p_2' - k)(p_2 - p_2')_\mu (p_2 - p_2')_\nu$$

$$= \left(g_{\mu\nu} + \frac{k_\mu k_\nu}{M^2} \right) \frac{1}{3} \frac{1}{(4\pi)^2} M^2 \left(1 - \frac{4m^2}{M^2} \right)^{3/2}. \tag{4-4.7}$$

The instant evaluation of the last integral conveys the recognition that it has already been encountered in the spin 0 vacuum polarization discussion [Eqs. (4–3.35, 19, 37)]. Let us also note that the current vector

$$j^\mu(x) = \phi_1(x) eq(1/i) \partial^\mu \phi_1(x), \tag{4-4.8}$$

evaluated in a region that is prior to the action of the detection source K_1, is given by

$$j^\mu(x) = \int d\omega_{p_1} d\omega_{p_1'} \, iK_1(-p_1) eq\tfrac{1}{2}(p_1 - p_1')^\mu iK_1(-p_1') \exp[-i(p_1 + p_1')x].$$

(4-4.9)

Accordingly, the second term of (4-4.5) is

$$i\,\frac{\alpha}{12\pi} \int \frac{dM^2}{M^2} \left(1 - \frac{4m^2}{M^2}\right)^{3/2} \int (dx)(dx') j^\mu(x) \left[i \int d\omega_k \exp[ik(x - x')]\right] J_\mu(x')$$

$$= i \int (dx)(dx') j^\mu(x) [\bar{D}_+(x - x') - D_+(x - x')] J_\mu(x'),$$

(4-4.10)

which is the recognition of the spin 0 form of the modified photon propagation function [Eq. (4-3.39)]. Here is the anticipated mechanism that substitutes $\bar{D}_+$ for D_+ in the vector potential of the primitive interaction.

Our attention is now concentrated on the first term of (4-4.5), and the integral (4-4.6). This vector must be a linear combination of the final momenta p_1^μ, $p_1'^\mu$. Now, the interchange of these momenta, combined with that of p_2 and p_2', reverses the sign of I^μ [recall that $(p_1' - p_2)^2 = (p_1 - p_2)^2$]. We conclude that

$$I^\mu = (p_1 - p_1')^\mu S(M^2),$$

(4-4.11)

where

$$S(M^2) = \int d\omega_{p_2} d\omega_{p_2'} (2\pi)^3 \delta(p_2 + p_2' - k) \frac{(p_1 + p_2)(p_1' + p_2')}{(p_1 - p_2)^2} \frac{(p_1 - p_1')(p_2 - p_2')}{(p_1 - p_1')^2}$$

$$\equiv \frac{1}{(4\pi)^2} \left(1 - \frac{4m^2}{M^2}\right)^{1/2} \left\langle \frac{(p_1 + p_2)(p_1' + p_2')}{(p_1 - p_2)^2} \frac{(p_1 - p_1')(p_2 - p_2')}{M^2 - 4m^2} \right\rangle.$$

(4-4.12)

In the rest frame of the vector k, all four particle energies equal $\tfrac{1}{2}M$, and the integration that evaluates (4-4.12) is extended over the scattering angle θ, that between $\mathbf{p}_1 = -\mathbf{p}_1'$ and $\mathbf{p}_2 = -\mathbf{p}_2'$:

$$\langle \cdots \rangle = \int_{-1}^{1} \frac{d(\cos\theta)}{2} \frac{-M^2 - (M^2 - 4m^2)\cos^2\tfrac{1}{2}\theta}{(M^2 - 4m^2)\sin^2\tfrac{1}{2}\theta} \cos\theta$$

$$= -2(M^2 - 2m^2) \int_{-1}^{1} \frac{d(\cos\theta)}{2} \frac{1}{(M^2 - 4m^2)\sin^2\tfrac{1}{2}\theta} + 4\frac{M^2 - 2m^2}{M^2 - 4m^2}.$$

(4-4.13)

At this point, we must acknowledge that the angular integral has a logarithmic singularity at $\theta = 0$. It is a consequence of the unlimited forward scattering that the long-range Coulomb potential produces. Here is the reminder by the formalism that the primitive interaction context of noninteracting particles cannot be entirely realized for charged particles. We are encountering another aspect of the infrared problem, since the zero mass of the photon is responsible for the unlimited range of the Coulomb potential. This suggests that the difficulty is only superficial and will disappear when additional soft photon processes are considered. For the moment, however, we choose to bypass the problem by imagining that the photon has a very small but finite mass μ. Recalling the origin of $(p_1 - p_2)^2$ in the structure of the photon propagation function, we recognize that the singular integral of (4–4.13) becomes

$$\int_{-1}^{1} \frac{d(\cos\theta)}{2} \frac{1}{(M^2 - 4m^2)\sin^2\tfrac{1}{2}\theta + \mu^2} = \frac{1}{M^2 - 4m^2} \log \frac{M^2 - 4m^2}{\mu^2} \quad (4–4.14)$$

and

$$S(M^2) = -\frac{2}{(4\pi)^2}\left(1 - \frac{4m^2}{M^2}\right)^{-1/2}\left(1 - \frac{2m^2}{M^2}\right)\left(\log\frac{M^2 - 4m^2}{\mu^2} - 2\right). \quad (4–4.15)$$

Inspecting the form of the first term, in (4–4.5), that results from (4–4.11) we see that the vector structure of the current (4–4.9) is again present, permitting this term to be displayed as

$$i\frac{\alpha}{2\pi}\int dM^2\left(1 - \frac{4m^2}{M^2}\right)^{-1/2}\left(1 - \frac{2m^2}{M^2}\right)\left(\log\frac{M^2 - 4m^2}{\mu^2} - 2\right)$$

$$\times \int (dx)(dx')j^\mu(x)\left[i\int d\omega_k \exp[ik(x - x')]\right]A_\mu(x'). \quad (4–4.16)$$

But, prior to performing the space-time extrapolation that now lies before us, we must notice one point concerning the gauge invariance of this expression. It is gauge invariant, indeed, under the causal conditions being considered, since

$$\partial_\mu j^\mu(x) = 0 \qquad [k(p_1 - p_1') = 0]. \quad (4–4.17)$$

Being rooted in the kinematics of free particles, however, this property will not be maintained after the space-time extrapolation is performed. Accordingly, we must rewrite (4–4.16) in a way that is without consequence for the causal situation but assures its gauge invariance in general. Returning to the momentum space for a moment, we observe that

$$\frac{1}{M^2}(k_\mu k_\nu - g_{\mu\nu}k^2)A^\nu(k) = A_\mu(k) + k_\mu\left(\frac{1}{M^2}k_\nu A^\nu(k)\right) \quad (4–4.18)$$

differs only by a gauge transformation from $A_\mu(k)$, and can replace it in (4–4.16). This is the substitution

$$A_\mu(x) \rightarrow -\frac{1}{M^2}\partial^\nu F_{\mu\nu}(x), \quad (4–4.19)$$

and the resulting space-time extrapolation of (4–4.16) is

$$- i \int_{(2m)^2}^{\infty} \frac{dM^2}{M^2} f(M^2) \int (dx)(dx') j^{\mu}(x) \Delta_+(x - x', M^2) \, \partial'^{\nu} F_{\mu\nu}(x'), \quad (4\text{–}4.20)$$

where we have defined

$$f(M^2) = \frac{\alpha}{2\pi} \left(1 - \frac{4m^2}{M^2} \right)^{-1/2} \left(1 - \frac{2m^2}{M^2} \right) \left(\log \frac{M^2 - 4m^2}{\mu^2} - 2 \right). \quad (4\text{–}4.21)$$

The action term that combines the primitive interaction with the two inter-action-induced modifications that have just been evaluated can be presented as

$$\int (dx)(dx') j^{\mu}(x) f_{\mu\nu}(x - x') \bar{A}^{\nu}(x'), \quad (4\text{–}4.22)$$

with

$$f_{\mu\nu}(x - x') = f_{\nu\mu}(x - x')$$

$$= g_{\mu\nu} \delta(x - x') - (\partial_\mu \partial_\nu - g_{\mu\nu} \partial^2) \int \frac{dM^2}{M^2} f(M^2) \Delta_+(x - x', M^2), \quad (4\text{–}4.23)$$

although this goes somewhat beyond the explicit calculations in its uniform use of the modified field

$$\bar{A}^{\mu}(x) = \int (dx') D_+(x - x') J^{\mu}(x'). \quad (4\text{–}4.24)$$

The four-dimensional momentum space equivalent of (4–4.23) is

$$f_{\mu\nu}(k) = g_{\mu\nu} + (k_\mu k_\nu - g_{\mu\nu} k^2) \int \frac{dM^2}{M^2} \frac{f(M^2)}{k^2 + M^2 - i\varepsilon}. \quad (4\text{–}4.25)$$

If the vector potential is restricted to a Lorentz gauge, and (4–4.24) is so written, the $k_\mu k_\nu$ or $- \partial_\mu \partial_\nu$ terms disappear and

$$f_{\mu\nu}(k) = g_{\mu\nu} F(k), \quad (4\text{–}4.26)$$

in which

$$F(k) = 1 - k^2 \int \frac{dM^2}{M^2} \frac{f(M^2)}{k^2 + M^2 - i\varepsilon}. \quad (4\text{–}4.27)$$

The corresponding form of the interaction (4–4.22) is

$$\int (dx)(dx') j^{\mu}(x) F(x - x') \bar{A}_{\mu}(x') \equiv \int (dx) j^{\mu}(x) \big|_{\text{eff.}} \bar{A}_{\mu}(x); \quad (4\text{–}4.28)$$

the latter way of writing it emphasizes that the current effective in interaction with the vector potential at a given point is a weighted average, of the local field structure j^μ, over all space-time. The term "form factor" applied to $F(x - -.x')$, or $F(k)$, describes the role of this function in producing an additional distribution or shape of the charge represented by $j^\mu(x)$. It should be kept in mind that the single form factor $F(k)$ is a simplified way of presenting the tensor form factor $f_{\mu\nu}(k)$.

Before discussing physical implications of the form factor, let us repeat the calculation for spin $\frac{1}{2}$ particles. The effective two-particle emission source associated with the electromagnetic vector potential [Eq. (4–3.5)] is conveyed by

$$i\eta_2(p)\eta_2(p')|_{\text{eff.}} = eq\gamma^\mu\gamma^0 A_\mu(k). \tag{4–4.29}$$

Since it is now quite clear that the annihilation mechanism of scattering produces the modified photon propagation function, we shall consider only the vacuum amplitude for the Coulomb deflection of the particles. To present this in a form useful for our purposes, it is simplest to return to the interaction expression

$$i\tfrac{1}{2}\int (dx)(dx')\psi_1(x)\gamma^0\gamma^\mu eq\psi_2(x)D_+(x - x')\psi_2(x')\gamma^0\gamma_\mu eq\psi_1(x') \tag{4–4.30}$$

and introduce the causal field structures

$$\psi_1(x)\gamma^0 = \int d\omega_p \, i\eta_1(-p)\gamma^0(m - \gamma p)\exp(-ipx),$$

$$\psi_1(x) = -\int d\omega_p \exp(-ipx)\,(-m - \gamma p)i\eta_1(-p), \tag{4–4.31}$$

and

$$\psi_2(x) = \int d\omega_p \exp(ipx)\,(m - \gamma p)i\eta_2(p),$$

$$\psi_2(x)\gamma^0 = -\int d\omega_p \, i\eta_2(p)\gamma^0(-m - \gamma p)\exp(ipx). \tag{4–4.32}$$

This gives the following vacuum amplitude term:

$$ie^2\tfrac{1}{2}\int d\omega_{p_1}\, d\omega_{p_1'}\, d\omega_{p_2}\, d\omega_{p_2'}\, (2\pi)^4\delta(p_1 + p_1' - p_2 - p_2')\frac{1}{(p_1 - p_2)^2}$$

$$\times \eta_1(-p_1)\gamma^0(m - \gamma p_1)\gamma^\mu q(m - \gamma p_2)\eta_2(p_2)\eta_2(p_2')\gamma^0$$

$$\times (-m - \gamma p_2')\gamma_\mu q(-m - \gamma p_1')\eta_1(-p_1'). \tag{4–4.33}$$

The insertion of the effective source (4–4.29) then produces

$$e^2 \tfrac{1}{2} \int d\omega_{p_1} d\omega_{p_1'} \cdot (2\pi)^4 \, \delta(p_1 + p_1' - k) dM^2 \, d\omega_k \, \eta_1(-p_1) \gamma^0 (m - \gamma p_1)$$

$$\times \, eq I^\mu(-m - \gamma p_1') \eta_1(-p_1') A_\mu(k), \tag{4-4.34}$$

with

$$I^\mu = \int d\omega_{p_2} d\omega_{p_2'} (2\pi)^3 \delta(p_2 + p_2' - k) \gamma^\nu (m - \gamma p_2) \gamma^\mu (-m - \gamma p_2') \gamma_\nu \frac{1}{(p_1 - p_2)^2}.$$
$$\tag{4-4.35}$$

The gauge invariance of this coupling, as expressed by

$$k_\mu I^\mu = 0, \tag{4-4.36}$$

is a consequence of the algebraic properties of the projection matrices $m - \gamma p_2$ and $-m - \gamma p_2'$:

$$(m - \gamma p_2)\gamma k(-m - \gamma p_2') = (m - \gamma p_2)[(\gamma p_2 + m) + (\gamma p_2' - m)](-m - \gamma p_2')$$

$$= 0. \tag{4-4.37}$$

A reduction of the matrices in I^μ is effected with the aid of the projection matrices that appear in (4-4.34), $m - \gamma p_1$ and $-m - \gamma p_1'$. They imply the equivalence of m with a matrix $-\gamma p_1$ that stands on the left of I^μ, and of m with a matrix $\gamma p_1'$ that occurs on the right. This is combined with the algebraic properties of the matrices in the following:

$$\gamma^\nu (m - \gamma p_2) \gamma^\mu (-m - \gamma p_2') \gamma_\nu$$

$$= [2p_2^\nu + (m + \gamma p_2)\gamma^\nu] \gamma^\mu [2p_{2\nu}' + \gamma_\nu(-m + \gamma p_2')]$$

$$\rightarrow [2p_2^\nu - \gamma(p_1 - p_2)\gamma^\nu] \gamma^\mu [2p_{2\nu}' - \gamma_\nu \gamma(p_1' - p_2')]$$

$$= 4p_2 p_2' \gamma^\mu + 2\gamma^\mu \gamma p_2 \gamma(p_1 - p_2) - 2\gamma(p_1 - p_2)\gamma p_2' \gamma^\mu$$

$$- 2\gamma(p_1 - p_2)\gamma^\mu \gamma(p_1 - p_2), \tag{4-4.38}$$

where we have also used the relations

$$p_1' - p_2' = -(p_1 - p_2), \qquad \gamma^\nu \gamma^\mu \gamma_\nu = 2\gamma^\mu. \tag{4-4.39}$$

Further reductions are indicated by

$$\gamma^\mu \gamma p_2 \gamma(p_1 - p_2) = \gamma^\mu \gamma p_1 \gamma(p_1 - p_2) + (p_1 - p_2)^2 \gamma^\mu$$

$$\rightarrow m\gamma^\mu \gamma(p_1 - p_2) - 2p_1^\mu \gamma(p_1 - p_2) + (p_1 - p_2)^2 \gamma^\mu, \tag{4-4.40}$$

and

$$- \gamma(p_1 - p_2)\gamma p_2' \gamma^\mu = - \gamma(p_1 - p_2)\gamma p_1' \gamma^\mu + (p_1 - p_2)^2 \gamma^\mu$$
$$\rightarrow m\gamma(p_1 - p_2)\gamma^\mu + 2p_1'^\mu \gamma(p_1 - p_2) + (p_1 - p_2)^2 \gamma^\mu. \quad (4\text{-}4.41)$$

These lead to a form suitable for integration,

$$\gamma^\nu(m - \gamma p_2)\gamma^\mu(-m - \gamma p_2')\gamma_\nu \rightarrow - 2(M^2 - 2m^2)\gamma^\mu + 2(p_1 - p_2)^2 \gamma^\mu$$
$$- 4m(p_1 - p_2)^\mu + 4(p_1' - p_2)^\mu \gamma(p_1 - p_2). \quad (4\text{-}4.42)$$

Using the notation indicated in (4-4.12) we now have

$$I^\mu = - \frac{2}{(4\pi)^2}\left(1 - \frac{4m^2}{M^2}\right)^{1/2}\left[(M^2 - 2m^2)\left\langle \frac{1}{(p_1 - p_2)^2}\right\rangle \gamma^\mu - \gamma^\mu \right.$$
$$\left. + 2m\left\langle \frac{(p_1 - p_2)^\mu}{(p_1 - p_2)^2}\right\rangle - 2\left\langle \frac{(p_1' - p_2)^\mu(p_1 - p_2)^\nu}{(p_1 - p_2)^2}\right\rangle \gamma_\nu \right]. \quad (4\text{-}4.43)$$

First, let us observe that

$$\left\langle \frac{(p_1 - p_2)^\mu}{(p_1 - p_2)^2}\right\rangle = a(p_1 - p_1')^\mu, \quad (4\text{-}4.44)$$

because the interchange of p_1 and p_1', combined with that of p_2 and p_2', reverses the sign of this vector.   Multiplication of both sides with $2p_{1\mu}$ gives

$$a = \frac{1}{M^2 - 4m^2}, \quad (4\text{-}4.45)$$

since

$$2p_1(p_1 - p_2) = (p_1 - p_2)^2,$$
$$2p_1(p_1 - p_1') = (p_1 - p_1')^2 = M^2 - 4m^2. \quad (4\text{-}4.46)$$

Then, using the reductions

$$(p_1 - p_1')^\mu = - \tfrac{1}{2}\{\gamma^\mu, \gamma p_1 - \gamma p_1'\}$$
$$\rightarrow - \tfrac{1}{2}\gamma^\mu(\gamma k - 2m) - \tfrac{1}{2}(-\gamma k - 2m)\gamma^\mu$$
$$= 2m\gamma^\mu + i\sigma^{\mu\nu}k_\nu, \quad (4\text{-}4.47)$$

we get

$$2m\left\langle \frac{(p_1 - p_2)^\mu}{(p_1 - p_2)^2}\right\rangle \rightarrow \frac{2m}{M^2 - 4m^2}(2m\gamma^\mu + i\sigma^{\mu\nu}k_\nu). \quad (4\text{-}4.48)$$

A related contribution can be separated from the last term of (4-4.43):

$$2(p_1 - p_1')^\mu\left\langle \frac{(p_1 - p_2)^\nu}{(p_1 - p_2)^2}\right\rangle \gamma_\nu = \frac{2}{M^2 - 4m^2}(p_1 - p_1')^\mu \gamma(p_1 - p_1')$$

$$\rightarrow -\frac{4m}{M^2 - 4m^2}(p_1 - p_1')^\mu$$

$$\rightarrow -\frac{4m}{M^2 - 4m^2}(2m\gamma^\mu + i\sigma^{\mu\nu}k_\nu). \tag{4-4.49}$$

Now consider

$$\left\langle \frac{(p_1 - p_2)^\mu (p_1 - p_2)^\nu}{(p_1 - p_2)^2} \right\rangle = b\left(g^{\mu\nu} + \frac{k^\mu k^\nu}{M^2}\right) + c(p_1 - p_1')^\mu (p_1 - p_1')^\nu, \tag{4-4.50}$$

where the right-hand side gives the general form of a symmetrical tensor that is unaltered by interchanging p_1 and p_1', and vanishes on multiplication with k_μ. Contracting indices on both sides supplies one relation between b and c,

$$1 = 3b + (M^2 - 4m^2)c. \tag{4-4.51}$$

The other is produced by multiplication with $(p_1 - p_1')_\mu = (2p_1 - k)_\mu$,

$$\tfrac{1}{2} = b + (M^2 - 4m^2)c, \tag{4-4.52}$$

which involves the observation that

$$\langle p_2^\nu \rangle = \tfrac{1}{2}(p_1 + p_1')^\nu. \tag{4-4.53}$$

The result,

$$b = \frac{1}{4}, \qquad c = \frac{1}{4}\frac{1}{M^2 - 4m^2}, \tag{4-4.54}$$

is displayed in

$$-2\left\langle \frac{(p_1 - p_2)^\mu (p_1 - p_2)^\nu}{(p_1 - p_2)^2} \right\rangle \gamma_\nu$$

$$= -\frac{1}{2}\left(\gamma^\mu + \frac{k^\mu}{M^2}\gamma k\right) - \frac{1}{2}\frac{1}{M^2 - 4m^2}(p_1 - p_1')^\mu \gamma(p_1 - p_1')$$

$$\rightarrow -\tfrac{1}{2}\gamma^\mu + \frac{m}{M^2 - 4m^2}(2m\gamma^\mu + i\sigma^{\mu\nu}k_\nu), \tag{4-4.55}$$

which also contains the remark:

$$\gamma k = (\gamma p_1 + m) + (\gamma p_1' - m) \rightarrow 0. \tag{4-4.56}$$

Putting these evaluations together gives

$$I^\mu = -\frac{2}{(4\pi)^2}\left(1 - \frac{4m^2}{M^2}\right)^{1/2}\left[(M^2 - 2m^2)\left\langle \frac{1}{(p_1 - p_2)^2} \right\rangle \gamma^\mu - \frac{3}{2}\gamma^\mu \right.$$

$$-\frac{m}{M^2 - 4m^2}(2m\gamma^\mu + i\sigma^{\mu\nu}k_\nu)\bigg].\tag{4-4.57}$$

As in the spin 0 discussion, the remaining integral is altered to

$$\left\langle\frac{1}{(p_1 - p_2)^2 + \mu^2}\right\rangle = \frac{1}{M^2 - 4m^2}\log\frac{M^2 - 4m^2}{\mu^2}\tag{4-4.58}$$

and we now get

$$e^2 I^\mu = -f_1(M^2)\gamma^\mu + (\tfrac{1}{2}g - 1)\frac{1}{2m}f_2(M^2)i\sigma^{\mu\nu}k_\nu,\tag{4-4.59}$$

where

$$f_1(M^2) = \frac{\alpha}{2\pi}\left(1 - \frac{4m^2}{M^2}\right)^{-1/2}\left[\left(1 - \frac{2m^2}{M^2}\right)\log\frac{M^2 - 4m^2}{\mu^2} - \frac{3}{2} + \frac{4m^2}{M^2}\right]\tag{4-4.60}$$

and

$$(\tfrac{1}{2}g - 1)f_2(M^2) = \frac{\alpha}{2\pi}\left(1 - \frac{4m^2}{M^2}\right)^{-1/2}\frac{2m^2}{M^2}.\tag{4-4.61}$$

Despite the apparent detailed dependence of I^μ [Eq. (4-4.35)] on p_1 and p_1', the kinematical reductions have produced a form [Eq. (4-4.59)] in which just the total momentum k is in evidence. Accordingly, the reference to the final particles in (4-4.34) occurs only in the two local field combinations

$$\tfrac{1}{2}\psi_1(x)\gamma^0 eq\gamma^\mu\psi_1(x) = \tfrac{1}{2}\int d\omega_{p_1}d\omega_{p_1'}\cdot\eta_1(-p_1)\gamma^0(m - \gamma p_1)eq\gamma^\mu(-m - \gamma p_1')$$

$$\times \eta_1(-p_1')\exp[-i(p_1 + p_1')x],\tag{4-4.62}$$

and

$$\tfrac{1}{2}\psi_1(x)\gamma^0 eq\sigma^{\mu\nu}\psi_1(x) = \tfrac{1}{2}\int d\omega_{p_1}d\omega_{p_1'}\cdot\eta_1(-p_1)\gamma^0(m - \gamma p_1)eq\sigma^{\mu\nu}(-m - \gamma p_1')$$

$$\times \eta_1(-p_1')\exp[-i(p_1 + p_1')x].\tag{4-4.63}$$

The coupling (4-4.34) thus appears as

$$-i\int\frac{dM^2}{M^2}f_1(M^2)\int(dx)(dx')\tfrac{1}{2}\psi(x)\gamma^0 eq\gamma^\mu\psi(x)$$

$$\times\left[i\int d\omega_k\exp[ik(x - x')]\right]\partial'^\nu F_{\mu\nu}(x')$$

$$+ i(\tfrac{1}{2}g - 1)\int dM^2 f_2(M^2)\int(dx)(dx')\tfrac{1}{2}\psi(x)\gamma^0\frac{eq}{2m}\sigma^{\mu\nu}\psi(x)$$

$$\times \left[i \int d\omega_k \exp[ik(x - x')] \right] \tfrac{1}{2} F_{\mu\nu}(x'), \tag{4-4.64}$$

where gauge invariance has been made explicit by the substitution (4–4.19). The space-time extrapolation of the additional coupling is combined with the primitive interaction to give the action term

$$\int (dx)(dx') \left[\tfrac{1}{2}\psi(x)\gamma^0 eq\gamma^\mu \psi(x) f_{\mu\nu}(x - x')\bar{A}^\nu(x') \right.$$

$$\left. + (\tfrac{1}{2}g - 1)\tfrac{1}{2}\psi(x)\gamma^0 \frac{eq}{2m} \sigma^{\mu\nu}\psi(x) F_2(x - x')\tfrac{1}{2}F_{\mu\nu}(x') \right], \tag{4-4.65}$$

in which

$$f_{\mu\nu}(k) = g_{\mu\nu} + (k_\mu k_\nu - g_{\mu\nu}k^2) \int \frac{dM^2}{M^2} \frac{f_1(M^2)}{k^2 + M^2 - i\varepsilon}. \tag{4-4.66}$$

In a Lorentz gauge, this tensor can be replaced by

$$f_{\mu\nu}(k) = g_{\mu\nu} F_1(k) \tag{4-4.67}$$

with the charge form factor

$$F_1(k) = 1 - k^2 \int \frac{dM^2}{M^2} \frac{f_1(M^2)}{k^2 + M^2 - i\varepsilon}. \tag{4-4.68}$$

Notice that

$$F_1(k^2 = 0) = 1, \tag{4-4.69}$$

which states that the primitive interaction is not modified for slowly varying fields ($|k^2| \ll m^2$). We shall give the same normalization to the second form factor,

$$F_2(k) = \int dM^2 \frac{f_2(M^2)}{k^2 + M^2 - i\varepsilon}, \qquad F_2(k^2 = 0) = 1. \tag{4-4.70}$$

According to the observation that

$$\int_{(2m)^2}^{\infty} \frac{dM^2}{M^2} \frac{2m^2}{M^2} \left(1 - \frac{4m^2}{M^2}\right)^{-1/2} = \int_{(2m)^2}^{\infty} d\left(1 - \frac{4m^2}{M^2}\right)^{1/2} = 1, \tag{4-4.71}$$

this is achieved with

$$f_2(M^2) = \frac{2m^2}{M^2} \left(1 - \frac{4m^2}{M^2}\right)^{-1/2}, \tag{4-4.72}$$

and then

$$\tfrac{1}{2}g - 1 = \frac{\alpha}{2\pi}.\qquad(4\text{–}4.73)$$

As the notation indicates, the last result is a rederivation of the additional magnetic moment appropriate to homogeneous fields [Eq. (4–2.36)]. The dynamical origin of this magnetic moment is evidenced here in the vanishing of the magnetic moment form factor $F_2(k)$ for $k^2 \to \infty$. An asymptotic statement is

$$k^2 \gg m^2: \quad F_2(k) \sim \frac{2m^2}{k^2} \log \frac{k^2}{m^2};\qquad(4\text{–}4.74)$$

it follows from the general form

$$F_2(k) = \int_0^1 dv \; \frac{1}{1 + \left(\dfrac{k^2}{4m^2}\right)(1 - v^2) - i\varepsilon}$$

$$= \frac{2m^2}{k^2}\left(1 + \frac{4m^2}{k^2}\right)^{-1/2} \log \frac{\left(1 + \dfrac{4m^2}{k^2}\right)^{1/2} + 1}{\left(1 + \dfrac{4m^2}{k^2}\right)^{1/2} - 1},\qquad(4\text{–}4.75)$$

where

$$v = \left(1 - \frac{4m^2}{M^2}\right)^{1/2}.\qquad(4\text{–}4.76)$$

For $-k^2 > 4m^2$, incidentally, the phase of the logarithm is appropriately chosen to give the imaginary part of the integral, $\pi i f_2(-k^2)$. The situation seems different for the charge form factor, where

$$\int \frac{dM^2}{M^2} \frac{f_1(M^2)}{k^2 + M^2 - i\varepsilon} = \frac{\alpha}{2\pi} \frac{1}{4m^2} \int_0^1 dv \; \frac{(1 + v^2) \log\left(\dfrac{4m^2}{\mu^2} \dfrac{v^2}{1 - v^2}\right) - 1 - 2v^2}{1 + \left(\dfrac{k^2}{4m^2}\right)(1 - v^2) - i\varepsilon}$$

$$\sim \frac{\alpha}{2\pi} \frac{1}{k^2} \log \frac{k^2}{m^2} \log \frac{k^2}{\mu^2},\qquad(4\text{–}4.77)$$

which exhibits only the dominant logarithmic terms. Since this function is multiplied by $-k^2$ in $F_1(k)$, the addition to the primitive interaction does not vanish as $k^2 \to \infty$. But another view of the matter appears on reinstating the tensor form factor $f_{\mu\nu}(k)$, for

$$f_{\mu\nu}\bar{A}^{\nu}(k) = \bar{A}_{\mu}(k) - \int \frac{dM^2}{M^2} \frac{f_1(M^2)}{k^2 + M^2 - i\varepsilon} \, \partial^\nu \bar{F}_{\mu\nu}(k)\qquad(4\text{–}4.78)$$

contains two different couplings. One is the primitive interaction, modified by vacuum polarization effects, and the other is a dynamically induced coupling with the field $\partial^\nu F_{\mu\nu}$, which does vanish as $k^2 \to \infty$.

The latter remark of principle notwithstanding, the practical consequence of the charge form factor structure [Eq. (4–4.68)] is a progressive weakening of the corresponding interaction with increasing k^2. This will be manifested in energy level displacements and in scattering cross section alterations. As a first indication of these effects, consider slowly varying fields for which one can simplify $F_1(k)$ to

$$F_1(k) \simeq 1 - k^2 \int_{(2m)^2}^{\infty} \frac{dM^2}{(M^2)^2} f_1(M^2). \tag{4–4.79}$$

The integral that appears here, according to (4–4.77), is

$$\frac{\alpha}{2\pi} \frac{1}{4m^2} \int_0^1 dv \left[(1 + v^2) \log\left(\frac{4m^2}{\mu^2} \frac{v^2}{1 - v^2} \right) - 1 - 2v^2 \right] = \frac{\alpha}{3\pi} \frac{1}{m^2} \left(\log \frac{m}{\mu} - \frac{3}{8} \right), \tag{4–4.80}$$

and

$$F_1(k) \simeq 1 - \frac{\alpha}{3\pi} \frac{k^2}{m^2} \left(\log \frac{m}{\mu} - \frac{3}{8} \right). \tag{4–4.81}$$

This effect can be compared to and combined with the vacuum polarization increase in coupling, which is described, in (4–3.76), for these circumstances by

$$\bar{A}_\mu(k) = \left(1 + \frac{\alpha}{15\pi} \frac{k^2}{m^2} \right) A_\mu(k). \tag{4–4.82}$$

Thus,

$$F_1(k) \bar{A}_\mu(k) = \left[1 - \frac{\alpha}{3\pi} \frac{k^2}{m^2} \left(\log \frac{m}{\mu} - \frac{3}{8} - \frac{1}{5} \right) \right] A_\mu(k), \tag{4–4.83}$$

which suggests that the form factor reduction quite overpowers the vacuum polarization strengthening, although the effective replacement for the transitional parameter μ must be clarified. In discussing energy level displacements, for example, it is quite plausible that some atomic excitation energy ΔE stands in the place of μ. The replacement $-\frac{3}{8} \to \log(m/\Delta E)$, which ignores additive constants, converts the vacuum polarization formula (4–3.57) into

$$\frac{\delta E_{ns}}{\left(\frac{1}{2} \frac{Z^2 \alpha^2}{n^2} m \right)} \sim \frac{8}{3\pi} \frac{Z^2 \alpha^3}{n} \log \frac{m}{\Delta E}. \tag{4–4.84}$$

This is indeed a correct indication, but we shall defer the more precise discussion, turning now to the consideration of scattering modifications.

We begin with spin 0 particles that are scattered by a Coulomb field. In deflecting a particle from momentum $\mathbf{p}_2$ to momentum $\mathbf{p}_1$, the Coulomb field supplies the momentum

$$\mathbf{q} = \mathbf{p}_1 - \mathbf{p}_2. \tag{4-4.85}$$

The effects we have been discussing modify the differential cross section by the factor

$$\left| \frac{F(\mathbf{q})D_+(\mathbf{q})}{D_+(\mathbf{q})} \right|^2$$

$$\cong 1 - 2q^2 \int dM^2 \frac{(1/M^2)f(M^2) - a(M^2)}{q^2 + M^2}$$

$$= 1 - \frac{\alpha}{4\pi} \frac{q^2}{m^2} \int_0^1 dv \frac{(1 + v^2)\left(\log\left(\dfrac{4m^2}{\mu^2} \dfrac{v^2}{1 - v^2}\right) - 2\right) - \dfrac{1}{3} v^4}{1 + (q^2/4m^2)(1 - v^2)}, \tag{4-4.86}$$

which contains only the spin 0 contribution to vacuum polarization; correction terms not linear in α have been discarded. To this elastic cross section must be added, at least, the cross section for scattering with emission of a soft photon having any energy less than $k^0_{\text{min.}}$, the minimum detectable photon energy in the experimental arrangement. This produces the physically meaningful cross section for essentially elastic scattering, which can always be supplemented by the cross section for radiative scattering with any degree of inelasticity.

In evaluating the soft-photon addition we must be consistent. The introduction of a photon mass requires that the photon be treated as a spin one particle, with its three polarization states. In practice, however, the only significant point is the altered relation between photon energy and momentum,

$$k^0 = (\mathbf{k}^2 + \mu^2)^{1/2}. \tag{4-4.87}$$

The transition matrix element for soft-photon emission during deflection in a Coulomb field appears in Eq. (3–14.61) as a product of two factors. One of these describes the elastic scattering process; the other is

$$(d\omega_k)^{1/2} eq e_{k\lambda}^* \left(\frac{p_1}{kp_1} - \frac{p_2}{kp_2} \right). \tag{4-4.88}$$

To be accurate, the denominators should be altered slightly, since

$$\frac{1}{(k + p_1)^2 + m^2} = \frac{1}{2kp_1 - \mu^2}, \qquad \frac{1}{(-k + p_2)^2 + m^2} = -\frac{1}{2kp_2 + \mu^2}. \tag{4-4.89}$$

However, the corrections are not larger than the ratio between μ and a particle energy, which is completely negligible, for we certainly require that

$$\mu \ll k^0_{\text{min.}}.$$ (4–4.90)

One may be given pause on noting that the polarization vector summation

$$\sum_\lambda e^\mu_{k\lambda} e^{\nu*}_{k\lambda} = g^{\mu\nu} + (k^\mu k^\nu / \mu^2)$$ (4–4.91)

contains the factor $1/\mu^2$, for, with the altered denominators, the effective photon source given in (4–4.88) is not conserved. Happily,

$$k\left(\frac{p_1}{kp_1 - \frac{1}{2}\mu^2} - \frac{p_2}{kp_2 + \frac{1}{2}\mu^2}\right) = \frac{1}{2}\mu^2\left(\frac{1}{kp_1 - \frac{1}{2}\mu^2} + \frac{1}{kp_2 + \frac{1}{2}\mu^2}\right)$$ (4–4.92)

has enough powers of μ in the numerator to save the day. And, of course, (4–4.88) is an approximation to an expression for which conservation holds exactly.

The relative soft photon emission probability is now obtained as

$$e^2 \int d\omega_k \left(\frac{p_1}{kp_1} - \frac{p_2}{kp_2}\right)^2 = 4\pi\alpha \int \frac{(dk)}{(2\pi)^3} \frac{1}{2k^0}\left[-\frac{2p_1p_2}{kp_1kp_2} - \frac{m^2}{(kp_1)^2} - \frac{m^2}{(kp_2)^2}\right],$$ (4–4.93)

or

$$\frac{\alpha}{4\pi^2} \int dk^0 |\mathbf{k}| \, d\Omega \left[(2m^2 + \mathbf{q}^2) \int_{-1}^1 \frac{dv}{2} \frac{1}{\left[k^0 p^0 - \mathbf{k}\cdot\left(\frac{1+v}{2}\mathbf{p}_1 + \frac{1-v}{2}\mathbf{p}_2\right)\right]^2}\right.$$

$$\left. - \frac{m^2}{(k^0 p^0 - \mathbf{k}\cdot\mathbf{p}_1)^2} - \frac{m^2}{(k^0 p^0 - \mathbf{k}\cdot\mathbf{p}_2)^2}\right],$$ (4–4.94)

which uses the combinatorial device introduced in (3–14.116). After carrying out the angular integrations this becomes

$$\frac{2\alpha}{\pi} \int_\mu^{k^0_{\text{min.}}} dk^0 |\mathbf{k}| \left[\left(1 + \frac{\mathbf{q}^2}{2m^2}\right)\int_0^1 dv \frac{1}{\left(1 + \frac{\mathbf{q}^2}{4m^2}(1 - v^2)\right)(k^0)^2 + \frac{\mu^2}{m^2}\left(\mathbf{p}^2 - \mathbf{q}^2\frac{1-v^2}{4}\right)}\right.$$

$$\left. - \frac{1}{(k^0)^2 + \frac{\mu^2}{m^2}\mathbf{p}^2}\right],$$ (4–4.95)

where

$$\mathbf{p}^2 = \mathbf{p}_1^2 = \mathbf{p}_2^2.$$ (4–4.96)

The remaining integral has the form $(k^0_{\text{min.}} \gg \mu)$

$$\int_\mu^{k^0_{\text{min.}}} dk^0((k^0)^2 - \mu^2)^{1/2} \frac{1}{(k^0)^2 + \lambda^2\mu^2}$$

$$= \log \frac{2k^0_{\min.}}{\mu} - \frac{1}{2} \left(1 + \frac{1}{\lambda^2}\right)^{1/2} \log \frac{\left(1 + \frac{1}{\lambda^2}\right)^{1/2} + 1}{\left(1 + \frac{1}{\lambda^2}\right)^{1/2} - 1}, \qquad (4\text{–}4.97)$$

and the resulting evaluation of (4–4.95) is given by

$$\frac{\alpha}{2\pi} \frac{q^2}{m^2} \log \frac{2k^0_{\min.}}{\mu} \int_0^1 dv \, \frac{1 + v^2}{1 + \frac{q^2}{4m^2}(1 - v^2)}$$

$$- \frac{\alpha}{\pi} \left[\int_0^1 dv \, \frac{1 + \frac{q^2}{2m^2}}{1 + \frac{q^2}{4m^2}(1 - v^2)} \frac{p^0}{\left(\mathbf{p}^2 - q^2 \frac{1 - v^2}{4}\right)^{1/2}} \right.$$

$$\left. \times \log \frac{p^0 + \left(\mathbf{p}^2 - q^2 \frac{1 - v^2}{4}\right)^{1/2}}{p^0 - \left(\mathbf{p}^2 - q^2 \frac{1 - v^2}{4}\right)^{1/2}} - \frac{p^0}{|\mathbf{p}|} \log \frac{p^0 + |\mathbf{p}|}{p^0 - |\mathbf{p}|} \right]. \qquad (4\text{–}4.98)$$

Although the v parameters of Eqs. (4–4.86) and (4–4.98) were introduced in very different ways, the two expressions can be added, and the fictitious mass μ does indeed cancel completely. Let us consider first the nonrelativistic situation, where $q^2 \ll m^2$, $\mathbf{p}^2 \ll m^2$, $p^0 \simeq m$. The elementary evaluations of the two terms are exhibited in

$$\left[1 - \frac{2\alpha}{3\pi} \frac{q^2}{m^2} \left(\log \frac{m}{\mu} - \frac{31}{40}\right)\right] + \left[\frac{2\alpha}{3\pi} \frac{q^2}{m^2} \left(\log \frac{2k^0_{\min.}}{\mu} - \frac{5}{6}\right)\right]$$

$$= 1 - \frac{2\alpha}{3\pi} \frac{q^2}{m^2} \left(\log \frac{m}{2k^0_{\min.}} + \frac{7}{120}\right). \qquad (4\text{–}4.99)$$

Also available is the nonrelativistic differential cross section for all inelastic processes, comprising photon energies that range from $k^0_{\min.}$ up to the kinetic energy T of the incident particle [Eq. (3–14.70)]. Presented as a fraction of the elastic differential cross section, this inelastic contribution is

$$\frac{2\alpha}{3\pi} \frac{q^2}{m^2} \left[\log \frac{4T}{k^0_{\min.}} - (\pi - \theta) \tan \tfrac{1}{2}\theta - \frac{\cos \theta}{\cos^2 \tfrac{1}{2}\theta} \log \frac{1}{\sin \tfrac{1}{2}\theta}\right]. \qquad (4\text{–}4.100)$$

The essentially elastic and inelastic cross sections combine to give the differential cross section for scattering without regard to the final energy of the particle. It is the following fraction of the Rutherford cross section:

$$1 - \frac{2\alpha}{3\pi} \frac{q^2}{m^2} \left[\log \frac{m}{8T} + \frac{7}{120} + (\pi - \theta) \tan \tfrac{1}{2}\theta + \frac{\cos \theta}{\cos^2 \tfrac{1}{2}\theta} \log \frac{1}{\sin \tfrac{1}{2}\theta} \right]. \quad (4\text{--}4.101)$$

These dynamical modifications of scattering cross sections remain quite small until we penetrate into the relativistic region. To facilitate the consideration of high energies, we introduce the symbols

$$\xi(v) = \frac{1}{p^0} \left(\mathbf{p}^2 - q^2 \frac{1 - v^2}{4} \right)^{1/2}, \qquad \beta = \frac{|\mathbf{p}|}{p^0}, \quad (4\text{--}4.102)$$

and also make the scattering angle θ explicit in

$$q^2 = 4\mathbf{p}^2 \sin^2 \tfrac{1}{2}\theta. \quad (4\text{--}4.103)$$

As v grows from 0 to 1, the new variable ξ changes from $\beta \cos \tfrac{1}{2}\theta$ to β. In the special situation of back scattering, $\theta = \pi$, we have $\xi(v) = \beta v$. With this notation the complicated v integral of the second term in (4–4.98) appears as

$$\left(1 + \frac{q^2}{2m^2}\right)\left(\frac{m}{p^0}\right)^2 \int_0^1 dv \, \frac{1}{1 - \xi^2} \frac{1}{\xi} \log \frac{1 + \xi}{1 - \xi}, \quad (4\text{--}4.104)$$

which also uses the relation

$$1 + \frac{q^2}{4m^2}(1 - v^2) = \left(\frac{p^0}{m}\right)^2 (1 - \xi^2). \quad (4\text{--}4.105)$$

We now observe that

$$\frac{1}{1 - \xi^2} \frac{1}{\xi} \log \frac{1 + \xi}{1 - \xi} + \frac{1}{1 - \xi^2} \log \frac{1 - \xi^2}{4} = \frac{1}{\xi} \left[\frac{\log \tfrac{1}{2}(1 + \xi)}{1 - \xi} - \frac{\log \tfrac{1}{2}(1 - \xi)}{1 + \xi} \right],$$

$$(4\text{--}4.106)$$

where the right-hand side is only weakly singular as $\xi \to 1$. A similar statement, with ξ replaced by β, is also useful. After defining the quantity

$$\frac{q^2}{2m^2} L = \left(1 + \frac{q^2}{2m^2}\right) \int_0^1 dv \, \frac{1}{\xi} \left[\frac{\log \tfrac{1}{2}(1 + \xi)}{1 - \xi} - \frac{\log \tfrac{1}{2}(1 - \xi)}{1 + \xi} \right]$$
$$- \frac{1}{\beta} \left(\frac{\log \tfrac{1}{2}(1 + \beta)}{1 - \beta} - \frac{\log \tfrac{1}{2}(1 - \beta)}{1 + \beta} \right), \quad (4\text{--}4.107)$$

one can present (4–4.98) in the following manner,

$$\frac{\alpha}{2\pi} \frac{q^2}{m^2} \log \left(\frac{k^0_{\text{min.}}}{\mu} \frac{m}{p^0} \right) \int_0^1 dv \, \frac{1 + v^2}{1 + \frac{q^2}{4m^2}(1 - v^2)} - \frac{\alpha}{2\pi} \frac{q^2}{(p^0)^2} L$$

$$+ \frac{\alpha}{\pi}\left(1 + \frac{q^2}{2m^2}\right)\int_0^1 dv \; \frac{\log\left(1 + \frac{q^2}{4m^2}(1 - v^2)\right)}{1 + \frac{q^2}{4m^2}(1 - v^2)}. \tag{4-4.108}$$

This is not the final version, however, for the last term of (4–4.108) has a more relevant form, which is the content of an identity:

$$\left(1 + \frac{q^2}{2m^2}\right)\int_0^1 dv \; \frac{\log\left(1 + \frac{q^2}{4m^2}(1 - v^2)\right)}{1 + \frac{q^2}{4m^2}(1 - v^2)} = \frac{q^2}{4m^2}\int_0^1 dv \; \frac{(1 + v^2)\log\frac{4v^2}{1 - v^2}}{1 + \frac{q^2}{4m^2}(1 - v^2)}; \tag{4-4.109}$$

it is stated for space-like q. The proof begins with this identity,

$$\left(1 + \frac{q^2}{2m^2}\right)\int_0^1 dv \; \frac{\log\left(1 + \frac{q^2}{4m^2}(1 - v^2)\right)}{1 + \frac{q^2}{4m^2}(1 - v^2)}$$

$$= \frac{q^2}{4m^2}\int_0^1 dv \int_0^1 \frac{du}{1 - u}\left[\frac{2 - (1 - v^2)u}{1 + \frac{q^2}{4m^2}(1 - v^2)u} - \frac{1 + v^2}{1 + \frac{q^2}{4m^2}(1 - v^2)}\right], \tag{4-4.110}$$

which is verified by performing the u integration. Now let us relabel the v parameter of the first term on the right side as v', and define

$$1 - v^2 = (1 - v'^2)u. \tag{4-4.111}$$

The new v variable also ranges from 0 to 1. The transformation implies that

$$v' \, dv' = \frac{v \, dv}{u}, \qquad dv' = \frac{v \, dv}{[u(u - (1 - v^2))]^{1/2}}, \tag{4-4.112}$$

which emphasizes the restriction $u \geqslant 1 - v^2$, for an assigned v. The new form of (4–4.110) is

$$\frac{q^2}{4m^2}\int_0^1 dv \; \frac{1 + v^2}{1 + \frac{q^2}{4m^2}(1 - v^2)}\left\{\int_{1-v^2}^1 \frac{du}{1 - u}\left[\frac{v}{[u(u - (1 - v^2))]^{1/2}} - 1\right] - \int_0^{1-v^2} \frac{du}{1 - u}\right\},$$

$$\tag{4-4.113}$$

and the identity (4–4.109) incorporates the value of the u integral, $\log(4v^2/1 - v^2)$.

Now we can effectively combine (4–4.108, 109) with (4–4.86) to produce the dynamical correction factor

$$1 - \frac{\alpha}{2\pi} \frac{q^2}{m^2} \left[\left(\log \frac{p^0}{k^0_{\text{min.}}} - 1 \right) \int_0^1 dv \, \frac{1 + v^2}{1 + \frac{q^2}{4m^2}(1 - v^2)} \right.$$

$$\left. - \frac{1}{6} \int_0^1 dv \frac{v^4}{1 + \frac{q^2}{4m^2}(1 - v^2)} + \left(\frac{m}{p^0} \right)^2 L \right]. \qquad (4\text{-}4.114)$$

The complexity of this structure resides entirely in the integral L, Eq. (4-4.107), with its somewhat involved relation between the integration variable v and

$$\xi(v) = \beta(1 - (1 - v^2) \sin^2 \tfrac{1}{2}\theta)^{1/2}. \qquad (4\text{-}4.115)$$

There are some simple limiting situations, however. At low energies,

$$p^0 \sim m: \quad L = \frac{4}{3}(1 - \log 2) + \frac{1}{9}, \qquad (4\text{-}4.116)$$

from which we recover (4-4.99). In the high energy limit, $\beta \to 1$, and

$$p^0 \gg m: \quad L = \frac{1}{\sin(\theta/2)} \phi(\cos \tfrac{1}{2}\theta), \qquad (4\text{-}4.117)$$

where

$$\phi(\cos \tfrac{1}{2}\theta) = \int_{\cos \frac{1}{2}\theta}^1 \frac{d\xi}{(\xi^2 - \cos^2 \tfrac{1}{2}\theta)^{1/2}} \left[\frac{\log \tfrac{1}{2}(1 + \xi)}{1 - \xi} - \frac{\log \tfrac{1}{2}(1 - \xi)}{1 + \xi} \right]. \quad (4\text{-}4.118)$$

The integral can be performed analytically for scattering at the angle $\theta = \pi$,

$$\phi(0) = \frac{\pi^2}{12}, \qquad (4\text{-}4.119)$$

which also continues to be quite accurate even at $\theta = \pi/2$. Close to the forward direction,

$$\phi(\cos \tfrac{1}{2}\theta \sim 1) \sim \tfrac{1}{2} \sin \tfrac{1}{2}\theta \left[\log \frac{1}{2(1 - \cos \tfrac{1}{2}\theta)} + 1 \right]. \qquad (4\text{-}4.120)$$

The accompanying high energy form of (4-4.114) is

$$1 - \frac{4\alpha}{\pi} \left[\left(\log \frac{p^0}{k^0_{\text{min.}}} - \frac{13}{12} \right) \left(\log \frac{2p^0}{m} \sin \tfrac{1}{2}\theta - \tfrac{1}{2} \right) + \frac{5}{72} + \frac{1}{2} \sin \frac{\theta}{2} \phi(\cos \tfrac{1}{2}\theta) \right],$$

$$(4\text{-}4.121)$$

which uses the high energy limits of the various integrals,

$$K_0 = \int_0^1 dv \frac{1}{1 + \lambda^2(1 - v^2)} = \frac{\log[\lambda + (\lambda^2 + 1)^{1/2}]}{\lambda(\lambda^2 + 1)^{1/2}},$$

$$K_1 = \int_0^1 dv \frac{v^2}{1 + \lambda^2(1 - v^2)} = \frac{(1 + \lambda^2)K_0 - 1}{\lambda^2},$$

$$K_2 = \int_0^1 dv \frac{v^4}{1 + \lambda^2(1 - v^2)} = \frac{(1 + \lambda^2)K_1 - \frac{1}{3}}{\lambda^2}. \qquad (4\text{-}4.122)$$

The scattering of spin $\frac{1}{2}$ particles introduces a new feature, for here the magnetic moment interaction also plays a role. Compared to the discussion of Section 3–14, the dynamical corrections have altered the coupling, as indicated by [Eqs. (4–4.65, 67)]

$$\gamma A(q) \rightarrow F_1(q)\gamma \bar{A}(q) + \frac{\alpha}{2\pi}\frac{1}{2m} F_2(q)\tfrac{1}{2}\sigma^{\mu\nu}\bar{F}_{\mu\nu}(q). \qquad (4\text{-}4.123)$$

When only the scalar potential appears, describing the modified Coulomb potential, this reduces to

$$\gamma^0 A^0(\mathbf{q}) \rightarrow \left[F_1(\mathbf{q})\gamma^0 + \frac{\alpha}{2\pi}\frac{1}{2m} F_2(\mathbf{q})i\gamma_5\boldsymbol{\sigma}\cdot\mathbf{q}\right]\bar{A}^0(\mathbf{q}). \qquad (4\text{-}4.124)$$

The transition matrix element, replacing (3–14.11), contains

$$u_{p_1\sigma_1}^* \left(F_1(\mathbf{q}) + \frac{\alpha}{2\pi}\frac{1}{2m} F_2(\mathbf{q})\gamma^0 i\gamma_5\boldsymbol{\sigma}\cdot\mathbf{q}\right) u_{p_2\sigma_2}\bar{A}^0(\mathbf{q}). \qquad (4\text{-}4.125)$$

When helicity states are used, as in (3–14.13), we now encounter

$$u_{p_1\sigma_1}^*\gamma^0 i\gamma_5\boldsymbol{\sigma}\cdot(\mathbf{p}_1 - \mathbf{p}_2)u_{p_2\sigma_2} = |\mathbf{p}|(\sigma_1 - \sigma_2)u_{p_1\sigma_1}^*\gamma^0 i\gamma_5 u_{p_2\sigma_2}, \qquad (4\text{-}4.126)$$

where, it is recalled [Eq. (2–6.90)],

$$u_{p_2\sigma_2} = \left[\left(\frac{p^0 + m}{2m}\right)^{1/2} + i\gamma_5\sigma_2\left(\frac{p^0 - m}{2m}\right)^{1/2}\right]v_{\sigma_2},$$

$$u_{p_1\sigma_1}^*\gamma^0 = v_{\sigma_1}^*\left[\left(\frac{p^0 + m}{2m}\right)^{1/2} - \left(\frac{p^0 - m}{2m}\right)^{1/2}i\gamma_5\sigma_1\right]. \qquad (4\text{-}4.127)$$

Accordingly,

$$u_{p_1\sigma_1}^*\gamma^0 i\gamma_5 u_{p_2\sigma_2} = (\sigma_2 - \sigma_1)\frac{|\mathbf{p}|}{2m}v_{\sigma_1}^*v_{\sigma_2} \qquad (4\text{-}4.128)$$

and

$$u_{p_1\sigma_1}^*\gamma^0 i\gamma_5\boldsymbol{\sigma}\cdot\mathbf{q}u_{p_2\sigma_2} = -(1 - \sigma_1\sigma_2)\frac{\mathbf{p}^2}{m}v_{\sigma_1}^*v_{\sigma_2}, \qquad (4\text{-}4.129)$$

which states that the magnetic moment interaction contributes only to transitions in which the helicity changes. On referring to Eq. (3–14.13), we recognize that the scattering intensity will now contain the factor

$$\left[\left(\frac{p^0}{m}\right)^2 \cos^2{\tfrac{1}{2}\theta} F_1(\mathbf{q})^2 + \sin^2{\tfrac{1}{2}\theta}\left(F_1(\mathbf{q}) - \frac{\alpha}{2\pi}\frac{\mathbf{p}^2}{m^2} F_2(\mathbf{q})\right)^2\right] \bar{A}^0(\mathbf{q})^2. \quad (4\text{–}4.130)$$

Relative to the differential cross section given in (3–14.14), this implies the following dynamical modification:

$$\left(F_1(\mathbf{q})^2 - \frac{\alpha}{\pi}\frac{\beta^2 \sin^2{\tfrac{1}{2}\theta}}{1 - \beta^2 \sin^2{\tfrac{1}{2}\theta}} F_2(\mathbf{q})\right)\left(\frac{D_+(\mathbf{q})}{D_+(\mathbf{q})}\right)^2$$

$$= 1 - 2\mathbf{q}^2 \int dM^2 \frac{(1/M^2)f_1(M^2) - a(M^2)}{\mathbf{q}^2 + M^2}$$

$$- \frac{\alpha}{\pi}\left(\frac{m}{p^0}\right)^2 \frac{1}{1 - \beta^2 \sin^2{\tfrac{1}{2}\theta}} \frac{\mathbf{q}^2}{4m^2} \int dM^2 \frac{f_2(M^2)}{\mathbf{q}^2 + M^2}, \quad (4\text{–}4.131)$$

where only terms linear in α have been retained.

The spin 0 discussion can be transferred, with minor changes, to spin $\tfrac{1}{2}$. In addition to the explicit magnetic moment term, it is only necessary to account for the slight difference between the charge form factors,

$$f_1(M^2) - f(M^2) = \frac{\alpha}{4\pi}\left(1 - \frac{4m^2}{M^2}\right)^{-1/2}, \quad (4\text{–}4.132)$$

and to substitute the spin $\tfrac{1}{2}$ description of vacuum polarization,

$$M^2 a(M^2)_{\text{spin }\frac{1}{2}} - M^2 a(M^2)_{\text{spin }0} = \frac{\alpha}{4\pi}\left(1 - \frac{4m^2}{M^2}\right)^{1/2}\left(1 + \frac{4m^2}{M^2}\right)$$

$$= \frac{\alpha}{4\pi}\left(1 - \frac{4m^2}{M^2}\right)^{-1/2}\left[1 - \left(\frac{4m^2}{M^2}\right)^2\right]. \quad (4\text{–}4.133)$$

The combination that appears in (4–4.131) is altered by

$$(f_1(M^2) - M^2 a(M^2)_{\text{spin }\frac{1}{2}}) - (f(M^2) - M^2 a(M^2)_{\text{spin }0}) = \frac{\alpha}{4\pi}\left(1 - \frac{4m^2}{M^2}\right)^{-1/2}\left(\frac{4m^2}{M^2}\right)^2,$$

$$(4\text{–}4.134)$$

which gives a rapidly converging contribution to the spectral integral. This is significant in high energy behavior for it means that all terms which increase logarithmically with energy are identical in the two situations. Indeed, apart from the last term of (4–4.131), the difference between that expression and (4–4.86) is

$$-\frac{\alpha}{2\pi}q^2\int_{(2m)^2}^{\infty}\frac{dM^2}{M^2}\frac{\left(1-\frac{4m^2}{M^2}\right)^{-1/2}\left(\frac{4m^2}{M^2}\right)^2}{q^2+M^2}=-\frac{\alpha}{\pi}\frac{q^2}{4m^2}\int_0^1 dv\frac{(1-v^2)^2}{1+\frac{q^2}{4m^2}(1-v^2)}$$

$$\rightarrow-\frac{2\alpha}{3\pi},\qquad q^2\gg m^2.\qquad(4\text{–}4.135)$$

Since the additional magnetic moment effect is negligible at high energies, owing to the $(m/p^0)^2$ factor, the spin $\frac{1}{2}$ analogue of (4–4.121) is immediately obtained:

$$1-\frac{4\alpha}{\pi}\left[\left(\log\frac{p^0}{k_{min.}^0}-\frac{13}{12}\right)\left(\log\frac{2p^0}{m}\sin\tfrac{1}{2}\theta-\tfrac{1}{2}\right)+\frac{17}{72}+\tfrac{1}{2}\sin\tfrac{1}{2}\theta\phi(\cos\tfrac{1}{2}\theta)\right].\quad(4\text{–}4.136)$$

In the nonrelativistic limit the value obtained from (4–4.135), $-(2\alpha/3\pi)$ $\times(q^2/m^2)\frac{1}{5}$, added to the magnetic moment contribution, $-(2\alpha/3\pi)(q^2/m^2)\frac{3}{8}$, converts (4–4.99) into

$$1-\frac{2\alpha}{3\pi}\frac{q^2}{m^2}\left(\log\frac{m}{2k_{min.}^0}+\frac{19}{30}\right),\qquad(4\text{–}4.137)$$

and the analogue of (4–4.101) is, correspondingly,

$$1-\frac{2\alpha}{3\pi}\frac{q^2}{m^2}\left[\log\frac{m}{8T}+\frac{19}{30}+(\pi-\theta)\tan\tfrac{1}{2}\theta+\frac{\cos\theta}{\cos^2\tfrac{1}{2}\theta}\log\frac{1}{\sin\tfrac{1}{2}\theta}\right].\quad(4\text{–}4.138)$$

The general spin $\frac{1}{2}$ formula, written in the notation of (4–4.122), is

$$1-\frac{\alpha}{2\pi}\frac{q^2}{m^2}\left[\left(\log\frac{p^0}{k_{min.}^0}-1\right)(K_0+K_1)+\tfrac{1}{2}K_0-K_1+\tfrac{1}{3}K_2\right.$$
$$\left.+\left(\frac{m}{p^0}\right)^2L+\frac{1}{2}\left(\frac{m}{p^0}\right)^2\frac{1}{1-\beta^2\sin^2\tfrac{1}{2}\theta}K_0\right].\qquad(4\text{–}4.139)$$

Concerning experimental verification of these spin $\frac{1}{2}$ predictions in electron scattering, we shall only remark that such agreement exists, and that the dynamical effects we have discussed are now routinely included in analyzing the very high energy electron scattering measurements that are exploring the inner structure of the proton.

With an historical gleam in his eye, Harold asks a question.

H.: I know that the spin $\frac{1}{2}$ results you have been describing were first derived by you many years ago. My question refers to a comment made shortly after and occasionally repeated since, that you had made two mistakes in that early calculation, which somehow managed to cancel and give the right answer. Would you comment on this remarkable achievement?

S.: I think I have been given too much credit. No one is clever enough to make two such complicated and exactly compensating errors. The real situation is far less spectacular. It is artificial to divide the essentially elastic scattering cross section into an elastic part and a soft photon addition. Neither component has physical significance, and alternative methods of calculation will give different expressions for the separate parts while agreeing on the sum. In particular, one can perform rearrangements in the two component terms, which may or may not be justified individually, but find their validity in the complete structure. Thus, there are several opportunities to exhibit different forms for the two pieces, while retaining a fixed answer for their sum. Perhaps I should confess that I was in no doubt about the correctness of the result because I had performed the calculation in two different ways, with concordant answers. I was, at the time, very insistent on working only with the physical massless photon because the operator method, then in use, alters discontinuously as the number of degrees of freedom changes with the introduction of a photon mass. Nevertheless, I carried out the calculation using a finite photon mass, but did not report it since it violated my self-imposed ground rules. No such inhibitions about the transitory use of a photon mass appear in these pages because we know that the description of a massive spin 1 particle with conserved sources is connected continuously with that of the photon. But, surely, the proper conclusion from all this history is that a better method must exist, one that gives the physical cross section directly, without the intervention of nonphysical distinctions. We shall describe such a method later.

4–5 FORM FACTORS II. SINGLE AND DOUBLE SPECTRAL FORMS

The primitive interactions we have been discussing, which are linear in the photon field and quadratic in charged particle fields, have two different applications where just one source operates in the extended sense to emit two noninteracting particles. The subsequent interaction of these particles modifies the effective ability of the source to emit, or absorb, the two particles. That has been considered for extended photon sources emitting a pair of charged particles. Now we turn to a particle source that emits a charged particle with an accompanying photon, and examine the modification produced by the scattering interaction of the particle and photon. The description of a charged particle source involves a choice of electromagnetic model. The simplest of these is the covariant model in which no charge acceleration radiation occurs. But, then we must pass over spin 0 and proceed directly to spin $\frac{1}{2}$, since no radiation appears in this spin 0 situation. We recall the effective two-particle source

$$iJ_2^\mu(k)\eta_2(p)|_{\text{eff.}} = eq[\gamma^\mu\psi_2(P) - if^\mu(k, P)\eta_2(P)], \tag{4–5.1}$$

where

$$if^\mu(k, P) = \frac{P^\mu}{kP} = \frac{2P^\mu}{P^2 + m^2}. \qquad (4\text{-}5.2)$$

The subsequent scattering process is described by the vacuum amplitude [Eq. (3-14.137)]

$$ie^2 \int d\omega_{p_1}\, d\omega_{k_1}\, d\omega_{p_2}\, d\omega_{k_2} (2\pi)^4 \delta(p_1 + k_1 - p_2 - k_2)$$

$$\times J_1^\mu(-k_1)\eta_1(-p_1)\gamma^0(m - \gamma p_1)\left[\gamma_\mu \frac{1}{\gamma(p_2 + k_2) + m}\gamma_\nu\right.$$

$$\left. + \gamma_\nu \frac{1}{\gamma(p_2 - k_1) + m}\gamma_\mu\right](m - \gamma p_2)\eta_2(p_2)J_2^\nu(k_2)|_{\text{eff.}}, \qquad (4\text{-}5.3)$$

or

$$\frac{\alpha}{4\pi} \int d\omega_{p_1}\, d\omega_{k_1}(2\pi)^4 \delta(p_1 + k_1 - P)\, dM^2\, d\omega_P \left(1 - \frac{m^2}{M^2}\right)$$

$$\times J_1^\mu(-k_1)\eta_1(-p_1)\gamma^0(m - \gamma p_1)eq \left[L_\mu \psi_2(P) + \frac{2}{M^2 - m^2} S_\mu \eta_2(P)\right]. \qquad (4\text{-}5.4)$$

Here, using the notation

$$\int d\omega_{k_2}\, d\omega_{p_2}(2\pi)^3 \delta(k_2 + p_2 - P)(\cdots) = \frac{1}{(4\pi)^2}\left(1 - \frac{m^2}{M^2}\right)\langle\cdots\rangle, \qquad (4\text{-}5.5)$$

we have defined

$$L_\mu = \left\langle\left(\gamma_\mu \frac{1}{\gamma P + m}\gamma_\nu + \gamma_\nu \frac{1}{\gamma(p_2 - k_1) + m}\gamma_\mu\right)(m - \gamma p_2)\gamma^\nu\right\rangle \qquad (4\text{-}5.6)$$

and

$$S_\mu = \left\langle\left(\gamma_\mu \frac{1}{\gamma P + m}\gamma_\nu + \gamma_\nu \frac{1}{\gamma(p_2 - k_1) + m}\gamma_\mu\right)(m - \gamma p_2)P^\nu\right\rangle. \qquad (4\text{-}5.7)$$

The conservation requirement on the effective photon emission sources is satisfied in the following way,

$$k_1^\mu\left(\gamma_\mu \frac{1}{\gamma P + m}\gamma_\nu + \gamma_\nu \frac{1}{\gamma(p_2 - k_1) + m}\gamma_\mu\right)(m - \gamma p_2)$$

$$= \left[-(\gamma p_1 + m)\frac{1}{\gamma P + m}\gamma_\nu + \gamma_\nu \frac{1}{\gamma(p_2 - k_1) + m}(\gamma p_2 + m)\right](m - \gamma p_2)$$

$$\to 0, \qquad (4\text{-}5.8)$$

owing to the explicit presence of the projection matrix $m - \gamma p_2$ and the appearance of $m - \gamma p_1$ in (4–5.4). We begin the calculation by introducing rearrangements that are designed to satisfy this requirement explicitly:

$$\left(\gamma_\mu \frac{1}{\gamma P + m}\gamma_\nu + \gamma_\nu \frac{1}{\gamma(p_2 - k_1) + m}\gamma_\mu\right)(m - \gamma p_2)$$

$$= -\left[\gamma_\mu \frac{m - \gamma P}{M^2 - m^2}\gamma_\nu + \gamma_\nu \frac{m - \gamma(p_2 - k_1)}{2p_2 k_1}\gamma_\mu\right](m - \gamma p_2)$$

$$\rightarrow -\left[\left(\frac{2p_{1\mu}}{M^2 - m^2} + \frac{p_{2\mu}}{p_2 k_1}\right)\gamma_\nu + \frac{\gamma_\nu \gamma k_1 \gamma_\mu}{2p_2 k_1} - \frac{\gamma_\mu \gamma k_1 \gamma_\nu}{M^2 - m^2}\right](m - \gamma p_2), \quad (4\text{–}5.9)$$

where expressions containing $\gamma p_1 + m$ on the left have been discarded. The three terms appearing in the final version of Eq. (4–5.9) satisfy the conservation condition individually, since

$$(\gamma k_1)^2 = 0, \qquad -2k_1 p_1 = M^2 - m^2. \tag{4–5.10}$$

There is, incidentally, no significant difference between $\gamma_\mu \gamma k_1$ and $-\gamma k_1 \gamma_\mu$, for any multiple of $k_{1\mu}$ can be omitted when applied in (4–5.4). Both of these matrix products can be replaced by $-i\sigma_{\mu\nu}k_1^\nu$.

The integrations involved in evaluating $\langle \cdots \rangle$ are illustrated by the rest frame calculation of $\langle 1/(-p_2 k_1)\rangle$,

$$\left\langle -\frac{1}{p_2 k_1}\right\rangle = \int_{-1}^{1} \frac{dz}{2} \frac{1}{M^2 - m^2 \left[\dfrac{M^2 + m^2}{2M} + \dfrac{M^2 - m^2}{2M}z\right]} = \left(\frac{2M}{M^2 - m^2}\right)^2 \frac{1}{2}\log\frac{M^2}{m^2},$$
$$\tag{4–5.11}$$

where z is the cosine of the scattering angle. Other such integrals contain the additional factors $p_{2\mu}$ and $p_{2\mu}p_{2\nu}$. The opportunity to omit multiples of $k_{1\mu}$ simplifies these calculations. Thus,

$$\left\langle \frac{p_{2\mu}}{p_2 k_1}\right\rangle + \frac{2p_{1\mu}}{M^2 - m^2} = ak_{1\mu} \rightarrow 0, \tag{4–5.12}$$

where a possible multiple of $p_{1\mu}$ does not appear since the left-hand side vanishes on multiplication with k_1^μ. A similar reduction gives

$$\left\langle \left(\frac{p_{2\mu}}{p_2 k_1} + \frac{2p_{1\mu}}{M^2 - m^2}\right)p_{2\nu}\right\rangle \rightarrow -\left(p_{1\mu}k_{1\nu} + \frac{M^2 - m^2}{2}g_{\mu\nu}\right)\frac{1}{M^2}\chi(M^2), \tag{4–5.13}$$

in which

$$\chi(M^2) = \left(\frac{M^2}{M^2 - m^2}\right)^2\left[\frac{M^2 + m^2}{M^2} - \frac{2m^2}{M^2 - m^2}\log\frac{M^2}{m^2}\right]. \tag{4–5.14}$$

This positive function varies from a value of $\frac{1}{3}$ at $M^2 = m^2$ to unity, for $M^2 \gg m^2$. The outcome of these calculations is given by

$$\left\langle \left(\gamma_\mu \frac{1}{\gamma P + m} \gamma_\nu + \gamma_\nu \frac{1}{\gamma(p_2 - k_1) + m} \gamma_\mu \right) (m - \gamma p_2) \right\rangle$$

$$= \gamma_\nu \gamma p_1 \gamma_\mu \gamma k_1 \frac{1}{2M^2} \chi(M^2) + \gamma_\nu \gamma_\mu \gamma k_1 \gamma P \left(\frac{1}{M^2 - m^2} - \frac{1}{2M^2} \chi(M^2) \right)$$

$$+ \frac{1}{2m} \gamma_\nu \gamma_\mu \gamma k_1 \left(\frac{M^2 - m^2}{M^2} \chi(M^2) - \frac{M^2 + m^2}{M^2 - m^2} \right)$$

$$+ \gamma_\mu \gamma k_1 \gamma_\nu \frac{1}{M^2 - m^2} \left(m - \frac{M^2 + m^2}{2M^2} \gamma P \right). \tag{4-5.15}$$

The quantities we need have additional factors of γ^ν or P^ν on the right-hand side. Only short reductions are required to produce the results:

$$L_\mu = i\sigma_{\mu\nu} k_1^\nu \left[\frac{m}{M^2} + \frac{1}{M^2} \left(\chi(M^2) + \frac{M^2 + m^2}{M^2 - m^2} \right) (\gamma P + m) \right],$$

$$S_\mu = -i\sigma_{\mu\nu} k_1^\nu \frac{m}{2M^2} \chi(M^2)(\gamma P + m). \tag{4-5.16}$$

The combination that appears in (4-5.4) is

$$L_\mu \psi_2(P) + \frac{2}{M^2 - m^2} S_\mu \eta_2(P)$$

$$= \frac{m}{M^2} i\sigma_{\mu\nu} k_1^\nu \psi_2(P) + i\sigma_{\mu\nu} k_1^\nu \frac{1}{M^2} \left(\chi(M^2) + \frac{M^2 + m^2}{M^2 - m^2} \right) \eta_2(P)$$

$$- i\sigma_{\mu\nu} k_1^\nu \frac{m}{M^2} \frac{1}{M^2 - m^2} \chi(M^2)(\gamma P + m)\eta_2(P). \tag{4-5.17}$$

Since the only reference to the individual momenta of the final particles is in the matrix $i\sigma_{\mu\nu} k_1^\nu$, all aspects of these particles in (4-5.4) can be combined into a local field product,

$$\int d\omega_{p_1} d\omega_{k_1} (2\pi)^4 \delta(p_1 + k_1 - P) J_1^\mu(-k_1)\eta_1(-p_1)\gamma^0(m - \gamma p_1)eq i\sigma_{\mu\nu} k_1^\nu$$

$$= -\int (dx)\psi_1(x)\gamma^0 eq \tfrac{1}{2}\sigma_{\mu\nu} F_1^{\mu\nu}(x) \exp(iPx). \tag{4-5.18}$$

When attention is confined to the term that involves particle fields explicitly, the vacuum amplitude (4-5.4) becomes

$$i\frac{\alpha}{2\pi}\int dM^2\frac{m^2}{M^2}\left(1-\frac{m^2}{M^2}\right)\int (dx)(dx')\psi_1(x)\gamma^0\frac{eq}{2m}\frac{1}{2}\sigma_{\mu\nu}F_1^{\mu\nu}(x)$$

$$\times\left[i\int d\omega_P\exp[iP(x-x')]\right]\psi_2(x').\qquad(4\text{–}5.19)$$

Its space-time extrapolation is stated as

$$i\frac{\alpha}{2\pi}\int (dx)(dx')\psi(x)\gamma^0\frac{eq}{2m}\frac{1}{2}\sigma_{\mu\nu}F^{\mu\nu}(x)G_2(x-x')\psi_{\text{ext.}}(x'),$$

$$G_2(p)=\int_{m^2}^{\infty}dM^2\frac{m^2}{M^2}\left(1-\frac{m^2}{M^2}\right)\frac{1}{p^2+M^2-i\varepsilon},\qquad(4\text{–}5.20)$$

where $\psi_{\text{ext.}}$ specifically distinguishes the field of the extended source.

This asymmetry must be removed, of course. Accordingly, let us note that

$$G_2(p)=1-(p^2+m^2)\int_{m^2}^{\infty}\frac{dM^2}{M^2}\frac{m^2}{M^2}\frac{1}{p^2+M^2-i\varepsilon},\qquad(4\text{–}5.21)$$

which uses the rearrangement

$$\frac{1}{p^2+M^2-i\varepsilon}=\frac{1}{M^2-m^2}-\frac{p^2+m^2}{M^2-m^2}\frac{1}{p^2+M^2-i\varepsilon}\qquad(4\text{–}5.22)$$

and the integral value

$$G_2(p^2+m^2=0)=\int_{m^2}^{\infty}\frac{dM^2}{M^2}\frac{m^2}{M^2}=1.\qquad(4\text{–}5.23)$$

When applied to the field $\psi_{\text{ext.}}$, the factor $m+\gamma p$ in $p^2+m^2=(m-\gamma p)(m+\gamma p)$ introduces the corresponding source. We shall group this with the other explicitly source dependent terms. Then the remaining field structure, stated as an action expression with a unified particle field, is

$$\int (dx)\tfrac{1}{2}\psi(x)\gamma^0\frac{\alpha}{2\pi}\frac{eq}{2m}\frac{1}{2}\sigma_{\mu\nu}F^{\mu\nu}(x)\psi(x);\qquad(4\text{–}5.24)$$

we have derived again the additional spin magnetic moment of $\alpha/2\pi$ magnetons.

Let us consider now a causal arrangement involving one extended particle source and one extended photon source, the latter providing space-like momenta. This time we shall work with spin 0 particles. The extended particle source K_2 emits a charged particle and a photon. The two particles travel to the vicinity of the extended photon source where a scattering process occurs, involving a virtual photon associated with the extended source, and the scattered charged particle goes on to be detected by the simple source K_1. The vacuum amplitude for the scattering process is given by [Eq. (3–12.92), but without the use of a Lorentz gauge]

$$\int d\omega_{p_1}(dk_1)d\omega_{p_2}d\omega_{k_2}\delta(p_1 + k_1 - p_2 - k_2)A^\mu(-k_1)K_1(-p_1)e^2 2V_{\mu\nu}K_2(p_2)J_2^\nu(k_2),$$

$$(4\text{–}5.25)$$

where

$$2V_{\mu\nu} = \frac{(2p_1 + k_1)_\mu(2p_2 + k_2)_\nu}{(p_2 + k_2)^2 + m^2} + \frac{(2p_2 - k_1)_\mu(2p_1 - k_2)_\nu}{(p_1 - k_2)^2 + m^2} - 2g_{\mu\nu} \quad (4\text{–}5.26)$$

and

$$iJ_2^\nu(k_2)K_2(p_2)\big|_{\text{eff.}} = eq[(2P - k_2)^\nu\phi_2(P) - if^\nu(k_2)K_2(P)]. \quad (4\text{–}5.27)$$

We shall only consider the ϕ_2 term in the effective source, which represents radiation by the charged particle rather than from the source. Harold interrupts.

H.: This question may be embarrassing, but I think it should be asked. I have heard that in your lectures on Source Theory you make frequent reference to space-time pictures of physical processes, which you call causal diagrams. Are these the same as the famous Feynman diagrams? And why have causal diagrams not been used in this book?

S.: The utility of a diagram as an instructional aid varies with the circumstances. In a lecture, where constant attention to the subject matter is required, such visual aids can be indispensable. Reading a book is a different activity, however. The reader has ample opportunity to supply his own additional material and should do so whenever this is helpful. I have preferred to put no diagrams in the text, both to emphasize that the analytical structure of the theory has priority, and to keep down the cost of the book. But perhaps I should use the causal situation now under discussion to illustrate causal diagrams. Three of them are needed to represent the three terms of the vacuum amplitude (4–5.25, 26, 27).

Causal Diagrams

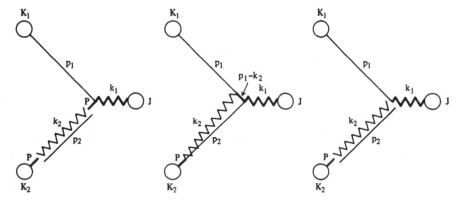

The following conventions are used: time is read vertically; circles represent sources; a thin straight line indicates the causal propagation of a real particle; a thin zigzag line correspondingly refers to a real photon; heavy straight and zigzag lines represent the noncausal propagation of a virtual particle and a virtual photon, respectively. The various lines can be labeled by the fields or propagation functions which they symbolize or, as in these pictures, by the momenta of the several real and virtual particles.

Causal diagrams are not Feynman diagrams. The latter do not involve a distinction between real and virtual particles; Feynman diagrams are noncausal.

Returning to the vacuum amplitude (4–5.25, 26, 27), we observe that the tensor $V_{\mu\nu}$ has the following gauge invariance properties

$$k_1^\mu 2V_{\mu\nu} = (2p_2 + k_2)_\nu - (2p_1 - k_2)_\nu - 2k_{1\nu}$$

$$= 2(p_2 + k_2 - p_1 - k_1)_\nu = 0, \qquad (4\text{–}5.28)$$

which uses the momentum relations

$$p_2 + k_2 = p_1 + k_1, \qquad p_1 - k_2 = p_2 - k_1, \qquad (4\text{–}5.29)$$

and similarly

$$V_{\mu\nu}k_2^\nu = 0. \qquad (4\text{–}5.30)$$

The vacuum amplitude contribution we are concerned with now appears as

$$- i2e^2 \int d\omega_{p_1}(dk_1)\, d\omega_P\, dM^2\, \delta(p_1 + k_1 - P) A^\mu(- k_1) K_1(- p_1) eq V_\mu \phi_2(P), \quad (4\text{–}5.31)$$

with

$$V_\mu = \int d\omega_{k_2} d\omega_{p_2} (2\pi)^3 \delta(p_2 + k_2 - P) V_{\mu\nu} 2P^\nu \qquad (4\text{–}5.32)$$

where

$$2V_{\mu\nu}P^\nu = (2p_1 + k_1)_\mu \left(\frac{M^2 + m^2}{M^2 - m^2} + \frac{1}{2} \right) + (2p_2 - k_1)_\mu$$

$$\times \frac{M^2 + m^2 + k_1{}^2 - \tfrac{1}{2}(M^2 - m^2)}{2p_1 k_2} - 2P_\mu. \qquad (4\text{–}5.33)$$

The kinematical relations that have been used in producing (4–5.33) are

$$- 2Pp_2 = M^2 + m^2, \qquad - 2Pk_2 = M^2 - m^2, \qquad - 2Pp_1 = M^2 + m^2 + k_1{}^2.$$

$$(4\text{–}5.34)$$

All such relations are based on the physical photon property, $k_2^2 = 0$. Again using expectation value notation for integrals of the type of (4–5.32), we present V_μ in the following way:

$$V_\mu = \frac{1}{(4\pi)^2}\left(1 - \frac{m^2}{M^2}\right)\left[(2p_1 + k_1)_\mu\left(\frac{M^2 + m^2}{M^2 - m^2} - \frac{1}{2}\right) - k_{1\mu}\right.$$

$$\left. + (M^2 + m^2 + k_1{}^2 - \tfrac{1}{2}(M^2 - m^2))\left\langle\frac{(p_2 - \tfrac{1}{2}k_1)_\mu}{p_1 k_2}\right\rangle\right]. \qquad (4\text{–}5.35)$$

The expectation value that occurs here has the vectorial form

$$\left\langle\frac{p_2 - \tfrac{1}{2}k_1}{p_1 k_2}\right\rangle = (2p_1 + k_1)a + k_1 b. \qquad (4\text{–}5.36)$$

Then, since

$$k_1(p_2 - \tfrac{1}{2}k_1) = k_2 p_1, \qquad (4\text{–}5.37)$$

we learn that

$$1 = -(M^2 - m^2)a + k_1{}^2 b. \qquad (4\text{–}5.38)$$

A second relation can be produced by multiplication with p_1, using

$$p_1(p_2 - \tfrac{1}{2}k_1) = -p_1 k_2 - \tfrac{1}{4}(M^2 + m^2 + k_1{}^2) - \tfrac{1}{2}m^2, \qquad (4\text{–}5.39)$$

which gives us

$$-1 + \frac{1}{2}\left(\frac{M^2 + m^2 + k_1{}^2}{2} + m^2\right)\left\langle-\frac{1}{p_1 k_2}\right\rangle$$

$$= -\left(\frac{M^2 + m^2 + k_1{}^2}{2} + m^2\right)a - \tfrac{1}{2}(M^2 - m^2 + k_1{}^2)b. \qquad (4\text{–}5.40)$$

The solution for a is expressed by

$$-\Delta a = M^2 - m^2 - k_1{}^2 + k_1{}^2\left(\frac{M^2 + m^2 + k_1{}^2}{2} + m^2\right)\left\langle-\frac{1}{p_1 k_2}\right\rangle, \qquad (4\text{–}5.41)$$

where

$$\Delta = (M^2 - m^2)^2 + 2(M^2 + m^2)k_1{}^2 + (k_1{}^2)^2$$

$$= (M^2 + m^2 + k_1{}^2)^2 - 4M^2 m^2. \qquad (4\text{–}5.42)$$

Conveniently evaluated in the P rest frame, the expectation value that appears in (4–5.41) is

$$\left\langle -\frac{1}{p_1 k_2} \right\rangle = \frac{2M}{M^2 - m^2} \int_{-1}^{1} \frac{dz}{2} \frac{1}{\dfrac{M^2 + m^2 + k_1{}^2}{2M} + z\left[\left(\dfrac{M^2 + m^2 + k_1{}^2}{2M}\right)^2 - m^2\right]^{1/2}}$$

$$= \frac{4M^2}{M^2 - m^2}(M^2 + m^2 + k_1{}^2) \int_0^1 dz \frac{1}{\Delta(1 - z^2) + 4M^2 m^2}. \qquad (4\text{-}5.43)$$

The vector V_μ is exhibited as a linear combination of the vectors $2p_1 + k_1$ and k_1. Apart from the factor $(4\pi)^{-2}[1 - (m/M)^2]$, the coefficient of $2p_1 + k_1$ is

$$\frac{M^2 + m^2}{M^2 - m^2} - \tfrac{1}{2} + (M^2 + m^2 + k_1{}^2 - \tfrac{1}{2}(M^2 - m^2))a = -k_1{}^2 I, \qquad (4\text{-}5.44)$$

where

$$\Delta I = \left[\Delta + 4M^2 m^2 - m^2(M^2 - m^2) + (M^2 + m^2 + k_1{}^2)\left(2m^2 - \frac{M^2 - m^2}{2}\right)\right]$$

$$\times \frac{1}{2}\left\langle -\frac{1}{p_1 k_2} \right\rangle - \left(\frac{M^2 + m^2}{M^2 - m^2} - \frac{1}{2}\right)(2(M^2 + m^2) + k_1{}^2)$$

$$- (M^2 + m^2 + k_1{}^2) + \tfrac{3}{2}(M^2 - m^2). \qquad (4\text{-}5.45)$$

After some rearrangement, I can be presented in the following way,

$$I(M^2, k_1{}^2) = \int_0^1 dz \frac{1}{\Delta(1 - z^2) + 4M^2 m^2}\left[(k_1{}^2 + 4m^2)\frac{2M^2}{M^2 - m^2}z^2\right.$$

$$\left. - \tfrac{1}{2}(M^2 - m^2 - k_1{}^2)(1 - z^2) + M^2\right]. \qquad (4\text{-}5.46)$$

The coefficient of k_1 is completely determined by the gauge invariance property

$$k_1{}^\mu V_\mu = 0, \qquad (4\text{-}5.47)$$

which specifies the combination

$$- k_1{}^2(2p_1 + k_1)_\mu + k_{1\mu}k_1(2p_1 + k_1) = -(k_1{}^2 g_{\mu\nu} - k_{1\mu}k_{1\nu})(2p_1 + k_1)^\nu. \qquad (4\text{-}5.48)$$

We can now recognize that (4–5.31) has the field structure

$$- i\frac{\alpha}{2\pi}\int dM^2\left(1 - \frac{m^2}{M^2}\right)\int (dx)(dx')I\partial_\nu F^{\mu\nu}(x)\phi_1(x)\left(\frac{1}{i}\partial_\mu^T + \frac{1}{i}\partial_\mu\right)$$

$$\times eq\left[i\int d\omega_P \exp[iP(x - x')]\right]\phi_2(x'), \qquad (4\text{-}5.49)$$

where

$$\phi_1(x)\partial_\mu^T = - \partial_\mu\phi_1(x) \tag{4-5.50}$$

and I appears as a spatially nonlocal but temporally localized operation on the electromagnetic field strengths. A space-time extrapolation of (4–5.49) is produced by introducing $\Delta_+(x - x', M^2)$ in place of its causal form for $x^0 > x^{0\prime}$. But, as in an earlier discussion of this section, a decomposition must be introduced [Eq. (4–5.22)] in order to give a unified presentation of the coupling that involves the fields ϕ_1 and ϕ_2. In combination with the primitive interaction, the resulting action term can be written as

$$\int (dx)(dx')j^\mu(x)f_{\mu\nu}(x - x')A^\nu(x'), \tag{4-5.51}$$

where

$$j^\mu(x) = \phi(x)eq(1/i)\partial^\mu\phi(x) \tag{4-5.52}$$

and

$$f_{\mu\nu}(k) = g_{\mu\nu} + (k_\mu k_\nu - g_{\mu\nu}k^2)\frac{\alpha}{2\pi}\int_{\to m^2}^\infty \frac{dM^2}{M^2} I(M^2, k^2). \tag{4-5.53}$$

The form factor arrived at in this way should be identical with (4–4.25, 21). Or rather, it should be had we introduced a photon mass to avoid the infrared singularity in the integral of (4–5.53). We shall be quite content to verify that the logarithmic infrared singularities of the two expressions are identical, and omit the additional considerations associated with a nonzero photon mass. The singular term in (4–5.46) is the one with $M^2 - m^2$ in the denominator. This contribution to the integral in (4–5.53) is

$$2(k^2 + 4m^2)\int_0^1 dz\, z^2 \int_{(m+\mu)^2}^\infty \frac{dM^2}{M^2 - m^2} \frac{1}{\Delta(1 - z^2) + 4M^2 m^2}$$

$$= 2\int_0^1 dz\, z^2 \frac{k^2 + 4m^2}{k^2(k^2 + 4m^2)(1 - z^2) + (2m^2)^2}\left(\log\frac{1}{\mu} + \cdots\right), \tag{4-5.54}$$

where the introduction of the photon mass in the lower limit of integration is an essential but incomplete statement of the modifications that are required; it suffices to indicate the photon mass dependence, however. What we are being asked to confirm is the following identity:

$$\int_0^1 dz\, z^2 \frac{k^2 + 4m^2}{k^2(k^2 + 4m^2)(1 - z^2) + (2m^2)^2} = \int_0^1 dv\, \frac{1 + v^2}{4m^2 + k^2(1 - v^2)}. \tag{4-5.55}$$

The two similar integrations can be performed, and one arrives at the common expression [in the notation of (4–4.122)]

$$\frac{1}{4m^2}(K_0(\lambda) + K_1(\lambda)), \qquad \lambda^2 = \frac{k^2}{4m^2}. \tag{4-5.56}$$

All this is not very elegant, of course. We should prefer to find a virtuoso transformation of integration variables that would directly interconnect the two equivalent expressions. That ideal will be approached more closely in our next exercise.

Continuing with the discussion of spin 0 particles, let us consider a causal arrangement of two extended particle sources, K_1 and K_2. Causally located between them is an extended photon source, which is capable of transmitting space-like momenta. The extended particle sources emit, or absorb, a photon and a charged particle. The photon propagates undisturbed between the two extended sources; the charged particle is deflected by the extended photon source. What we are describing is a generalization of the arrangement used in Section 4–2 for the first derivation of the additional spin magnetic moment of spin $\frac{1}{2}$ particles. The homogeneous magnetic field has been replaced by an arbitrary field. The vacuum amplitude that describes the history of the two particles is

$$\int (dx) \cdots (d\xi') i J_1^\mu(\xi) K_1(x)|_{\text{eff.}} D_+(\xi - \xi') \Delta_+^A(x, x') i J_{2\mu}(\xi') K_2(x')|_{\text{eff.}}, \tag{4-5.57}$$

where the effective sources are illustrated in (4–5.27). Only the term linear in A will be retained of the expansion [Eq. (3–12.28)]

$$\Delta_+^A = \Delta_+ + \Delta_+ eq(pA + Ap)\Delta_+ + \cdots. \tag{4-5.58}$$

And, when just the field-dependent parts of the effective sources are used, the resulting vacuum amplitude term is

$$- ie^2 \int d\omega_k\, d\omega_{p_1} d\omega_{p_2} (2P_1 - k)(2P_2 - k)\phi_1(- P_1) eq(P_1 + P_2 - 2k) A(K)\phi_2(P_2), \tag{4-5.59}$$

where

$$K = P_1 - P_2 \tag{4-5.60}$$

and

$$P_1 = p_1 + k, \qquad P_2 = p_2 + k. \tag{4-5.61}$$

It is convenient to introduce these kinematical relations by writing (the positiveness of all energies is understood)

$$\int d\omega_{p_1}\, d\omega_{p_2} = \int \frac{(dp_1)}{(2\pi)^3} \frac{(dp_2)}{(2\pi)^3} \delta(p_1{}^2 + m^2)\delta(p_2{}^2 + m^2)$$

$$= \int \frac{(dP_1)}{(2\pi)^3} \frac{(dP_2)}{(2\pi)^3} \delta((P_1 - k)^2 + m^2)\delta((P_2 - k)^2 + m^2), \tag{4-5.62}$$

which converts (4–5.59) into

$$ie^2 \int dM_1^2\, dM_2^2\, d\omega_{P_1} d\omega_{P_2} d\omega_k\, \delta(2kP_1 + M_1^2 - m^2)\delta(2kP_2 + M_2^2 - m^2)$$

$$\times\, (2K^2 + M_1^2 + M_2^2 + 2m^2)\phi_1(-P_1)eq(P_1 + P_2 - 2k)A(K)\phi_2(P_2).$$

$$(4\text{–}5.63)$$

The basic invariant integral that appears here,

$$\int d\omega_k\, \delta(2kP_1 + M_1^2 - m^2)\delta(2kP_2 + M_2^2 - m^2), \qquad (4\text{–}5.64)$$

can be evaluated in the rest frame of either momentum. Using that of P_1 gives us

$$\frac{1}{8\pi^2} \int k^0\, dk^0\, dz\, \delta(-2k^0 M_1 + M_1^2 - m^2)\delta(-2k^0 P_2^0 + 2k^0|\mathbf{P}_2|z + M_2^2 - m^2),$$

$$(4\text{–}5.65)$$

where z is the cosine of the angle between the vectors $\mathbf{k}$ and $\mathbf{P}_2$. The invariant meaning of the P_2 components is supplied by

$$P_2^0 = -\frac{P_1 P_2}{M_1} = \frac{K^2 + M_1^2 + M_2^2}{2M_1},$$

$$|\mathbf{P}_2| = \left[\left(\frac{P_1 P_2}{M_1}\right)^2 - M_2^2\right]^{1/2} = \frac{\Delta^{1/2}}{2M_1}, \qquad (4\text{–}5.66)$$

in which

$$\Delta = (2P_1 P_2)^2 - 4M_1^2 M_2^2 = (K^2 + M_1^2 + M_2^2)^2 - 4M_1^2 M_2^2. \qquad (4\text{–}5.67)$$

If the z integral is not to vanish, it is necessary that

$$\left| M_2^2 - m^2 - \frac{M_1^2 - m^2}{M_1}\frac{K^2 + M_1^2 + M_2^2}{2M_1} \right| < \frac{M_1^2 - m^2}{M_1}\frac{\Delta^{1/2}}{2M_1}, \qquad (4\text{–}5.68)$$

or that

$$(M_1^2 - M_2^2)^2 < \frac{K^2}{m^2}(M_1^2 - m^2)(M_2^2 - m^2). \qquad (4\text{–}5.69)$$

Under these conditions, the value of the integral (4–5.65) is stated as

$$\int d\omega_k\, \delta(2kP_1 + M_1^2 - m^2)\delta(2kP_2 + M_2^2 - m^2) = \frac{1}{(4\pi)^2}\frac{1}{\Delta^{1/2}}. \qquad (4\text{–}5.70)$$

It is useful to satisfy the inequality (4–5.69) by introducing new variables, x and v, according to

$$M_1{}^2 = m^2 + \tfrac{1}{2}x[K^2 + 4m^2 + v(K^2(K^2 + 4m^2))^{1/2}]$$

$$M_2{}^2 = m^2 + \tfrac{1}{2}x[K^2 + 4m^2 - v(K^2(K^2 + 4m^2))^{1/2}]. \qquad (4\text{-}5.71)$$

The inequality now reads

$$v^2 < 1; \qquad (4\text{-}5.72)$$

the domain of x extends from 0 to ∞. We also find that

$$\varDelta = K^2(K^2 + 4m^2)(1 + 2x + x^2v^2), \qquad (4\text{-}5.73)$$

while

$$dM_1{}^2\, dM_2{}^2 = \tfrac{1}{2}(K^2 + 4m^2)[K^2(K^2 + 4m^2)]^{1/2}x\, dx\, dv. \qquad (4\text{-}5.74)$$

The effective replacement for $P_1 + P_2 - 2k$ in (4-5.63) is of the form

$$\langle P_1 + P_2 - 2k \rangle = (P_1 + P_2)\alpha + (P_1 - P_2)\beta. \qquad (4\text{-}5.75)$$

Multiplication by $P_1 + P_2$ and $P_1 - P_2$, respectively, implies the equations

$$K^2 + M_1{}^2 + M_2{}^2 + 2m^2 = (K^2 + 2M_1{}^2 + 2M_2{}^2)\alpha + (M_1{}^2 - M_2{}^2)\beta,$$

$$0 = -(M_1{}^2 - M_2{}^2)\alpha + K^2\beta. \qquad (4\text{-}5.76)$$

The result is

$$\langle P_1 + P_2 - 2k \rangle_\mu = \frac{K^2 + M_1{}^2 + M_2{}^2 + 2m^2}{\varDelta} (K^2 g_{\mu\nu} - K_\mu K_\nu)(P_1 + P_2)^\nu, \quad (4\text{-}5.77)$$

where

$$\frac{K^2 + M_1{}^2 + M_2{}^2 + 2m^2}{\varDelta} = \frac{1}{K^2}\frac{1 + x}{1 + 2x + x^2v^2}. \qquad (4\text{-}5.78)$$

We also note that

$$2K^2 + M_1{}^2 + M_2{}^2 + 2m^2 = K^2 + (K^2 + 4m^2)(1 + x). \qquad (4\text{-}5.79)$$

All these relations are combined in the vacuum amplitude

$$i\,\frac{\alpha}{4\pi}\int d\omega_{P_1}\, d\omega_{P_2}\, x\, dx\, \frac{dv}{2}\,(K^2 + 4m^2)\frac{1 + x}{(1 + 2x + x^2v^2)^{3/2}}\left(2 + x + \frac{4m^2}{K^2}(1 + x)\right)$$

$$\times\ \phi_1(-P_1)eq\phi_2(P_2)(P_1 + P_2)(K^2A(K) - KKA(K)). \qquad (4\text{-}5.80)$$

We now observe that

$$\phi_1(-P_1)A(K)\phi_2(P_2) = \int (dx)\phi_1(x)\exp(iP_1x)\int (d\xi)\exp[-i(P_1 - P_2)\xi]$$

$$\times A(\xi) \int (dx') \exp(-iP_2 x')\, \phi_2(x'), \qquad (4\text{-}5.81)$$

and therefore

$$-\int d\omega_{P_1} d\omega_{P_2} \phi_1(-P_1) A(K)\phi_2(P_2)$$

$$= \int (dx)(d\xi)(dx')\phi_1(x) \left[i \int d\omega_{P_1} \exp[iP_1(x-\xi)] \right]$$

$$\times A(\xi) \left[i \int d\omega_{P_2} \exp[iP_2(\xi-x')] \right] \phi_2(x'), \qquad (4\text{-}5.82)$$

in which the causal forms of two propagation functions are evident. In carrying out the space-time extrapolations it is convenient to introduce the substitutions

$$i \int d\omega_{P_1} \exp[iP_1(x-\xi)] \rightarrow \int \frac{(dp_1)}{(2\pi)^4} \frac{\exp[ip_1(x-\xi)]}{p_1^2 + M_1^2 - i\varepsilon},$$

$$i \int d\omega_{P_2} \exp[iP_2(\xi-x')] \rightarrow \int \frac{(dp_2)}{(2\pi)^4} \frac{\exp[ip_2(\xi-x')]}{p_2^2 + M_2^2 - i\varepsilon}, \qquad (4\text{-}5.83)$$

and then return to four-dimensional momentum space, as in

$$-\int d\omega_{P_1}\, d\omega_{P_2}\, \phi_1(-P_1) A(K)\phi_2(P_2)$$

$$\rightarrow \int \frac{(dp_1)}{(2\pi)^4} \frac{(dp_2)}{(2\pi)^4} \phi_1(-p_1) \frac{1}{p_1^2 + M_1^2 - i\varepsilon} A(k) \frac{1}{p_2^2 + M_2^2 - i\varepsilon} \phi_2(p_2), \quad (4\text{-}5.84)$$

where

$$k = p_1 - p_2. \qquad (4\text{-}5.85)$$

One must not forget the possibility of adding contact terms in each of the two spectral forms. Keeping this in mind we write the space-time extrapolation of (4-5.80) as the double spectral form

$$-i\frac{\alpha}{4\pi} \int \frac{(dp_1)}{(2\pi)^4} \frac{(dp_2)}{(2\pi)^4} \phi_1(-p_1)eq\phi_2(p_2)(p_1+p_2)k^2 A(k)(k^2 + 4m^2)$$

$$\times \int x\, dx\, \frac{dv}{2} \frac{1+x}{(1+2x+x^2v^2)^{3/2}} \left(2 + x + \frac{4m^2}{k^2}(1+x)\right)$$

$$\times \frac{1}{p_1^2 + M_1^2 - i\varepsilon} \frac{1}{p_2^2 + M_2^2 - i\varepsilon} \qquad (4\text{-}5.86)$$

where, for simplicity, we have adopted a Lorentz gauge,

$$kA(k) = 0. \tag{4-5.87}$$

The most elementary situation to which this vacuum amplitude can be applied is one where no field is present. In a Lorentz gauge, the proportionality of the vectors A and k demands that $k^2 = 0$. Note that (4-5.86) does not vanish for this circumstance, owing to the term containing $1/k^2$; there is a potential-dependent vacuum amplitude. According to (4-5.71), with $K^2 \to k^2$, we have

$$k^2 = 0: \quad M_1{}^2 = M_2{}^2 = m^2(1 + 2x) = M^2. \tag{4-5.88}$$

The v integration in (4-5.86) can then be performed,

$$\int_{-1}^{1} \frac{dv}{2} \frac{1}{(1 + 2x + x^2 v^2)^{3/2}} = \frac{1}{(1 + x)(1 + 2x)}, \tag{4-5.89}$$

and (4-5.86) reduces to

$$-i \frac{\alpha}{2\pi} \int \frac{(dp_1)}{(2\pi)^4} \frac{(dp_2)}{(2\pi)^4} \phi_1(-p_1) eq \phi_2(p_2)(p_1 + p_2) A(k)$$

$$\times \int \frac{dM^2}{M^2} \frac{M^2 + m^2}{M^2 - m^2} \left(\frac{M^2 - m^2}{p_1{}^2 + M^2 - i\varepsilon} - 1 \right) \left(\frac{M^2 - m^2}{p_2{}^2 + M^2 - i\varepsilon} - 1 \right), \tag{4-5.90}$$

where we have also indicated the contact terms that are necessary physically. They serve to replace fields with sources, according to

$$\left(\frac{M^2 - m^2}{p_1{}^2 + M^2 - i\varepsilon} - 1 \right) \left(\frac{M^2 - m^2}{p_2{}^2 + M^2 - i\varepsilon} - 1 \right) = \frac{p_1{}^2 + m^2}{p_1{}^2 + M^2 - i\varepsilon} \frac{p_2{}^2 + m^2}{p_2{}^2 + M^2 - i\varepsilon},$$

$$\tag{4-5.91}$$

which converts (4-5.90) into

$$-i \frac{\alpha}{2\pi} \int \frac{dM^2}{M^2} \frac{M^2 + m^2}{M^2 - m^2} \frac{(dp_1)}{(2\pi)^4} \frac{(dp_2)}{(2\pi)^4}$$

$$\times K_1(-p_1) \frac{1}{p_1{}^2 + M^2 - i\varepsilon} eq(p_1 + p_2) A(k) \frac{1}{p_2{}^2 + M^2 - i\varepsilon} K_2(p_2). \tag{4-5.92}$$

The correctness of this procedure will be evident on comparison with (4-1.65), which gives the corresponding modified propagation function in the absence of an electromagnetic field, and in the gauge $A = 0$. A more general description of the field free situation is produced by the matrix substitution $p \to p - eqA$:

$$\frac{1}{(p - eqA)^2 + M^2 - i\varepsilon} = \frac{1}{p^2 + M^2 - i\varepsilon} + \frac{1}{p^2 + M^2 - i\varepsilon}$$

$$\times\, eq(pA + Ap)\,\frac{1}{p^2 + M^2 - i\varepsilon} + \cdots. \tag{4-5.93}$$

The additional term, linear in A, for the coupling of two sources is precisely (4–5.92). This is what the normalization requirements of the phenomenological theory demand.

Now let us return to (4–5.86) and extract the particle field terms, rejecting all explicit source terms, by means of the rearrangements illustrated in

$$\frac{1}{p_1{}^2 + M_1{}^2 - i\varepsilon} = \frac{1}{M_1{}^2 - m^2} - \frac{p_1{}^2 + m^2}{M_1{}^2 - m^2}\frac{1}{p_1{}^2 + M_1{}^2 - i\varepsilon}. \tag{4-5.94}$$

Effectively, this is the substitution

$$\frac{1}{p_1{}^2 + M_1{}^2 - i\varepsilon}\frac{1}{p_2{}^2 + M_2{}^2 - i\varepsilon} \to \frac{1}{M_1{}^2 - m^2}\frac{1}{M_2{}^2 - m^2}$$

$$= \frac{1}{k^2 + 4m^2}\frac{1}{m^2 + (k^2/4)(1 - v^2)}\frac{1}{x^2}. \tag{4-5.95}$$

In addition, we shall write

$$\frac{4m^2}{k^2}\frac{1}{m^2 + (k^2/4)(1 - v^2)} = \frac{4}{k^2} - \frac{1 - v^2}{m^2 + (k^2/4)(1 - v^2)} \tag{4-5.96}$$

and omit the $1/k^2$ term since, as we have just shown, its function is to introduce the gauge covariant combination $p - eqA$ in the modified propagation function. The result can be presented as a field dependent action term. Combined with the primitive interaction, it is

$$\int (dx)(dx')\phi(x)eq(1/i)\,\partial_\mu\phi(x)F(x - x')A^\mu(x'), \tag{4-5.97}$$

with

$$F(k) = 1 - \frac{k^2}{4m^2}\int_0^1 2v\,dv\,\frac{f(v)}{1 + \dfrac{k^2}{4m^2}(1 - v^2)} \tag{4-5.98}$$

and

$$vf(v) = \frac{\alpha}{2\pi}\int_{\to 0}^{\infty}\frac{dx}{x}\frac{1 + x}{(1 + 2x + x^2v^2)^{3/2}}(1 + (1 + x)v^2). \tag{4-5.99}$$

We have met the inevitable logarithmic singularity at the lower integration limit. This time it is worth while to introduce the finite photon mass μ for a detailed comparison with earlier results. The significant effect of the photon mass appears in the inequality (4–5.68) which now reads (k^2 appears in place of K^2)

$$\left| M_2{}^2 + \mu^2 - m^2 - \frac{M_1{}^2 + \mu^2 - m^2}{M_1} \frac{k^2 + M_1{}^2 + M_2{}^2}{2M_1} \right|$$

$$< \left[\left(\frac{M_1{}^2 + \mu^2 - m^2}{2M_1} \right)^2 - \mu^2 \right]^{1/2} \frac{\Delta^{1/2}}{M_1}, \tag{4–5.100}$$

or

$$x^2(1 - v^2)\left(m^2 + \frac{k^2}{4} \right) > \mu^2(1 + x). \tag{4–5.101}$$

It suffices to introduce the following lower limit in the integral of (4–5.99):

$$x_{\text{lower lim.}} = \frac{\mu}{m}(1 - v^2)^{-1/2}\left(1 + \frac{k^2}{4m^2} \right)^{-1/2}. \tag{4–5.102}$$

After performing the x integration we find

$$vf(v) = \frac{\alpha}{2\pi} \frac{1 + v^2}{2} \left[\log\left(\frac{4m^2}{\mu^2} \frac{1 - v^2}{(1 + v)^2} \right) - 2 + \frac{2v}{1 + v^2} + \log\left(1 + \frac{k^2}{4m^2} \right) \right]. \tag{4–5.103}$$

But there is an identity, somewhat reminiscent of (4–4.109):

$$\int_0^1 dv(1 + v^2) \frac{\log\left(1 + \frac{k^2}{4m^2} \right)}{1 + \frac{k^2}{4m^2}(1 - v^2)} = \int_0^1 dv(1 + v^2) \frac{\log\left(\frac{(1 + v)^2}{1 - v^2} \frac{v^2}{1 - v^2} \right) - \frac{2v}{1 + v^2}}{1 + \frac{k^2}{4m^2}(1 - v^2)},$$

$$\tag{4–5.104}$$

which can be verified by algebraic rearrangement, or by regarding $- k^2/4m^2$ as a complex variable. In the latter approach, both sides of (4–5.104) represent functions of a complex variable that vanish at infinity and have branch lines extending from 1 to ∞ along the real axis. On observing that the two imaginary parts are equal on that interval, we confirm (4–5.104). Now we have, effectively,

$$vf(v) = \frac{\alpha}{2\pi} \frac{1 + v^2}{2} \left[\log\left(\frac{4m^2}{\mu^2} \frac{v^2}{1 - v^2} \right) - 2 \right], \tag{4–5.105}$$

which correctly identifies $f(v)$ with the $f(M^2)$ of Eq. (4–4.21).

There are some subtleties in the treatment of finite photon mass that would bear further examination, but this will be deferred until the corresponding spin $\frac{1}{2}$ discussion is before us.

4–6 FORM FACTORS III. SPIN ½

The spin ½ counterpart of these considerations uses the two-particle amplitude

$$\int (dx) \cdots (d\xi') i J_1{}^\mu(\xi)\eta_1(x)\gamma^0|_{\text{eff.}} D_+(\xi - \xi')G_+^A(x, x')i J_{2\mu}(\xi')\eta_2(x')|_{\text{eff.}}, \quad (4\text{–}6.1)$$

retaining only the field dependent part of the source:

$$i J^\mu(\xi)\eta(x)|_{\text{eff.}} = eq\, \delta(x - \xi)\gamma^\mu\psi(x) + \cdots, \quad (4\text{–}6.2)$$

and considers the linear field term of the expansion

$$G_+^A = G_+ + G_+ eq\gamma A G_+ + \cdots. \quad (4\text{–}6.3)$$

The corresponding vacuum amplitude is

$$- ie^2 \int dM_1{}^2 dM_2{}^2 d\omega_{P_1} d\omega_{P_2} d\omega_k\, \delta(p_1{}^2 + m^2)\delta(p_2{}^2 + m^2)\psi_1(- P_1)\gamma^0\gamma^\nu(m - \gamma p_1)$$

$$\times\, eq\gamma A(K)(m - \gamma p_2)\gamma_\nu\psi_2(P_2), \quad (4\text{–}6.4)$$

where

$$p_1 = P_1 - k, \qquad p_2 = P_2 - k. \quad (4\text{–}6.5)$$

We shall first carry out an algebraic simplification of the matrix

$$M^\mu = \gamma^\nu(m - \gamma p_1)\gamma^\mu(m - \gamma p_2)\gamma_\nu, \quad (4\text{–}6.6)$$

based only on the kinematical relations

$$- p_1{}^2 = - p_2{}^2 = m^2. \quad (4\text{–}6.7)$$

It is designed to make explicit the gauge invariance property

$$K_\mu M^\mu = 0, \quad (4\text{–}6.8)$$

which is a consequence of the equality

$$\gamma K = (\gamma p_1 + m) - (\gamma p_2 + m). \quad (4\text{–}6.9)$$

The first stage of the simplification is displayed in

$$M^\mu = (2p_1{}^\nu + (m + \gamma p_1)\gamma^\nu)\gamma^\mu(2p_{2\nu} + \gamma_\nu(m + \gamma p_2))$$

$$= - 2(K^2 + 2m^2)\gamma^\mu + 2(m + \gamma p_1)\gamma p_2\gamma^\mu + 2\gamma^\mu\gamma p_1(m + \gamma p_2)$$

$$+ 2(m + \gamma p_1)\gamma^\mu(m + \gamma p_2). \quad (4\text{–}6.10)$$

The next step involves a systematic use of the connection

$$K = p_1 - p_2, \tag{4-6.11}$$

as illustrated by

$$(m + \gamma p_1)\gamma p_2 = (m + \gamma p_1)(\gamma p_1 - \gamma K) = m(m + \gamma p_1) - (m + \gamma p_1)\gamma K, \tag{4-6.12}$$

which leads finally to

$$M^\mu = - K^2 \gamma^\mu + K^\mu \gamma K - 4m(p_1 + p_2)^\mu - \tfrac{1}{2}\gamma(p_1 + p_2)i\sigma^{\mu\nu}K_\nu$$
$$- i\sigma^{\mu\nu}K_\nu \tfrac{1}{2}\gamma(p_1 + p_2) - (p_1 + p_2)^\mu(p_1 + p_2)^\nu \gamma_\nu. \tag{4-6.13}$$

There are four sets of terms here that individually obey the property (4-6.8). Two of them involve identities and the other two depend upon the orthogonality property

$$K(p_1 + p_2) = 0. \tag{4-6.14}$$

The integration process in (4-6.4) implies an average value of the vector $p_1 + p_2$ which, according to (4-6.14), must have a form that is indicated in matrix notation as

$$\langle p_1 + p_2 \rangle = \alpha \left(1 - \frac{KK}{K^2} \right)(P_1 + P_2). \tag{4-6.15}$$

The algebraic property

$$(P_1 + P_2)(p_1 + p_2) = (P_1 + P_2)^2 - 2k(P_1 + P_2)$$
$$= - (K^2 + M_1{}^2 + M_2{}^2 + 2m^2), \tag{4-6.16}$$

combined with the recognition that

$$- (P_1 + P_2)(K^2 - KK)(P_1 + P_2) = \Delta, \tag{4-6.17}$$

gives

$$\alpha = \frac{K^2(K^2 + M_1{}^2 + M_2{}^2 + 2m^2)}{\Delta}, \tag{4-6.18}$$

which repeats the result of (4-5.77). The other expectation value that we need is given by the symmetrical matrix

$$\langle (p_1 + p_2)(p_1 + p_2) \rangle = a \left(1 - \frac{KK}{K^2} \right) + b \left(1 - \frac{KK}{K^2} \right)(P_1 + P_2)(P_1 + P_2)\left(1 - \frac{KK}{K^2} \right). \tag{4-6.19}$$

The trace of this matrix and the result of multiplication on both sides with the vector $P_1 + P_2$ supply the necessary information to determine a and b:

$$- (K^2 + 4m^2) = 3a - b(\Delta/K^2),$$

$$(K^2 + M_1{}^2 + M_2{}^2 + 2m^2)^2 = - a(\Delta/K^2) + b(\Delta/K^2)^2. \qquad (4\text{--}6.20)$$

The solution is

$$a = \frac{2}{\Delta} [K^2(M_1{}^2 - m^2)(M_2{}^2 - m^2) - m^2(M_1{}^2 - M_2{}^2)^2],$$

$$b = \frac{K^2(K^2 + 4m^2)}{\Delta} + \frac{6K^2}{\Delta} [K^2(M_1{}^2 - m^2)(M_2{}^2 - m^2) - m^2(M_1{}^2 - M_2{}^2)^2].$$

$$(4\text{--}6.21)$$

When the variables of (4–5.71) are introduced, the three parameters α, a, b acquire the following form:

$$\alpha = \frac{1 + x}{\delta},$$

$$a = (K^2 + 4m^2)\, \frac{x^2(1 - v^2)}{2\delta},$$

$$b = \frac{1}{\delta} + \frac{3}{2} \frac{x^2(1 - v^2)}{\delta^2}, \qquad (4\text{--}6.22)$$

which uses the abbreviation

$$\delta = 1 + 2x + v^2x^2. \qquad (4\text{--}6.23)$$

Note that only a involves K^2, and in a linear way.

Now that just the vectors P_1, P_2 and their difference K appear in $\langle M_\mu \rangle$, we proceed to the final rearrangement. It utilizes identities, such as

$$(P_1 + P_2)^\mu = - (\gamma P_1 \gamma^\mu + \gamma^\mu \gamma P_2) + i\sigma^{\mu\nu}K_\nu \qquad (4\text{--}6.24)$$

and the relatively complicated one

$$(P_1 + P_2)^\mu \left[\gamma(P_1 + P_2) + \frac{M_1{}^2 - M_2{}^2}{K^2}\gamma K \right]$$

$$= - 2\gamma P_1 \gamma^\mu \gamma P_2 - \left(M_1{}^2 + M_2{}^2 + \frac{(M_1{}^2 - M_2{}^2)^2}{K^2} \right)\gamma^\mu$$

$$+ \left(1 + \frac{M_1{}^2 - M_2{}^2}{K^2} \right)\gamma P_1 i\sigma^{\mu\nu}K_\nu + \left(1 + \frac{M_2{}^2 - M_1{}^2}{K^2} \right)i\sigma^{\mu\nu}K_\nu \gamma P_2, \quad (4\text{--}6.25)$$

which are designed to express $\langle M^\mu \rangle$ in terms of the basic matrix vectors γ^μ, $i\sigma^{\mu\nu}K_\nu$, together with possible factors of γP_1 on the left and γP_2 on the right. In doing this we make the one simplification of introducing a Lorentz gauge, so that

$\gamma^{\mu} - (K^{\mu}\gamma K/K^2)$ is replaced by γ^{μ}. This is no loss of generality since the projection matrix $1 - (KK/K^2)$ induces a gauge transformation on the vector potential that puts it into a Lorentz gauge. The final result is

$$\langle M^{\mu}\rangle = \left[-(1+\alpha)K^2 - a + b\left(M_1{}^2 + M_2{}^2 + \frac{(M_1{}^2 - M_2{}^2)^2}{K^2}\right)\right]\gamma^{\mu}$$

$$+ 4m\alpha(\gamma P_1\gamma^{\mu} + \gamma^{\mu}\gamma P_2) + 2b\gamma P_1\gamma^{\mu}\gamma P_2 - 4m\alpha i\sigma^{\mu\nu}K_{\nu}$$

$$-\left[\alpha + b\left(1 + \frac{M_1{}^2 - M_2{}^2}{K^2}\right)\right]\gamma P_1 i\sigma^{\mu\nu}K_{\nu}$$

$$-\left[\alpha + b\left(1 + \frac{M_2{}^2 - M_1{}^2}{K^2}\right)\right]i\sigma^{\mu\nu}K_{\nu}\gamma P_2. \tag{4-6.26}$$

It enters the vacuum amplitude (4–6.4), which is now written as

$$-i\frac{\alpha}{4\pi}\int d\omega_{P_1}\,d\omega_{P_2}\frac{dM_1{}^2\,dM_2{}^2}{\Delta^{1/2}}\,\psi_1(-P_1)\gamma^0 eq\langle M^{\mu}\rangle A_{\mu}(K)\psi_2(P_2). \tag{4-6.27}$$

The space-time extrapolation of this vacuum amplitude, performed without regard to contact terms as in (4–5.84), is given by the double spectral form

$$i\frac{\alpha}{4\pi}\int\frac{(dp_1)}{(2\pi)^4}\frac{(dp_2)}{(2\pi)^4}\frac{dM_1{}^2\,dM_2{}^2}{\Delta^{1/2}}\,\psi_1(-p_1)\gamma^0 eq\langle M^{\mu}\rangle A_{\mu}(k)\psi_2(p_2)$$

$$\times\frac{1}{p_1{}^2 + M_1{}^2 - i\varepsilon}\frac{1}{p_2{}^2 + M_2{}^2 - i\varepsilon}, \tag{4-6.28}$$

where we recall that

$$\frac{dM_1{}^2\,dM_2{}^2}{\Delta^{1/2}} = \frac{1}{2}\frac{x\,dx\,dv}{\delta^{1/2}}(k^2 + 4m^2). \tag{4-6.29}$$

The electromagnetic field-free situation is considered first. This circumstance is conveyed in (4–6.26) by omitting all $\sigma^{\mu\nu}k_{\nu}$ and k^2 terms, which leaves

$$\langle M^{\mu}\rangle = 2m^2\beta\gamma^{\mu} + 4m\alpha(\gamma p_1\gamma^{\mu} + \gamma^{\mu}\gamma p_2) + 2b\gamma p_1\gamma^{\mu}\gamma p_2, \tag{4-6.30}$$

where

$$\beta = (1 + 2x + 2v^2x^2)b - \frac{x^2(1 - v^2)}{\delta}. \tag{4-6.31}$$

The following v integrals are required:

$$\int_{-1}^{1}\frac{1}{2}\,dv\,\frac{1}{\delta^{3/2}} = \frac{1}{(1+x)(1+2x)}, \qquad \frac{3}{2}\int_{-1}^{1}\frac{1}{2}\,dv\,\frac{1-v^2}{\delta^{5/2}} = \frac{1}{(1+x)(1+2x)^2},$$

$$x^3 \int_{-1}^{1} \frac{1}{2} \, dv \, \frac{v^2}{\delta^{3/2}} = \frac{1}{2} \log(1 + 2x) - \frac{x}{1 + x},$$

$$x^5 \int_{-1}^{1} \frac{1}{2} \, dv \, \frac{v^2(1 - v^2)}{\delta^{5/2}} = -\frac{1}{2} \log(1 + 2x) + \frac{x}{1 + x} + \frac{1}{3} \frac{x^3}{(1 + x)(1 + 2x)}. \quad (4\text{–}6.32)$$

They supply these evaluations,

$$\int_{-1}^{1} \frac{1}{2} \, dv \, \frac{\alpha}{\delta^{1/2}} = \frac{1}{1 + 2x}, \qquad \int_{-1}^{1} \frac{1}{2} \, dv \, \frac{\beta}{\delta^{1/2}} = \frac{1 + x}{1 + 2x},$$

$$\int_{-1}^{1} \frac{1}{2} \, dv \, \frac{b}{\delta^{1/2}} = \frac{1 + x}{(1 + 2x)^2}. \quad (4\text{–}6.33)$$

Accordingly, the effective value of $\langle M^\mu \rangle$ is

$$\langle M^\mu \rangle \to \frac{m^2}{M^4} [(M^2 + m^2)(M^2\gamma^\mu + \gamma p_1 \gamma^\mu \gamma p_2) + 4mM^2(\gamma p_1 \gamma^\mu + \gamma^\mu \gamma p_2)], \quad (4\text{–}6.34)$$

which introduces the single spectral mass parameter

$$M^2 = m^2(1 + 2x). \quad (4\text{–}6.35)$$

Apart from a factor of $(m^2/2M^4)$, the result we have just obtained can also be written as

$$[(M - m)^2 - 2mM](M - \gamma p_1)\gamma^\mu(M - \gamma p_2) + [(M + m)^2 + 2mM]$$

$$\times (M + \gamma p_1)\gamma^\mu(M + \gamma p_2). \quad (4\text{–}6.36)$$

This enables the unified version of the vacuum amplitude (4–6.28) to be displayed in the following form:

$$i \frac{\alpha}{4\pi} \int \frac{(dp_1)}{(2\pi)^4} \frac{(dp_2)}{(2\pi)^4} \int_{\to m}^{\infty} \frac{dM}{M} \left(1 - \frac{m^2}{M^2}\right) \frac{1}{2} \psi(-p_1)\gamma^0 \left\{[(M - m)^2 - 2mM]\right.$$

$$\times \frac{1}{\gamma p_1 + M - i\varepsilon} eq\gamma A(k) \frac{1}{\gamma p_2 + M - i\varepsilon}$$

$$+ [(M + m)^2 + 2mM] \frac{1}{\gamma p_1 - M + i\varepsilon} eq\gamma A(k) \frac{1}{\gamma p_2 - M + i\varepsilon}$$

$$\left. - 2\left(1 - \frac{4m^2M^2}{(M^2 - m^2)^2}\right) eq\gamma A(k)\right\} \psi(P_2). \quad (4\text{–}6.37)$$

A completely local contact term has also been added, in order to satisfy the physical normalization conditions. They are implied by the gauge covariant generalization of the modified propagation function. It is most convenient, and natural, to use the structure described in the action contribution (4–1.55, 56) as amended by the contact modifications (4–1.59, 60), without the algebraic recombinations that are also stated in the latter equation. The generalization given by ($\pm i\varepsilon$ is omitted)

$$\frac{1}{\gamma p \pm M} \rightarrow \frac{1}{\gamma(p - eqA) \pm M} = \frac{1}{\gamma p \pm M} + \frac{1}{\gamma p \pm M} eq\gamma A \frac{1}{\gamma p \pm M} + \cdots \quad (4\text{–}6.38)$$

and

$$\gamma p \rightarrow \gamma p - eq\gamma A \quad (4\text{–}6.39)$$

leads immediately to (4–6.37).

Incidentally, we might have proceeded similarly with spin 0, retaining fields and adding a suitable contact interaction in (4–1.64) to obtain the vacuum amplitude

$$-i\frac{\alpha}{2\pi}\int \frac{(dp)}{(2\pi)^4} \frac{dM^2}{M^2} (M^2 + m^2)(M^2 - m^2)\phi_1(-p)\left[\frac{1}{p^2 + M^2 - i\varepsilon} - \frac{1}{M^2 - m^2}\right.$$

$$\left. + \frac{p^2 + m^2}{(M^2 - m^2)^2}\right] \phi_2(p). \quad (4\text{–}6.40)$$

The gauge covariant substitution then implies the additional coupling term

$$-i\frac{\alpha}{2\pi}\int \frac{(dp_1)}{(2\pi)^4} \frac{(dp_2)}{(2\pi)^4} \phi_1(-p_1)eq\phi_2(p_2)(p_1 + p_2)A(k)$$

$$\times \int \frac{dM^2}{M^2} \frac{M^2 + m^2}{M^2 - m^2}\left(\frac{M^2 - m^2}{p_1^2 + M^2 - i\varepsilon}\frac{M^2 - m^2}{p_2^2 + M^2 - i\varepsilon} - 1\right). \quad (4\text{–}6.41)$$

It reproduces the double spectral form of (4–5.90), but replaces the additional contact terms appearing there, which involve single spectral forms, by one completely local contact term. The two expressions differ only in their treatment of couplings that depend explicitly upon sources, rather than fields. What we have now done for both spins 0 and $\frac{1}{2}$ has the advantage of indicating that no single spectral forms are needed.

Let us return to the double spectral form (4–6.28) and consider the magnetic moment terms separately. We shall apply them to three situations where single spectral forms are already known. One of these, described in Eq. (4–5.20), gives the field dependence for coupling of an extended particle source with a simple particle source and a simple photon source ($k^2 = 0$). Correspondingly, terms that are explicitly dependent upon sources are discarded in (4–6.26, 28) by the replacements $\gamma p_1 \rightarrow -m$, $\gamma p_2 \rightarrow -m$ and, if ψ_1 is identified as the field of the simple particle source, by $(p_1^2 + M^2)^{-1} \to (M^2 - m^2)^{-1}$. The resulting coefficient of $-2mi\sigma^{\mu\nu}k_\nu$ in (4-6.26) is $\alpha - b$, which enters the integral [cf. Eq. (4-6.33)]

$$\int_{-1}^{1} \frac{1}{2} \, dv \, \frac{(\alpha - b)}{\delta^{1/2}} = \frac{x}{(1 + 2x)^2} = m^2 \, \frac{M^2 - m^2}{2M^4} . \qquad (4\text{-}6.42)$$

The derived single spectral form,

$$i \, \frac{\alpha}{2\pi} \int \frac{(dp_1)}{(2\pi)^4} \, \frac{(dp_2)}{(2\pi)^4} \, dM^2 \, \frac{m^2}{M^2} \left(1 - \frac{m^2}{M^2}\right) \psi(-p_1)\gamma^0 \, \frac{eq}{2m} \, \sigma F(k)\psi_{\text{ext.}}(p_2) \frac{1}{p_2^2 + M^2 - i\varepsilon} , \qquad (4\text{-}6.43)$$

is the momentum space equivalent of Eq. (4-5.20).

The single spectral form stated in Eqs. (4-4.65, 73, 75) refers entirely to simple particle sources. The denominators of the double spectral form now reduce to

$$\frac{1}{M_1^2 - m^2} \frac{1}{M_2^2 - m^2} = \frac{1}{x^2 \, k^2 + 4m^2} \frac{1}{m^2 + (k^2/4)(1 - v^2)} . \qquad (4\text{-}6.44)$$

This time it is the x integration that must be performed. Using the integrals

$$\int_0^\infty \frac{dx}{\delta^{3/2}} = \frac{1}{1 + |v|} , \qquad 3 \int_0^\infty \frac{x \, dx}{\delta^{5/2}} = \frac{1}{(1 + |v|)^2} , \qquad (4\text{-}6.45)$$

we get

$$\int_0^\infty \frac{dx}{x} \frac{\alpha - b}{\delta^{1/2}} = \int_0^\infty dx \left[\frac{1}{\delta^{3/2}} - \frac{3}{2} \frac{x(1 - v^2)}{\delta^{5/2}}\right] = \frac{1}{2} . \qquad (4\text{-}6.46)$$

The derived single spectral form,

$$i \, \frac{\alpha}{2\pi} \int \frac{(dp_1)}{(2\pi)^4} \, \frac{(dp_2)}{(2\pi)^4} \left[\int_0^1 \frac{dv}{1 + \frac{k^2}{4m^2}(1 - v^2)}\right] \frac{1}{2} \, \psi(-p_1)\gamma^0 \, \frac{eq}{2m} \, \sigma F(k)\psi(p_2), \qquad (4\text{-}6.47)$$

is the expected one.

The third application is that discussed in Section 4-2. The electromagnetic field here is a uniform one, so that the relevant terms of $\langle M^\mu \rangle A_\mu(k)$ become

$$\frac{1}{(2\pi)^4} \, \delta(p_1 - p_2)[4m\alpha\sigma F + 2(\alpha + b)\gamma p\sigma F], \qquad (4\text{-}6.48)$$

where $p = p_1 = p_2$, and symmetrized multiplication between γp and σF is understood. The v integration is performed:

$$\int_{-1}^{1} \frac{1}{2} \, dv \, \frac{1}{\delta^{1/2}} [4m\alpha\sigma F + 2(\alpha + b)\gamma p\sigma F] = 4 \, \frac{m^2}{M^2}(m + \gamma p)\sigma F - \frac{m^2}{M^2}\left(1 - \frac{m^2}{M^2}\right)\gamma p\sigma F. \qquad (4\text{-}6.49)$$

The resulting vacuum amplitude,

$$i\,\frac{\alpha}{2\pi}\int\frac{(dp)}{(2\pi)^4}\frac{dM^2}{(p^2+M^2-i\varepsilon)^2}\frac{1}{2}\,\psi(-p)\gamma^0\left[2\left(1-\frac{m^2}{M^2}\right)(m+\gamma p)\right.$$

$$\left.-\frac{1}{2}\left(1-\frac{m^2}{M^2}\right)^2\gamma p\right]eq\sigma F\psi(p),\tag{4-6.50}$$

can be exhibited as a conventional single spectral integral by carrying out a partial integration with respect to M^2. Since there are no contributions at the endpoints, m^2 and ∞, this gives the action term (i is omitted)

$$-\frac{\alpha}{2\pi}\int\frac{(dp)}{(2\pi)^4}\frac{dM}{M}\frac{m^2}{M^2}\frac{1}{2}\,\psi(-p)\gamma^0\left[-4(m+\gamma p)+2\left(1-\frac{m^2}{M^2}\right)\gamma p\right]eq\sigma F\psi(p)$$

$$\times\frac{1}{p^2+M^2-i\varepsilon},\tag{4-6.51}$$

where we proceed to write

$$\frac{-4(m+\gamma p)+2(1-m^2/M^2)\,\gamma p}{p^2+M^2-i\varepsilon}=\frac{(1-m/M)^2}{\gamma p+M-i\varepsilon}+\frac{(1+m/M)^2}{\gamma p-M+i\varepsilon}.\tag{4-6.52}$$

Apart from the appearance of p rather than Π, this is the electromagnetic field dependent term of Eqs. (4-2.30, 31).

Our final task here is to derive, from the double spectral form, the known single spectral form of the spin $\frac{1}{2}$ charge form factor. The substitutions $\gamma p_1\to-m$, $\gamma p_2\to-m$, which reject explicit source terms, produce the following coefficient of γ^μ in $\langle M^\mu\rangle$:

$$-\left(1+\frac{1+x}{\delta}\right)k^2-8m^2\frac{1+x}{\delta}-\frac{x^2(1-v^2)}{2\delta}\,(k^2+4m^2)+4m^2\left(\frac{1}{\delta}+\frac{3}{2}\frac{x^2(1-v^2)}{\delta^2}\right)$$

$$+\left(\frac{1}{\delta}+\frac{3}{2}\frac{x^2(1-v^2)}{\delta^2}\right)(x+x^2v^2)(k^2+4m^2).\tag{4-6.53}$$

As in the derivation of the magnetic moment form factor in this circumstance $(-p_1{}^2,-p_2{}^2\to m^2)$, we also have

$$\frac{dM_1{}^2\,dM_2{}^2}{\varDelta^{1/2}}\frac{1}{(M_1{}^2-m^2)(M_2{}^2-m^2)}=\frac{1}{2}\,dv\,\frac{dx}{x\delta^{1/2}}\frac{1}{m^2+(k^2/4)(1-v^2)}.\tag{4-6.54}$$

Since the vector potential term derived for $k^2=0$ should be incorporated in the modified propagation function, we remove it by the effective substitution [Eq. (4-5.96)]

$$4m^2\to-(1-v^2)k^2.\tag{4-6.55}$$

The resulting coefficient of $-k^2\gamma^\mu$ inferred from (4–6.53) can be arranged as

$$\frac{(1+x)(1+v^2(1+x))}{\delta} + \frac{x^2v^2(1-v^2)}{2\delta} + \frac{3}{2}\frac{x^2(1-v^2)^2}{\delta^2}$$

$$-\left(\frac{1}{\delta} + \frac{3}{2}\frac{x^2(1-v^2)}{\delta^2}\right)(x+x^2v^2)v^2. \tag{4–6.56}$$

One recognizes in the first term the complete structure of the spin 0 form factor, as given in (4–5.99). This immediately supplies the relation

$$vf_1(v) - vf(v) = \frac{\alpha}{2\pi}\int_0^\infty dx \left[\frac{xv^2(1-v^2)}{2\delta^{3/2}} + \frac{3}{2}\frac{x(1-v^2)^2}{\delta^{5/2}}\right.$$

$$\left. -\left(\frac{1}{\delta^{3/2}} + \frac{3}{2}\frac{x^2(1-v^2)}{\delta^{5/2}}\right)(1+xv^2)v^2\right]. \tag{4–6.57}$$

The convergence of this integral at the lower limit means that photon mass considerations are the same for both spins and need not be repeated. The integrand can be simplified somewhat to become

$$\frac{3}{2}\frac{x(1-v^2)(1+xv^2)}{\delta^{5/2}} - \frac{(1+x)v^2}{\delta^{3/2}}. \tag{4–6.58}$$

Then, using the integrals ($v > 0$)

$$\int_0^\infty dx\,\frac{1+x}{\delta^{3/2}} = \frac{1}{v}, \qquad 3\int_0^\infty dx\,\frac{x(1+xv^2)}{\delta^{5/2}} = \frac{1}{1+v}, \tag{4–6.59}$$

we get

$$vf_1(v) - vf(v) = \frac{\alpha}{4\pi}(1-3v). \tag{4–6.60}$$

But, according to Eq. (4–4.132),

$$vf_1(v) - vf(v) = \frac{\alpha}{4\pi}. \tag{4–6.61}$$

It would seem that we have failed.

That is too somber a conclusion. But certainly here is a warning that we have ignored some subtlety in the calculation. Perhaps it is time to emphasize that the business we are about is more artistic than scientific. It is the attempt to exploit one double spectral form, without using additional single spectral forms, in order to reproduce the known single spectral form results. Whether or not this can be accomplished depends upon the organization of the calculation. We have succeeded with the spin $\frac{1}{2}$ magnetic moment effects, doubtless because, as a dynamically induced phenomenon, there no contact terms are involved. The contrast between the charge form factors of spins 0 and $\frac{1}{2}$ possibly stems from the presence of two kinds of propagation functions in the latter circumstance (characterized by opposite intrinsic parity) which are unmixed in the absence of an electromagnetic field [Eq. (4–6.37)], but become coupled together by a field in a way that need not have been properly considered. This raises the question whether the difficulty could be bypassed by organizing the spin $\frac{1}{2}$ calculation in a manner resembling that of spin 0.

We are already familiar with such a procedure. It is the replacement of the propagation function G_+^A [Eq. (4–2.1)] by the propagation function Δ_+^A, which obeys [Eq. (4–2.4)]

$$[\Pi^2 - eq\sigma F + m^2]\Delta_+^A = 1, \tag{4–6.62}$$

a form that differs from the spin 0 structure only in the additional $eq\sigma F$ term. In effect, the linear coupling with the electromagnetic field is changed according to

$$eq\gamma A \to \frac{eq}{2m}[pA + Ap + \sigma F], \tag{4–6.63}$$

while $G_+ \to \Delta_+$. To see the workings of this type of calculation, let us return to the considerations of Section 4–4, which examined an extended photon source emitting a pair of charged particles that subsequently interacted, and proceed to introduce the interaction combination, (4–6.63), by algebraic rearrangement of the known spin $\frac{1}{2}$ structure. We refer to Eq. (4–4.35), which contains the following matrix product:

$$\gamma^\nu(m - \gamma p_2)\gamma^\mu(-m - \gamma p_2')\gamma_\nu. \tag{4–6.64}$$

In view of the projection matrices that flank γ^μ, one can introduce the substitution

$$\gamma^\mu \to \frac{1}{2}\left[\left(-\frac{1}{m}\gamma p_2\right)\gamma^\mu + \gamma^\mu\left(\frac{1}{m}\gamma p_2'\right)\right]$$

$$= \frac{1}{2m}[(p_2 - p_2')^\mu - i\sigma^{\mu\rho}k_\rho]. \tag{4–6.65}$$

Then, recalling the additional projection matrices that appear in Eq. (4–4.34), namely $m - \gamma p_1$ and $-m - \gamma p_1'$, we write

$$\gamma^\nu(m - \gamma p_2) = (m + \gamma p_2)\gamma^\nu + 2p_2{}^\nu \to \gamma(p_2 - p_1)\gamma^\nu + 2p_2{}^\nu$$

$$= (p_1 + p_2)^\nu - i\sigma^{\nu\kappa}(p_1 - p_2)_\kappa, \tag{4–6.66}$$

and

$$(- m - \gamma p_2')\gamma_\nu = \gamma_\nu(- m + \gamma p_2') + 2p_{2\nu}' \to \gamma_\nu \gamma(p_2' - p_1') + 2p_{2\nu}',$$

$$= (p_1' + p_2')_\nu + i\sigma_{\nu\lambda}(p_1' - p_2')^\lambda. \tag{4-6.67}$$

The result is to replace (4-6.64) with

$$\frac{1}{2m}[(p_1 + p_2)^\nu - i\sigma^{\nu\kappa}(p_1 - p_2)_\kappa][(p_2 - p_2')^\mu - i\sigma^{\mu\rho}k_\rho][(p_1' + p_2')_\nu + i\sigma_{\nu\lambda}(p_1' - p_2')^\lambda], \tag{4-6.68}$$

which is the statement of the substitution (4-6.63). (Apart from powers of $2m$, which would be supplied by external factors.) On omitting the spin terms, we recognize the spin 0 structure contained in Eq. (4-4.6).

In working out the product (4-6.68), it is helpful to note that $[p_1' - p_2' = - (p_1 - p_2)]$

$$[- i\sigma^{\nu\kappa}(p_1 - p_2)_\kappa][- i\sigma_{\nu\lambda}(p_1 - p_2)^\lambda]$$

$$= [- \gamma(p_1 - p_2)\gamma^\nu - (p_1 - p_2)^\nu][\gamma_\nu\gamma(p_1 - p_2) + (p_1 - p_2)_\nu]$$

$$= - 3(p_1 - p_2)^2, \tag{4-6.69}$$

while

$$[- i\sigma^{\nu\kappa}(p_1 - p_2)_\kappa]\sigma^{\mu\rho}[- i\sigma_{\nu\lambda}(p_1 - p_2)^\lambda]$$

$$= [- \gamma(p_1 - p_2)\gamma^\nu - (p_1 - p_2)^\nu]\sigma^{\mu\rho}[\gamma_\nu\gamma(p_1 - p_2) \dotplus (p_1 - p_2)_\nu]$$

$$= (p_1 - p_2)^2\sigma^{\mu\rho}, \tag{4-6.70}$$

since

$$\gamma^\nu\sigma^{\mu\rho}\gamma_\nu = 0. \tag{4-6.71}$$

At this stage the integral of (4-4.35) becomes, apart from simple factors,

$$\left\langle \frac{(p_1 + p_2)(p_1' + p_2')}{(p_1 - p_2)^2} (p_2 - p_2')^\mu \right\rangle - i\sigma^{\mu\nu}k_\nu\left[\left\langle \frac{(p_1 + p_2)(p_1' + p_2')}{(p_1 - p_2)^2} \right\rangle + 1\right]$$

$$- 2k^\nu i\sigma_{\nu\kappa}\left\langle \frac{(p_1 - p_2)^\kappa(p_2 - p_2')^\mu}{(p_1 - p_2)^2} \right\rangle + i\sigma^{\mu\rho}k_\rho i\sigma_{\nu\lambda}\left\langle \frac{(p_1 + p_2)^\nu(p_1 - p_2)^\lambda}{(p_1 - p_2)^2} \right\rangle$$

$$+ i\sigma^{\nu\kappa}\left\langle \frac{(p_1' + p_2')_\nu(p_1 - p_2)_\kappa}{(p_1 - p_2)^2} \right\rangle i\sigma^{\mu\rho}k_\rho, \tag{4-6.72}$$

which use the fact that

$$\langle (p_2 - p_2') \rangle = 0. \tag{4-6.73}$$

We already know the integral [Eq. (4–4.13)]

$$\left\langle \frac{(p_1 + p_2)(p_1' + p_2')}{(p_1 - p_2)^2} (p_2 - p_2')^\mu \right\rangle = (p_1 - p_1')^\mu \left[-2(M^2 - 2m^2) \left\langle \frac{1}{(p_1 - p_2)^2} \right\rangle \right.$$

$$\left. + 4 \frac{M^2 - 2m^2}{M^2 - 4m^2} \right], \qquad (4\text{–}6.74)$$

and a similar evaluation gives

$$\left\langle \frac{(p_1 + p_2)(p_1' + p_2')}{(p_1 - p_2)^2} \right\rangle + 1 = -2(M^2 - 2m^2) \left\langle \frac{1}{(p_1 - p_2)^2} \right\rangle$$

$$+ 4 \frac{M^2 - 2m^2}{M^2 - 4m^2} - 2 \frac{M^2}{M^2 - 4m^2}. \quad (4\text{–}6.75)$$

Another familiar integral, Eq. (4–4.44, 45), is encountered in the last two terms of (4–6.72) where, since only the antisymmetrical parts of the tensors are required,

$$\left\langle \frac{(p_1 + p_2)^\nu (p_1 - p_2)^\lambda}{(p_1 - p_2)^2} \right\rangle \to 2p_1^\nu \left\langle \frac{(p_1 - p_2)^\lambda}{(p_1 - p_2)^2} \right\rangle = 2p_1^\nu (p_1 - p_1')^\lambda \frac{1}{M^2 - 4m^2}$$

$$(4\text{–}6.76)$$

and

$$\left\langle \frac{(p_1' + p_2')_\nu (p_1 - p_2)_\kappa}{(p_1 - p_2)^2} \right\rangle \to 2p_1'{}_\nu (p_1 - p_1')_\kappa \frac{1}{M^2 - 4m^2}. \quad (4\text{–}6.77)$$

This leaves the integral

$$\left\langle \frac{(p_1 - p_2)^\kappa (p_2 - p_2')^\mu}{(p_1 - p_2)^2} \right\rangle = a \left(g^{\kappa\mu} + \frac{k^\kappa k^\mu}{M^2} \right) + b(p_1 - p_1')^\kappa (p_1 - p_1')^\mu, \quad (4\text{–}6.78)$$

where the form of the right-hand side indicates that the integral is symmetrical under an interchange of p_1 and p_1', and vanishes on multiplication with k_κ or k_μ. The two scalar combinations produced by the trace and by multiplication with $(p_1 - p_1')_\kappa (p_1 - p_1')_\mu$ supply the information that

$$-1 = 3a + (M^2 - 4m^2)b, \qquad 0 = a + (M^2 - 4m^2)b, \qquad (4\text{–}6.79)$$

and therefore

$$a = -\frac{1}{2}, \qquad b = \frac{1}{2} \frac{1}{M^2 - 4m^2}. \qquad (4\text{–}6.80)$$

On combining these evaluations, one can present (4–6.72) as

$$[(p_1 - p_1')^\mu - i\sigma^{\mu\nu}k_\nu] \left[-2(M^2 - 2m^2) \left\langle \frac{1}{(p_1 - p_2)^2} \right\rangle + 4 \frac{M^2 - 2m^2}{M^2 - 4m^2} \right]$$

$$+ i\sigma^{\mu\nu}k_\nu \frac{M^2 + 4m^2}{M^2 - 4m^2} + \frac{1}{M^2 - 4m^2}[- k^\nu i\sigma_{\nu\lambda}(p_1 - p_1')^\lambda(p_1 - p_1')^\mu$$

$$+ i\sigma^{\mu\rho}k_\rho k^\nu i\sigma_{\nu\lambda}(p_1 - p_1')^\lambda + k^\nu i\sigma_{\nu\lambda}(p_1 - p_1')^\lambda i\sigma^{\mu\rho}k_\rho]. \qquad (4\text{-}6.81)$$

To simplify the last spin combination, we write it as

$$\gamma k\gamma(p_1 - p_1')(p_1 - p_1')^\mu + (\gamma^\mu\gamma k + k^\mu)\gamma k\gamma(p_1 - p_1') + \gamma(p_1 - p_1')\gamma k(\gamma k\gamma^\mu + k^\mu)$$

$$= - (\gamma k\gamma p_1' + \gamma p_1\gamma k)(p_1 - p_1')^\mu - M^2(p_1 - p_1')^\mu \rightarrow - M^2(p_1 - p_1')^\mu, \qquad (4\text{-}6.82)$$

where the final step is the projection matrix reduction $\gamma p_1 \rightarrow - m$, $\gamma p_1' \rightarrow m$. The result,

$$[(p_1 - p_1')^\mu - i\sigma^{\mu\nu}k_\nu]\left[- 2(M^2 - 2m^2)\left\langle\frac{1}{(p_1 - p_2)^2}\right\rangle + 4\frac{M^2 - 2m^2}{M^2 - 4m^2}\frac{M^2}{M^2 - 4m^2}\right]$$

$$+ i\sigma^{\mu\nu}k_\nu \frac{4m^2}{M^2 - 4m^2}, \qquad (4\text{-}6.83)$$

multiplied by the factors $(1/2m)$ and $(1/4\pi)^2(1 - (4m^2/M^2))^{1/2}$, is in complete agreement with (4-4.57). It cannot be said that this calculation is appreciably shorter than the earlier one, but certainly the related spin 0 result is constantly in evidence.

Now we turn to the causal arrangement that produced the double spectral form, and consider [Eq. (4-6.4)]

$$\gamma^\nu(m - \gamma(P_1 - k))\gamma^\mu(m - \gamma(P_2 - k))\gamma_\nu. \qquad (4\text{-}6.84)$$

Here the substitutions are

$$\gamma^\mu \rightarrow \frac{1}{2}\left[\left(- \frac{1}{m}\gamma(P_1 - k)\right)\gamma^\mu + \gamma^\mu\left(- \frac{1}{m}\gamma(P_2 - k)\right)\right]$$

$$= \frac{1}{2m}[(P_1 + P_2 - 2k)^\mu - i\sigma^{\mu\rho}K_\rho], \qquad (4\text{-}6.85)$$

and

$$\gamma^\nu[m - \gamma(P_1 - k)] = (m + \gamma P_1)\gamma^\nu + (2P_1 - k)^\nu - i\sigma^{\nu\kappa}k_\kappa,$$

$$[m - \gamma(P_2 - k)]\gamma_\nu = \gamma_\nu(m + \gamma P_2) + (2P_2 - k)_\nu + i\sigma_{\nu\lambda}k^\lambda. \qquad (4\text{-}6.86)$$

The action of the matrix factors $m + \gamma P_1$ and $m + \gamma P_2$ converts fields into sources. If these terms are removed, (4-6.84) is replaced with

$$\frac{1}{2m}[(2P_1 - k)^\nu - i\sigma^{\nu\kappa}k_\kappa][(P_1 + P_2 - 2k)^\mu - i\sigma^{\mu\rho}K_\rho][(2P_2 - k)_\nu + i\sigma_{\nu\lambda}k^\lambda].$$

$$(4\text{-}6.87)$$

On omitting the spin terms we recognize the spin 0 structure in Eq. (4–5.59).

Since the momentum $p_1 - p_2$ in Eqs. (4–6.69, 70) has been replaced with a real photon momentum ($-k^2 = 0$), there is no contribution from these terms and (4–6.87) becomes (apart from the factor $1/2m$)

$$(2P_1 - k)(2P_2 - k)((P_1 + P_2 - 2k)^\mu - i\sigma^{\mu\nu}K_\nu) + (P_1 + P_2 - 2k)^\mu 2K^\nu i\sigma_{\nu\lambda}k^\lambda$$

$$- i\sigma^{\mu\rho}K_\rho 2P_1{}^\nu i\sigma_{\nu\lambda}k^\lambda + 2P_{2\nu}i\sigma^{\nu\kappa}k_\kappa i\sigma^{\mu\rho}K_\rho, \tag{4–6.88}$$

in which

$$(2P_1 - k)(2P_2 - k) = -(2K^2 + M_1{}^2 + M_2{}^2 + 2m^2). \tag{4–6.89}$$

As described in Eqs. (4–6.15) and (4–6.19), the effect of the k integration is to produce the substitutions

$$P_1 + P_2 - 2k \rightarrow \alpha\left(1 - \frac{KK}{K^2}\right)(P_1 + P_2) \tag{4–6.90}$$

and

$$(P_1 + P_2 - 2k)(P_1 + P_2 - 2k)$$

$$\rightarrow a\left(1 - \frac{KK}{K^2}\right) + b\left(1 - \frac{KK}{K^2}\right)(P_1 + P_2)(P_1 + P_2)\left(1 - \frac{KK}{K^2}\right), \tag{4–6.91}$$

where the various coefficients are summarized in Eq. (4–6.22). This converts (4–6.88) into a form that, in a Lorentz gauge, can be presented as

$$(P_1 + P_2)^\mu[-\alpha(2K^2 + M_1{}^2 + M_2{}^2 + 2m^2) + (\alpha - b)2P_1{}^\nu i\sigma_{\nu\lambda}P_2{}^\lambda]$$

$$+ i\sigma^{\mu\nu}K_\nu[2K^2 + M_1{}^2 + M_2{}^2 + 2m^2 + a]$$

$$- (1 - \alpha)\{i\sigma^{\mu\rho}K_\rho, P_1{}^\nu i\sigma_{\nu\lambda}P_2{}^\lambda\} - \frac{M_1{}^2 - M_2{}^2}{K^2}\alpha[i\sigma^{\mu\rho}K_\rho, P_1{}^\nu i\sigma_{\nu\lambda}P_2{}^\lambda]. \tag{4–6.92}$$

Now we observe that

$$i\sigma^{\mu\rho}K_\rho P_1{}^\nu i\sigma_{\nu\lambda}P_2{}^\lambda = \gamma^\mu\gamma K(\gamma K\gamma P_2 + KP_2) = -K^2\gamma^\mu\gamma P_2 - KP_2 i\sigma^{\mu\nu}K_\nu,$$

$$P_1{}^\nu i\sigma_{\nu\lambda}P_2{}^\lambda i\sigma^{\mu\rho}K_\rho = (\gamma P_1\gamma K + KP_1)\gamma K\gamma^\mu = -K^2\gamma P_1\gamma^\mu + KP_1 i\sigma^{\mu\nu}K_\nu,$$

$$\tag{4–6.93}$$

from which we get

$$\{i\sigma^{\mu\rho}K_\rho, P_1{}^\nu\sigma_{\nu\lambda}P_2{}^\lambda\} = K^2(P_1 + P_2)^\mu \tag{4–6.94}$$

and

$$[i\sigma^{\mu\rho}K_\rho, P_1{}^\nu i\sigma_{\nu\lambda}P_2{}^\lambda] = K^2(\gamma P_1\gamma^\mu - \gamma^\mu\gamma P_2) + (M_1{}^2 - M_2{}^2)i\sigma^{\mu\nu}K_\nu. \tag{4–6.95}$$

The following is also needed:

$$2P_1{}^\nu i\sigma_{\nu\lambda}P_2{}^\lambda = \gamma P_1 \gamma K - \gamma K \gamma P_2 + K^2. \tag{4-6.96}$$

A further reduction, in which explicit source terms are removed, disposes of the combinations

$$(\gamma P_1 + m)\gamma^\mu - \gamma^\mu(\gamma P_2 + m), \qquad (\gamma P_1 + m)\gamma K - \gamma K(\gamma P_2 + m). \tag{4-6.97}$$

The outcome for (4-6.92) is this expression,

$$(P_1 + P_2)^\mu[-\alpha(2K^2 + M_1{}^2 + M_2{}^2 + 2m^2) + (\alpha - b)K^2 - (1 - \alpha)K^2]$$
$$+ i\sigma^{\mu\nu}K_\nu \left[2K^2 + M_1{}^2 + M_2{}^2 + 2m^2 + a - \alpha\frac{(M_1{}^2 - M_2{}^2)^2}{K^2}\right], \tag{4-6.98}$$

or, introducing the explicit forms of α, a, b [Eq. (4-6.22), and also (4-5.79)],

$$((P_1 + P_2)^\mu - i\sigma^{\mu\nu}K_\nu)\left[-\left(\frac{1 + x + v^2(1 + x)^2}{\delta} + \frac{x^2v^2}{\delta} + \frac{3}{2}\frac{x^2(1 - v^2)}{\delta^2}\right)K^2\right.$$
$$- \frac{(1 + x)^2}{\delta}(4m^2 + K^2(1 - v^2))\bigg] + i\sigma^{\mu\nu}K_\nu\left[-4m^2\left(\frac{x}{\delta} - \frac{3}{2}\frac{x^2(1 - v^2)}{\delta^2}\right)\right.$$
$$+ \left.\left(\frac{x(1 + x)}{\delta} + \frac{x^2(1 - v^2)}{2\delta} + \frac{x}{\delta} - \frac{3}{2}\frac{x^2(1 - v^2)}{\delta^2}\right)(K^2 + 4m^2)\right]. \tag{4-6.99}$$

In the earlier spin 0 and spin $\frac{1}{2}$ discussions we have omitted terms proportional to $4m^2 + K^2(1 - v^2)$, after introducing the specialization $p_1{}^2$, $p_2{}^2 = -m^2$, with the argument that the appropriate contact term produces the necessary cancellation. This tactic is somewhat unsatisfactory, however, since the individual terms are infrared divergent, and may not be handled in the same manner when a finite photon mass is introduced. Let us instead proceed in this way. The relation

$$(K^2 + 4m^2)(4m^2 + K^2(1 - v^2)) = \frac{4}{x^2}(M_1{}^2 - m^2)(M_2{}^2 - m^2) \tag{4-6.100}$$

concentrates all the K^2 dependence of such contributions to a double spectral form in the factor

$$\frac{M_1{}^2 - m^2}{p_1{}^2 + M_1{}^2 - i\varepsilon}\frac{M_2{}^2 - m^2}{p_2{}^2 + M_2{}^2 - i\varepsilon} - 1, \tag{4-6.101}$$

where the appropriate K^2-independent contact term has already been added. This combination, which is a generalization of that in (4-6.41), vanishes on placing p_1^2 and p_2^2 equal to $-m^2$.

We have only to divide (4-6.99) by $2m$ in order to produce the γ^μ and $i\sigma^{\mu\nu}k_\nu$ couplings that are to be compared with the previous calculation. Thus, replacing (4-6.56) as the coefficient of $-\gamma^\mu k^2$ is

$$\frac{(1 + x)(1 + v^2(1 + x))}{\delta} + \frac{x^2 v^2}{\delta} + \frac{3}{2}\frac{x^2(1 - v^2)}{\delta^2}, \tag{4–6.102}$$

where the first term is again the corresponding spin 0 result. The relation (4–6.57) is altered to

$$v f_1(v) - v f(v) = \frac{\alpha}{2\pi}\int_0^\infty dx \left[\frac{x v^2}{\delta^{3/2}} + \frac{3}{2}\frac{x(1 - v^2)}{\delta^{5/2}}\right], \tag{4–6.103}$$

where [Eqs. (4–6.45, 59)]

$$\int_0^\infty dx\,\frac{x}{\delta^{3/2}} = \frac{1}{v(1 + v)}, \qquad 3\int_0^\infty dx\,\frac{x}{\delta^{5/2}} = \frac{1}{(1 + v)^2} \tag{4–6.104}$$

yields

$$v f_1(v) - v f(v) = \frac{\alpha}{4\pi}. \tag{4–6.105}$$

In contrast with (4–6.60), we have now realized the desired result.

But, on turning to the magnetic form factor, we find that the situation has reversed. The first term in the coefficient of $i\sigma^{\mu\nu}k_\nu$ is the expected one. The second contribution involves the x integral

$$\int_0^\infty dx \left[\frac{1 + x}{\delta^{3/2}} + \frac{x(1 - v^2)}{2\delta^{3/2}} + \frac{1}{\delta^{3/2}} - \frac{3}{2}\frac{x(1 - v^2)}{\delta^{5/2}}\right] = \frac{3}{2}\frac{1}{v}, \tag{4–6.106}$$

and the attempt to reproduce the magnetic form factor seems to have failed. There is, however, a common pattern to these unwanted additional terms which suggests that both failures are only apparent. The additions to the charge and magnetic form factors in the two calculations are proportional to

$$\frac{k^2}{4m^2}\int dv\,\frac{2v}{1 + \dfrac{k^2}{4m^2}(1 - v^2)} = k^2\int\frac{dM^2}{M^2}\frac{1}{k^2 + M^2} = \int dM^2\left(\frac{1}{M^2} - \frac{1}{k^2 + M^2}\right) \tag{4–6.107}$$

and

$$\frac{k^2 + 4m^2}{4m^2}\int dv\,\frac{(2/v)}{1 + \dfrac{k^2}{4m^2}(1 - v^2)} = (k^2 + 4m^2)\int\frac{dM^2}{M^2 - 4m^2}\frac{1}{k^2 + M^2}$$

$$= \int dM^2\left(\frac{1}{M^2 - 4m^2} - \frac{1}{k^2 + M^2}\right), \tag{4–6.108}$$

respectively, where the initial space-like limitation on k^2, $k^2 > 0$, is still retained. But what are the domains of integration? They are ordinarily fixed by a square root factor that indicates the threshold of the multiparticle exchange processes. There are no threshold factors here. As we shall see again in the next section, the generalization from the initial causal situation also implies the removal of initial mass restrictions, subject only to constraints imposed by threshold factors. In this situation, space-time extrapolation extends the domain of M^2 without limit. The variable M^2 ranges from $-\infty$ to ∞, for such are the values assumed by the momentum structure $-p^2$ in the four-dimensional momentum space. The two terms on the right-hand side of Eq. (4–6.107) have singularities at the real values $M^2 = 0$ and $M^2 = -k^2$, and similarly there are singularities at $M^2 = 4m^2$ and $M^2 = -k^2$ in (4–6.108). But these terms are also related to each other by a finite translation of M^2. A finite but infinitely remote interval gives no contribution to the individual integrals. Accordingly, the two terms cancel, and the value that is thus assigned to (4–6.107), and to (4–6.108), is zero.

Harold interrupts.

H.: I have a helpful comment and a question. During the spin 0 discussion that used one extended particle source, the problem arose of establishing equivalence with known results. This was simplified to confirming the identity (4–5.55), and that was done by independent calculation of the two sides. But you expressed the wish to find a suitable transformation between the differently appearing forms. This is the transformation you wanted:

$$x + (x^2 - 1)^{1/2}z = \frac{(x + 1)^{1/2} + (x - 1)^{1/2}v}{(x + 1)^{1/2} - (x - 1)^{1/2}v},\qquad (4\text{–}6.109)$$

where both variables, z and v, range from -1 to $+1$. Since

$$dz = \frac{2dv}{((x + 1)^{1/2} - (x - 1)^{1/2}v)^2},\qquad (4\text{–}6.110)$$

we find that

$$\int_{-1}^{1} \frac{dz}{x + (x^2 - 1)^{1/2}z} = \int_{-1}^{1} \frac{dv}{1 + \frac{1}{2}(x - 1)(1 - v^2)}\qquad (4\text{–}6.111)$$

or, subtracting $2/x$ from both sides and rearranging,

$$2(x + 1)\int_{0}^{1} dz \frac{z^2}{1 + (x^2 - 1)(1 - z^2)} = \int_{0}^{1} dv \frac{1 + v^2}{1 + \frac{1}{2}(x - 1)(1 - v^2)}.\qquad (4\text{–}6.112)$$

The choice

$$x = 1 + \frac{k^2}{2m^2}\qquad (4\text{–}6.113)$$

then gives the desired result.

Here is the question. I am a little confused by what you have been doing. The same dynamical modifications of electromagnetic properties have now been derived in a number of different ways. Why is that important? Wouldn't one derivation suffice?

S.: Thank you for the transformation. As to your question, it is precisely the agreement of those various derivations that is significant since the consistency of the principles of causality and space-time uniformity was being tested. Different causal arrangements of sources, which were sometimes operated as simple sources, sometimes as extended sources, all led to the same conclusions. And, in the process, we have discussed two extended particle sources, which results are particularly useful in various applications. But first, we have more to learn about single and double spectral forms.

4-7   FORM FACTORS IV.   THE DEUTERON

The problems to which we now direct our attention are, strictly speaking, outside the framework of pure electrodynamics. They lie in the realm of low energy nuclear physics. Yet no explicit account of nuclear forces will appear in this discussion. Besides the photon, the particles of interest are the neutron, proton, and deuteron. The latter, in particular, is described phenomenologically although we have no doubts about the composite nature of this particle. For simplicity, all these particles are treated as though they were spinless objects. We shall not distinguish between the neutron and proton mass, and denote the common value by m. The deuteron mass will be written as

$$m_D = 2m - \varepsilon, \tag{4-7.1}$$

where ε, the deuteron binding energy, is quite small in comparison with m. The physical relation between the deuteron and the neutron and proton is introduced through the extended source concept. Given enough energy relative to the momentum—a sufficient excess of mass—the source that ordinarily emits a deuteron can emit a neutron and a proton. This is expressed by the primitive interaction

$$W_{Dpn} = 4\pi f \int (dx)\phi_D(x)\phi_p(x)\phi_n(x), \tag{4-7.2}$$

which involves a scalar product in the charge space common to the proton and the deuteron.

We first consider the modification in the deuteron propagation function that is implied by the primitive interaction, which portrays the deuteron field as an effective two-particle source:

$$iK_p(x)K_n(x')|_{\text{eff.}} = 4\pi f \phi_D(x)\delta(x - x'). \tag{4-7.3}$$

The resulting coupling between two causally arranged extended deuteron sources
is deduced from the vacuum amplitude

$$\int (dx) \cdots (dy') i K_{p1}(x) K_{n1}(y)|_{\text{eff.}} \Delta_p(x - x') \Delta_n(y - y') i K_{p2}(x') K_{n2}(y')|_{\text{eff.}}; \quad (4\text{–}7.4)$$

it is

$$(4\pi f)^2 \int (dx)(dx') \phi_{D1}(x) \Delta_p(x - x') \Delta_n(x - x') \phi_{D2}(x'). \quad (4\text{–}7.5)$$

In view of the causal arrangement, the product of propagation functions is
evaluated as

$$x^0 > x^{0'}: \qquad \Delta_p(x - x') \Delta_n(x - x')$$

$$= -\int d\omega_p \, d\omega_n \, \exp[iP(x - x')]$$

$$= -\int dM^2 \frac{1}{(4\pi)^2} \left(1 - \frac{4m^2}{M^2}\right)^{1/2} d\omega_P \exp[iP(x - x')]$$

$$= \frac{i}{(4\pi)^2} \int_{(2m)^2}^{\infty} dM^2 \left(1 - \frac{4m^2}{M^2}\right)^{1/2} \Delta_+(x - x', M^2), \quad (4\text{–}7.6)$$

which is an application of the kinematical integral (4–1.24). Contact terms must
be added to this expression in order to satisfy the physical normalization conditions. They demand that the additional coupling refer to sources and not the
deuteron field. Otherwise the initial description of the deuteron that is contained
in its propagation function would be altered. The needed supplementary terms
are indicated by

$$\frac{1}{p^2 + M^2 - i\varepsilon} \rightarrow \frac{1}{p^2 + M^2 - i\varepsilon} - \frac{1}{M^2 - m_D^2} + \frac{p^2 + m_D^2}{(M^2 - m_D^2)^2}$$

$$= \frac{(p^2 + m_D^2)^2}{(M^2 - m_D^2)^2} \frac{1}{p^2 + M^2 - i\varepsilon}. \quad (4\text{–}7.7)$$

The additional action term obtained in this way is

$$f^2 \int \frac{(dp)}{(2\pi)^4} \frac{1}{2} \phi_D(-p)(p^2 + m_D^2)^2 \phi_D(p) \int_{(2m)^2}^{\infty} \frac{dM^2}{(M^2 - m_D^2)^2} \left(1 - \frac{4m^2}{M^2}\right)^{1/2} \frac{1}{p^2 + M^2 - i\varepsilon}.$$

$$(4\text{–}7.8)$$

When it is added to the initial action expression,

$$\int \frac{(dp)}{(2\pi)^4} [K_D(p)\phi_D(-p) - \tfrac{1}{2}\phi_D(-p)(p^2 + m_D^2)\phi_D(p)], \quad (4\text{–}7.9)$$

the application of the stationary action principle supplies modified field equations that are solved by a modified propagation function,

$$\Delta_D(p) = \frac{1}{p^2 + m_D^2 - (p^2 + m_D^2)^2 f^2 \int_{(2m)^2}^{\infty} \frac{dM^2}{(M^2 - m_D^2)^2} \left(1 - \frac{4m^2}{M^2}\right)^{1/2} \frac{1}{p^2 + M^2 - i\varepsilon}} \cdot$$

(4–7.10)

Although we have used relativistic methods to derive it, the essential domain of application for this result is a nonrelativistic one. This restriction is introduced by writing

$$p^0 = 2m + E + \frac{\mathbf{p}^2}{4m}, \qquad M = 2m + W, \qquad E, W, \frac{\mathbf{p}^2}{4m}, \varepsilon \ll m. \qquad (4\text{–}7.11)$$

The limiting form of the propagation function is

$$\Delta_D(p) \to -\frac{1}{4m} G(E), \qquad\qquad (4\text{–}7.12)$$

with

$$G(E) = \frac{1}{E + \varepsilon - (E + \varepsilon)^2 \frac{f^2}{4m^{3/2}} \int_0^{\infty} dW \frac{W^{1/2}}{(W + \varepsilon)^2} \frac{1}{E + i\eta - W}} \cdot \qquad (4\text{–}7.13)$$

To avoid confusion between ε, the deuteron binding energy, and $\varepsilon \to +0$, the latter has been denoted by $\eta \to +0$. The integral that appears here can be evaluated by contour integration methods, either applied directly, or in simplified form by introducing a new integration variable, $W^{1/2} = x$:

$$\int_0^{\infty} dW \frac{W^{1/2}}{(W + \varepsilon)^2} \frac{1}{E + i\eta - W} = \frac{d}{d\varepsilon} \int_{-\infty}^{\infty} dx\, x^2 \frac{1}{x^2 + \varepsilon} \frac{1}{x^2 - (E + i\eta)}$$

$$= \frac{\pi}{2} \varepsilon^{-1/2} \frac{E - \varepsilon}{(E + \varepsilon)^2} - \pi i \frac{E^{1/2}}{(E + \varepsilon)^2} \cdot \qquad (4\text{–}7.14)$$

This gives

$$G(E) = \frac{1}{\left(1 - \frac{\pi}{8} \frac{f^2}{m\gamma}\right)(E + \varepsilon) + \frac{\pi}{4} \frac{f^2}{m\gamma}(\varepsilon + i(E\varepsilon)^{1/2})}, \qquad (4\text{–}7.15)$$

in which we have used the symbol

$$\gamma = (m\varepsilon)^{1/2}. \qquad\qquad (4\text{–}7.16)$$

For large values of E, $E \gg \varepsilon$, this function has the asymptotic behavior

$$G(E) \rightarrow \frac{1}{1 - (\pi/8)(f^2/my)} \frac{1}{E}. \tag{4-7.17}$$

The primitive interaction with the electromagnetic field is

$$\int (dx) j^\mu(x) A_\mu(x) = \int (dx) \left[\phi_D(x) eq \frac{1}{i} \partial^\mu \phi_D(x) A_\mu(x) + \phi_p(x) eq \frac{1}{i} \partial^\mu \phi_p(x) A_\mu(x) \right]. \tag{4-7.18}$$

In order to determine the dynamical modification of the deuteron electromagnetic properties, we consider the following causal arrangement. It involves two extended deuteron sources, and an extended photon source with space-like momenta. The virtual deuteron emitted by an extended source decays into a neutron and a proton. The proton is scattered by the photon source and later recombines with the neutron to form a virtual deuteron, which is detected by the other deuteron source. This is the arrangement discussed in previous sections, but with the exchanged photon replaced by a neutron. The corresponding vacuum amplitude is (4-7.5), where the proton propagation function is changed to

$$\Delta_p^A = \Delta_p + \Delta_p eq(pA + Ap)\Delta_p + \cdots. \tag{4-7.19}$$

Inserting the causal forms of the three propagation functions, we find that the vacuum amplitude for the process of interest becomes

$$- i(4\pi f)^2 \int d\omega_{p_1} d\omega_{p_2} d\omega_p \phi_{D1}(- P_1) eq(p_1 + p_2) A(K) \phi_{D2}(P_2)$$

$$= - i(4\pi f)^2 \int dM_1{}^2 dM_2{}^2 d\omega_{P_1} d\omega_{P_2} d\omega_p \delta((P_1 - p)^2 + m^2) \delta((P_2 - p)^2 + m^2)$$

$$\times \phi_{D1}(- P_1) eq \phi_{D2}(P_2)(P_1 + P_2 - 2p) A(K). \tag{4-7.20}$$

We have designated the neutron momentum by p; other symbols are chosen as before. The essential change from the earlier discussion is the neutron energy-momentum relation,

$$p^2 + m^2 = 0, \tag{4-7.21}$$

replacing that of the photon. Thus, the basic integral to be evaluated now reads

$$\int d\omega_p \delta(2pP_1 + M_1{}^2) \delta(2pP_2 + M_2{}^2)$$

$$= \frac{1}{8\pi^2} \int |\mathbf{p}| \, dp^0 \, dz \, \delta(- 2p^0 M_1 + M_1{}^2) \delta(- 2p^0 P_2{}^0 + 2|\mathbf{p}| \, |\mathbf{P}_2| z + M_2{}^2)$$

$$= \frac{1}{(4\pi)^2} \frac{1}{\varDelta^{1/2}} , 0. \qquad (4\text{-}7.22)$$

These are the same values as before, with

$$\varDelta = K^4 + 2K^2(M_1{}^2 + M_2{}^2) + (M_1{}^2 - M_2{}^2)^2, \qquad (4\text{-}7.23)$$

but the condition for the nonvanishing of the integral is different. It is deduced, from the requirement

$$\left| M_2{}^2 - M_1 \frac{K^2 + M_1{}^2 + M_2{}^2}{2M_1} \right| < (M_1{}^2 - 4m^2)^{1/2} \frac{\varDelta^{1/2}}{2M_1}, \qquad (4\text{-}7.24)$$

to be

$$\varDelta < \frac{K^2}{m^2} M_1{}^2 M_2{}^2, \qquad (4\text{-}7.25)$$

or

$$m^2(M_1{}^2 - M_2{}^2)^2 < K^2[(M_1{}^2 - 2m^2)(M_2{}^2 - 2m^2) - m^2(K^2 + 4m^2)]. \qquad (4\text{-}7.26)$$

A choice of variables that is consistent with this inequality is given by

$$M_1{}^2 - 2m^2 = m(K^2 + 4m^2)^{1/2}x + mK(x^2 - 1)^{1/2}v,$$

$$M_2{}^2 - 2m^2 = m(K^2 + 4m^2)^{1/2}x - mK(x^2 - 1)^{1/2}v, \qquad (4\text{-}7.27)$$

where

$$v^2 < 1, \quad x > 1. \qquad (4\text{-}7.28)$$

Some other useful quantities appear in these variables as

$$\varDelta = K^2[K^2 + 8m^2 + 4m(K^2 + 4m^2)^{1/2}x + 4m^2(x^2 - 1)v^2],$$

$$dM_1{}^2 \, dM_2{}^2 = 2m^2 K(K^2 + 4m^2)^{1/2}(x^2 - 1)^{1/2} dx \, dv. \qquad (4\text{-}7.29)$$

The algebraic property

$$K(P_1 + P_2 - 2p) = (p_1 - p_2)(p_1 + p_2) = 0 \qquad (4\text{-}7.30)$$

again fixes the form of the vector expectation value

$$\langle P_1 + P_2 - 2p \rangle = \alpha\left(1 - \frac{KK}{K^2}\right)(P_1 + P_2), \qquad (4\text{-}7.31)$$

where

$$\alpha = -\frac{K^2}{\varDelta}(P_1 + P_2)(P_1 + P_2 - 2p) = \frac{K^2(K^2 + M_1{}^2 + M_2{}^2)}{\varDelta}. \qquad (4\text{-}7.32)$$

Apart from contact terms, and expressed in a Lorentz gauge, the space-time extrapolation of the vacuum amplitude is then given by the double spectral integral

$$
if^2 \int \frac{(dp_1)}{(2\pi)^4} \frac{(dp_2)}{(2\pi)^4} \frac{dM_1^2 \, dM_2^2}{\Delta^{3/2}} (k^2 + M_1^2 + M_2^2)
$$

$$
\times \phi_{\mathrm{D1}}(-p_1) eq(p_1 + p_2) k^2 A(k) \phi_{\mathrm{D2}}(p_2) \frac{1}{p_1^2 + M_1^2 - i\varepsilon} \frac{1}{p_2^2 + M_2^2 - i\varepsilon}.
$$

$$(4\text{--}7.33)$$

In the field-free situation expressed by $k^2 = 0$, we have

$$
M_1^2 = M_2^2 = M^2 = 2m^2(1 + x) \tag{4--7.34}
$$

and

$$
k^2 \frac{dM_1^2 \, dM_2^2}{\Delta^{3/2}} = \frac{1}{2} \frac{(x^2 - 1)^{1/2} \, dx \, dv}{[2(1 + x) + (x^2 - 1)v^2]^{3/2}}. \tag{4--7.35}
$$

Using the integral

$$
\tfrac{1}{2} \int_{-1}^1 dv \, \frac{1}{[2(1 + x) + (x^2 - 1)v^2]^{3/2}} = \frac{1}{2} \frac{1}{(1 + x)^2}, \tag{4--7.36}
$$

we find that (4–7.33) becomes, correctly,

$$
if^2 \int \frac{(dp_1)}{(2\pi)^4} \frac{(dp_2)}{(2\pi)^4} \phi_{\mathrm{D1}}(-p_1) eq(p_1 + p_2) A(k) \phi_{\mathrm{D2}}(p_2) \, dM^2 \left(1 - \frac{4m^2}{M^2}\right)^{1/2}
$$

$$
\times \left(\frac{1}{p_1^2 + M^2 - i\varepsilon} \frac{1}{p_2^2 + M^2 - i\varepsilon} - \frac{1}{(M^2 - m_{\mathrm{D}}^2)^2}\right), \tag{4--7.37}
$$

in which we have inserted the purely local contact term that is implied by the gauge covariant generalization of Eqs. (4–7.7, 8).

The double spectral integral (4–7.33) will now be applied to deuteron fields that are associated with simple sources. The action term inferred through the replacement $p_1^2, p_2^2 \rightarrow -m_{\mathrm{D}}^2$ is

$$
f^2 \int \frac{(dp_1)}{(2\pi)^4} \frac{(dp_2)}{(2\pi)^4} \tfrac{1}{2} \phi_{\mathrm{D}}(-p_1) eq(p_1 + p_2) k^2 A(k) \phi_{\mathrm{D}}(p_2)
$$

$$
\times \int \frac{dM_1^2 \, dM_2^2}{\Delta^{3/2}} \frac{k^2 + M_1^2 + M_2^2}{(M_1^2 - m_{\mathrm{D}}^2)(M_2^2 - m_{\mathrm{D}}^2)}, \tag{4--7.38}
$$

where

$$(M_1{}^2 - m_D{}^2)(M_2{}^2 - m_D{}^2) = (m(k^2 + 4m^2)^{1/2}x + 2m^2 - m_D{}^2)^2 - m^2k^2(x^2 - 1)v^2.$$

$$(4-7.39)$$

At this point we confine ourselves to the nonrelativistic situation, expressed by such restrictions and simplifications as

$$k^2 \ll m^2 \tag{4-7.40}$$

and

$$m_D{}^2 = (2m - \varepsilon)^2 \cong 4m^2 - 4\gamma^2. \tag{4-7.41}$$

Thus,

$$(M_1{}^2 - m_D{}^2)(M_2{}^2 - m_D{}^2) \cong 4m^4\left[\left(x - 1 + \frac{2\gamma^2}{m^2} + \frac{k^2}{8m^2}x\right)^2 - \frac{k^2}{4m^2}(x^2 - 1)v^2\right],$$

$$(4-7.42)$$

which indicates that the important values of $x - 1$ are small, of the order of k^2/m^2 and γ^2/m^2. We therefore introduce the simplification $x^2 - 1 \cong 2(x - 1)$, and also write

$$x - 1 = y^2. \tag{4-7.43}$$

The spectral integral of (4–7.38) then becomes (including a factor of k^2)

$$k^2 \int \frac{dM_1{}^2 dM_2{}^2}{\varDelta^{3/2}} \frac{k^2 + M_1{}^2 + M_2{}^2}{(M_1{}^2 - m_D{}^2)(M_2{}^2 - m_D{}^2)} \cong \frac{2^{-3/2}}{m^2} \int \frac{y^2\, dy\, dv}{\left(y^2 + \dfrac{k^2}{8m^2} + \dfrac{2\gamma^2}{m^2}\right)^2 - \dfrac{k^2}{2m^2}v^2y^2}.$$

$$(4-7.44)$$

The variable y ranges from 0 to ∞. But an equivalent version of (4–7.44), apart from the multiplier $2^{-3/2}/m^2$, is

$$\int_{-1}^1 \frac{dv}{v} \frac{1}{2} \int_{-\infty}^\infty dy \, \frac{y}{y^2 + \dfrac{k^2}{8m^2} + \dfrac{2\gamma^2}{m^2} - \dfrac{k}{2^{1/2}m}vy} \left(\frac{2^{1/2}m}{k}\right), \tag{4-7.45}$$

and the translation

$$y \to y + \frac{k}{2^{3/2}m}v \tag{4-7.46}$$

now gives

$$\frac{1}{2}\int_0^1 dv \int_{-\infty}^\infty dy \, \frac{1}{y^2 + \dfrac{2\gamma^2}{m^2} + \dfrac{k^2}{8m^2}(1 - v^2)} = \frac{\pi}{2}\int_0^1 dv \, \frac{1}{\left[\dfrac{2\gamma^2}{m^2} + \dfrac{k^2}{8m^2}(1 - v^2)\right]^{1/2}}$$

$$= \frac{\pi}{2} \frac{2^{3/2}m}{k} \tan^{-1} \frac{k}{4\gamma} . \tag{4-7.47}$$

The k dependence of this factor appears in spectral form on noting that

$$\frac{1}{k} \tan^{-1} \frac{k}{4\gamma} = \int_{4\gamma}^{\infty} \frac{dM}{k^2 + M^2} . \tag{4-7.48}$$

The vector potential coupling deduced for $k^2 = 0$, which is to be associated with the modified propagation function, must still be removed. (This is automatically produced by the contact term.) Accordingly, the spectral integral actually occurs in the form

$$\int_{4\gamma}^{\infty} dM \left[\frac{1}{k^2 + M^2} - \frac{1}{M^2} \right] = - k^2 \int_{4\gamma}^{\infty} \frac{dM}{M^2} \frac{1}{k^2 + M^2} . \tag{4-7.49}$$

Adding the resulting action term to the primitive electromagnetic interaction of the deuteron, we obtain the deuteron form factor:

$$F(k) = 1 - k^2 \frac{\pi}{2} \frac{f^2}{m} \int_{4\gamma}^{\infty} \frac{dM}{M^2} \frac{1}{k^2 + M^2 - i\varepsilon} . \tag{4-7.50}$$

For large values of k, $k \gg 4\gamma$, the form factor approaches a constant limit,

$$F(k) \to 1 - \frac{\pi}{8} \frac{f^2}{m\gamma} . \tag{4-7.51}$$

We have met this combination before, in the asymptotic behavior of the deuteron propagation function, Eq. (4-7.17). Evidently there is a particular significance to a zero value for the combination, fixing the coupling constant f relative to the deuteron binding energy:

$$\frac{\pi}{8} \frac{f^2}{m\gamma} = 1. \tag{4-7.52}$$

The effective electromagnetic interaction of the deuteron with high frequency photons would be zero, and the characteristic particle behavior of the deuteron propagation function would disappear. Both remarks make clear that this is a composite deuteron, which is dissociated with certainty if probed with a sufficiently high energy disturbance. Under these circumstances, the two terms of the form factor (4-7.50) can be united to give

$$F(k) = 4\gamma \int_{4\gamma}^{\infty} \frac{dM}{k^2 + M^2 - i\varepsilon} . \tag{4-7.53}$$

For spatial values of k one can provide this formula with a conventional interpretation in terms of a charge distribution. On recalling that

$$\frac{1}{k^2 + M^2} = \int d(\tfrac{1}{2}\mathbf{r}) \exp(- i\mathbf{k} \cdot \tfrac{1}{2}\mathbf{r}) \frac{1}{4\pi} \frac{\exp(- M\tfrac{1}{2}r)}{\tfrac{1}{2}r}, \tag{4–7.54}$$

which uses $\tfrac{1}{2}\mathbf{r}$ as a position vector, we get

$$F(k) = \int (d\mathbf{r}) \exp(- i\mathbf{k} \cdot \tfrac{1}{2}\mathbf{r}) \left[\left(\frac{\gamma}{2\pi}\right)^{1/2} \frac{\exp(- \gamma r)}{r} \right]^2. \tag{4–7.55}$$

Exhibited here is the deuteron wave function as a function of the neutron-proton distance; the vector $\tfrac{1}{2}\mathbf{r}$ locates the proton relative to the center of mass of the deuteron. This wave function is familiar as the zero effective range limit. It is a solution of the Schrödinger equation for internal energy $- \varepsilon$, with no interaction energy at any finite distance between the particles.

We shall do better if we recognize that the limiting behavior (4–7.51) for k large compared to γ must refer to values of k that are small still in comparison with the momenta of the virtual particles that are exchanged between neutron and proton. The deuteron is not yet completely dissociated and the limiting value (4–7.51) need not be zero. Let us designate it as $- \gamma r_e/(1 - \gamma r_e)$, so that

$$\frac{\pi}{8} \frac{f^2}{m\gamma} = 1 + \frac{\gamma r_e}{1 - \gamma r_e} = \frac{1}{1 - \gamma r_e}. \tag{4–7.56}$$

Now the form factor becomes

$$F(k) = - \frac{\gamma r_e}{1 - \gamma r_e} + \int (d\mathbf{r}) \exp(- i\mathbf{k} \cdot \tfrac{1}{2}\mathbf{r}) |\psi_0(\mathbf{r})|^2, \tag{4–7.57}$$

where

$$\psi_0(\mathbf{r}) = \left(\frac{\gamma}{2\pi} \frac{1}{1 - \gamma r_e}\right)^{1/2} \frac{\exp(- \gamma r)}{r}. \tag{4–7.58}$$

We still describe the charge distribution in terms of the motion of proton and neutron, with the exchanged particles implicit in the neutron-proton force. This means that the charge density is deduced from a wave function $\psi(\mathbf{r})$:

$$F(k) = \int (d\mathbf{r}) \exp(- i\mathbf{k} \cdot \tfrac{1}{2}\mathbf{r}) |\psi(\mathbf{r})|^2$$

$$= \int (d\mathbf{r}) \exp(- i\mathbf{k} \cdot \tfrac{1}{2}\mathbf{r}) (|\psi(\mathbf{r})|^2 - |\psi_0(\mathbf{r})|^2) + \int (d\mathbf{r}) \exp(- i\mathbf{k} \cdot \tfrac{1}{2}\mathbf{r}) |\psi_0(\mathbf{r})|^2. \tag{4–7.59}$$

The comparison of the two expressions, (4–7.57) and (4–7.59), has this interpretation. The wave functions $\psi(\mathbf{r})$ and $\psi_0(\mathbf{r})$ coincide except for such values of r that kr is small in the restricted range of k being considered. That provides the identification

$$\int (d\mathbf{r})(|\psi_0(\mathbf{r})|^2 - |\psi(\mathbf{r})|^2) = \frac{\gamma r_e}{1 - \gamma r_e}. \tag{4-7.60}$$

This relation is more recognizable if we introduce radial wave functions $u(r)$ and $u_0(r)$ by writing

$$\psi(\mathbf{r}) = \left(\frac{\gamma}{2\pi} \frac{1}{1 - \gamma r_e}\right)^{1/2} \frac{u(r)}{r}, \qquad \psi_0(\mathbf{r}) = \left(\frac{\gamma}{2\pi} \frac{1}{1 - \gamma r_e}\right)^{1/2} \frac{u_0(r)}{r}, \tag{4-7.61}$$

in which

$$u_0(r) = \exp(-\gamma r), \qquad u_0(0) = 1. \tag{4-7.62}$$

Now (4-7.60) reads

$$\int_0^\infty dr(u_0^2 - u^2) = \tfrac{1}{2}r_e, \tag{4-7.63}$$

which is the conventional definition of the effective range r_e. The usual boundary condition, $u(0) = 0$, suggests, correctly, that r_e is a positive length.

It is interesting to use our present standpoint in order to derive two familiar sets of results involving the effective range. One of these refers to cross sections for the photodisintegration of the deuteron. The primitive interaction is enlarged by including the effect of an electromagnetic field on the charged particles. It is not the influence of the photon on the motion of the deuteron that is important, but the disturbance of the internal state through the action on the proton. Accordingly the relevant interaction term, derived from

$$4\pi f \int (dx)\phi_D(x) \left[\int (dx')\Delta_p^A(x, x')K_p(x')\right] \phi_n(x), \tag{4-7.64}$$

is

$$4\pi f \int (dx)(dx')\phi_D(x)\Delta_p(x - x')eq(pA + Ap)(x')\phi_p(x')\phi_n(x). \tag{4-7.65}$$

The T matrix element, the coefficient of the source products $iK_{p_pq}^* iK_{p_n}^* iK_{p_Dq} iJ_{k\lambda}$, is stated as $(q = +1)$

$$\langle p_p p_n | T | p_D k\lambda \rangle = 4\pi f e (d\omega_{p_p} d\omega_{p_n} d\omega_{p_D} d\omega_k)^{1/2} \frac{2e_{k\lambda}p_p}{(p_D - p_n)^2 + m^2}, \tag{4-7.66}$$

with momentum conservation expressed by

$$P = p_p + p_n = p_D + k. \tag{4-7.67}$$

We use the gauge in which the polarization vector has no time component in the rest frame of P. It validates the omission of the electromagnetic interaction with the deuteron. The invariant flux factor is [Eq. (3–12.70)]

$$2(M^2 - m_D^2)d\omega_{p_D}d\omega_k, \tag{4-7.68}$$

and the final state integration in the center of mass frame for prescribed solid angle $d\Omega$ is given by [Eq. (3–12.75)]

$$\int d\omega_{p_p} d\omega_{p_n} (2\pi)^4 \delta(p_p + p_n - P) = \frac{1}{32\pi^2}\left(1 - \frac{4m^2}{M^2}\right)^{1/2} d\Omega. \tag{4-7.69}$$

We deduce the differential cross section

$$\frac{d\sigma}{d\Omega} = 4\pi\alpha f^2\left(1 - \frac{4m^2}{M^2}\right)^{1/2}\frac{1}{M^2 - m_D^2}\left[\frac{e \cdot p_p}{2p_D p_n + m_D^2}\right]^2. \tag{4-7.70}$$

This formula will be applied in the nonrelativistic regime where

$$M \cong 2m + E = 2m - \varepsilon + k^0 \tag{4-7.71}$$

and thus

$$k^0 = E + \varepsilon. \tag{4-7.72}$$

The kinetic energy of neutron and proton,

$$E = \frac{p^2}{m}, \tag{4-7.73}$$

is computed from the relative momentum of the particles in the center of mass frame,

$$p_p = -p_n = p. \tag{4-7.74}$$

We also have

$$-2p_D p_n - m_D^2 \cong m_D(2p_n^0 - m_D) \cong 2m(E + \varepsilon), \tag{4-7.75}$$

leading to the nonrelativistic expression

$$\frac{d\sigma}{d\Omega} = 2\alpha\frac{\gamma}{1 - \gamma r_e}\frac{p}{(p^2 + \gamma^2)^3}(e \cdot p)^2, \tag{4-7.76}$$

in which (4–7.56) has been used to eliminate f^2. Averaging over the initial polarization and integrating over the final angular distribution gives us the total cross section

$$\sigma = \frac{8\pi}{3}\alpha\frac{\gamma}{1 - \gamma r_e}\frac{p^3}{(p^2 + \gamma^2)^3}. \tag{4-7.77}$$

Its maximum value, at $p = \gamma$, is

$$\sigma_{\text{max.}} = \frac{\pi}{3}\,\alpha\,\frac{1}{1 - \gamma r_e}\,\frac{1}{\gamma^2}\,. \qquad (4\text{-}7.78)$$

These are well-known results of effective range theory.

The second example is neutron-proton scattering. This takes place through the exchange of a deuteron as described by the modified propagation function. The implied interaction term is

$$(4\pi f)^2 \tfrac{1}{2} \int (dx)(dx')\phi_{\mathrm{p}}(x)\phi_{\mathrm{n}}(x)\varDelta_{\mathrm{D}}(x - x')\phi_{\mathrm{p}}(x')\phi_{\mathrm{n}}(x'), \qquad (4\text{-}7.79)$$

and the corresponding T matrix element is deduced as

$$\langle p_{\mathrm{p1}}p_{\mathrm{n1}}|T|p_{\mathrm{p2}}p_{\mathrm{n2}}\rangle = (4\pi f)^2 (d\omega_{p_{\mathrm{p1}}} d\omega_{p_{\mathrm{n1}}} d\omega_{p_{\mathrm{p2}}} d\omega_{p_{\mathrm{n2}}})^{1/2}[\varDelta_{\mathrm{D}}(P) + \varDelta_{\mathrm{D}}(p_{\mathrm{n2}} - p_{\mathrm{p1}})], \qquad (4\text{-}7.80)$$

where

$$P = p_{\mathrm{p1}} + p_{\mathrm{n1}} = p_{\mathrm{p2}} + p_{\mathrm{n2}} \qquad (4\text{-}7.81)$$

and

$$p_{\mathrm{n2}} - p_{\mathrm{p1}} = p_{\mathrm{n1}} - p_{\mathrm{p2}}. \qquad (4\text{-}7.82)$$

Since this is an elastic collision, similar kinematical factors involved in the cross section definition cancel, giving

$$\frac{d\sigma}{d\Omega} = \frac{(2\pi f^2)^2}{M^2}|\varDelta_{\mathrm{D}}(P) + \varDelta_{\mathrm{D}}(p_{\mathrm{n2}} - p_{\mathrm{p1}})|^2 \cong \left|-\frac{\gamma}{1 - \gamma r_e}\frac{2}{m}G(E)\right|^2. \qquad (4\text{-}7.83)$$

The second version is the nonrelativistic one. In this circumstance the $\varDelta_{\mathrm{D}}(p_{\mathrm{n2}} - p_{\mathrm{p1}})$ term, necessary to satisfy crossing symmetry, is negligible. We have also used the coupling constant relation (4-7.56). When the latter is employed in (4-7.15) we find that

$$-\frac{\gamma}{1 - \gamma r_e}\frac{2}{m}G(E) = \frac{1}{-\gamma + \frac{1}{2}r_e(p^2 + \gamma^2) - ip} = \frac{1}{p\cot\delta - ip} = \frac{1}{p}\sin\delta\exp(i\delta), \qquad (4\text{-}7.84)$$

where the real angle δ is determined by

$$p\cot\delta = -\gamma + \tfrac{1}{2}r_e(p^2 + \gamma^2). \qquad (4\text{-}7.85)$$

The differential cross section

$$\frac{d\sigma}{d\Omega} = \frac{1}{p^2}\sin^2\delta \qquad (4\text{-}7.86)$$

is the standard one, expressed in terms of the s-wave phase shift δ, and the phase shift formula (4–7.85) is the well-known effective range result. We also recognize, in (4–7.84), the usual complex form of the scattering amplitude in terms of the phase shift. This means that the unitarity requirement has been satisfied automatically.

Wearing a serious look, Harold speaks up.

H.: I think that scientists, like generals and statesmen, should write their memoirs. However influenced by a specific viewpoint and the natural human tendency to self-aggrandizement, the report of an individual who took part in events is irreplaceable. Surely, an appreciable fraction of the scientific history that is mechanically repeated in papers and books written by people who were not there, so to speak, is fictional or, in the phrase of Josephine Tey (*The Daughter of Time*), is Tonypandy. I say all this in connection with the effective range phase shift formula. Although there doubtless were qualitative precursors, I believe that you were the first to appreciate its significance and give a derivation in which the effective range, as you called it, had a precise meaning that could be transferred to other problems. That derivation used a variational technique. Later, other people produced more elementary derivations and one of these can be found described in textbooks as though its author had originated the formula. Since it is hard to believe that you were unaware of the possibility of an elementary derivation, why did you prefer to use the unfamiliar variational approach?

S.: It is an interesting question in motivations. Perhaps I should point out that both the variational method for scattering and the effective range formula had their origin in the electromagnetic problems of wave guide propagation. Some of that history is described in *Discontinuities in Waveguides*, Gordon and Breach, 1968. One will find there formulas for the frequency derivatives of certain electromagnetic quantities which also have stationary properties. These frequency derivatives are given as the difference between total and asymptotic expressions for energy. I was quite familiar with the Schrödinger equation analogue, in which a trigonometric function of the phase shift replaces the electromagnetic quantities, and probability appears in the place of electromagnetic energy to determine an energy derivative. What was needed, however, was not an exact formula with uncertain variability in energy of a parameter, but an approximation valid over a limited energy region. For that reason it seemed preferable to use the stationary property of the phase shift. Incidentally, the effective range formula was an early embodiment of the nonspeculative viewpoint that later found its full expression in source theory. It is rather pleasing, therefore, to derive the effective range results once again by using source theory.

In previous sections we have deduced electromagnetic form factors by considering extended photon sources that emit time-like momenta. Let us apply that approach to the deuteron form factor. A virtual photon decays into a proton and an antiproton. These particles later interact by exchanging a virtual neutron to form a deuteron-antideuteron pair. The scattering reaction is described by the interaction term

$$(4\pi f)^2 \tfrac{1}{2} \int (dx)(dx')\phi_D(x)\phi_p(x)\Delta_n(x - x')\phi_p(x')\phi_D(x'), \tag{4–7.87}$$

where the effective two-particle proton source is represented by

$$iK_p(p_2)K_p(p_2')|_{\text{eff.}} = eq(p_2 - p_2')A(k), \tag{4–7.88}$$

and

$$k = p_2 + p_2'. \tag{4–7.89}$$

The resulting probability amplitude can be presented as

$$(4\pi f)^2 \tfrac{1}{2} \int d\omega_{p_1} d\omega_{p_1'} (2\pi)^4 \delta(p_1 + p_1' - k)\, dM^2\, d\omega_k\, K_D(- p_1)eqK_D(- p_1')I^\mu A_\mu(k), \tag{4–7.90}$$

with

$$I^\mu = \int d\omega_{p_2} d\omega_{p_2'} (2\pi)^3 \delta(p_2 + p_2' - k)\, \frac{1}{(p_1 - p_2)^2 + m^2}(p_2 - p_2')^\mu$$
$$= (p_1 - p_1')^\mu S(M^2). \tag{4–7.91}$$

The scalar function is evaluated in terms of an expectation value:

$$S(M^2) = \int d\omega_{p_2} d\omega_{p_2'} (2\pi)^3 \delta(p_2 + p_2' - k)\, \frac{1}{(p_1 - p_2)^2 + m^2}\, \frac{(p_1 - p_1')(p_2 - p_2')}{(p_1 - p_1')^2}$$
$$= \frac{1}{(4\pi)^2}\left(1 - \frac{4m^2}{M^2}\right)^{1/2}\left\langle \frac{1}{(p_1 - p_2)^2 + m^2}\, \frac{(p_1 - p_1')(p_2 - p_2')}{(p_1 - p_1')^2} \right\rangle. \tag{4–7.92}$$

All this is quite similar to what was done in Section 4–4. But here it is necessary to distinguish between the initial particles of the collision, which are protons, and the final particles, deuterons. That distinction appears in the magnitudes assigned to the spatial momenta of the particles in the center of mass frame,

$$p_1^0 = p_1^{0\prime} = \tfrac{1}{2}M, \qquad |\mathbf{p}_1| = |\mathbf{p}_1'| = (\tfrac{1}{4}M^2 - m_D^2)^{1/2},$$
$$p_2^0 = p_2^{0\prime} = \tfrac{1}{2}M, \qquad |\mathbf{p}_2| = |\mathbf{p}_2'| = (\tfrac{1}{4}M^2 - m^2)^{1/2}. \tag{4–7.93}$$

Accordingly,

$$\langle \cdots \rangle = \int_{-1}^{1} \frac{1}{2} \, dz \, \frac{z}{\frac{1}{2}M^2 - m_D^2 - 2z(\frac{1}{4}M^2 - m_D^2)^{1/2}(\frac{1}{4}M^2 - m^2)^{1/2}} \frac{(\frac{1}{4}M^2 - m^2)^{1/2}}{(\frac{1}{4}M^2 - m_D^2)^{1/2}},$$

(4–7.94)

where z is the cosine of the scattering angle.

Before discussing this integral in detail, let us complete the formal space-time extrapolation. We write the scalar function $S(M^2)$ as

$$S(M^2) = \frac{1}{(4\pi)^2} \frac{1}{M} \sigma(M) \tag{4–7.95}$$

and identify the deuteron current vector to obtain the vacuum amplitude

$$if^2 \int \frac{dM^2}{M^2} \frac{\sigma(M)}{M} \int (dx)(dx') j_D^\mu(x) \left[i \int d\omega_k \exp[ik(x - x')] \right] (-\partial^{\nu'} F_{\mu\nu}(x')), \quad (4–7.96)$$

where the gauge invariant substitution (4–4.19) has also been introduced. The space-time extrapolation is performed, as always, by replacing the causal form of the propagation function with $\Delta_+(x - x', M^2)$. After adding the primitive interaction of the deuteron, the Lorentz gauge form of the deuteron form factor is obtained as

$$F(k) = 1 - k^2 f^2 \int \frac{dM^2}{M^2} \frac{\sigma(M)}{M} \frac{1}{k^2 + M^2 - i\varepsilon}. \tag{4–7.97}$$

The explicit expression for $\sigma(M)$ is

$$\sigma(M) = \frac{(M^2 - 4m^2)^{1/2}}{M^2 - 4m_D^2} \left[\frac{1}{\mu} \log \frac{1 + \mu}{1 - \mu} - 2 \right], \tag{4–7.98}$$

where

$$\mu = \frac{(M^2 - 4m^2)^{1/2}(M^2 - 4m_D^2)^{1/2}}{M^2 - 2m_D^2}. \tag{4–7.99}$$

Since a pair of deuterons are created, the threshold would seem to be at $M = 2m_D$, where $\mu = 0$. But the function of μ that appears in (4–7.98) is even and, for small μ,

$$\frac{1}{\mu} \log \frac{1 + \mu}{1 - \mu} - 2 = \tfrac{2}{3}\mu^2 + \cdots, \tag{4–7.100}$$

giving

$$M \rightarrow 2m_D \pm 0: \quad \sigma(M) = \frac{4}{3} \frac{(m_D^2 - m^2)^{3/2}}{m_D^4}. \tag{4–7.101}$$

There is no threshold here. As M becomes less than $2m_D$, μ turns imaginary,

$$\mu = i\,\frac{(M^2 - 4m^2)^{1/2}(4m_D{}^2 - M^2)^{1/2}}{M^2 - 2m_D{}^2} = i\nu, \qquad (4\text{-}7.102)$$

but $\sigma(M)$ remains real. For that reason there is no distinction between $i\nu$ and $-i\nu$. If we write

$$\frac{1 + i\nu}{1 - i\nu} = \exp[i\phi(M)] \qquad (4\text{-}7.103)$$

where

$$\phi = 2\tan^{-1}\nu, \qquad \phi(2m_D) = 0, \qquad (4\text{-}7.104)$$

we now have

$$M < 2m_D: \quad \sigma(M) = \frac{(M^2 - 4m^2)^{1/2}}{M^2 - 4m_D{}^2}\left(\frac{\phi}{\nu} - 2\right). \qquad (4\text{-}7.105)$$

A pair of protons is also created, and the next threshold approached with decreasing M is at $M = 2m$. If it had happened that

$$m_D < 2^{1/2}m, \qquad (4\text{-}7.106)$$

which represents a very large binding energy, greater than $(2 - 2^{1/2})m$, the denominator of ν, $M^2 - 2m_D^2$, would remain positive throughout and the angle ϕ would return to zero at $M = 2m$. More precisely, we have

$$M \to 2m + 0: \quad \frac{\phi}{\nu} - 2 \to -\tfrac{2}{3}\nu^2, \quad \sigma \to \frac{1}{6}\frac{(M^2 - 4m^2)^{3/2}}{(2m^2 - m_D{}^2)^2}, \qquad (4\text{-}7.107)$$

which is normal threshold behavior.

But the inequality (4-7.106) is not obeyed. At a value of M intermediate between $2m_D$ and $2m$, namely $M = 2^{1/2}m_D$, ν is infinite and $\phi = \pi$. Then, when $M \to 2m + 0$, ν approaches zero through negative values and $\phi \to 2\pi$. The limiting value is

$$M \to 2m + 0: \quad \sigma(M) = \frac{\pi}{2}\,\frac{m_D{}^2 - 2m^2}{(m_D{}^2 - m^2)^{3/2}}. \qquad (4\text{-}7.108)$$

Can we proceed below $M = 2m$? Yes. In this region,

$$(M^2 - 4m^2)^{1/2} = \pm i(4m^2 - M^2)^{1/2} \qquad (4\text{-}7.109)$$

and

$$i\nu = \pm\mu, \qquad \mu = \frac{(4m^2 - M^2)^{1/2}(4m_D{}^2 - M^2)^{1/2}}{2m_D{}^2 - M^2}, \qquad (4\text{-}7.110)$$

while

$$\phi = 2\pi \pm \frac{1}{i} \log \frac{1+\mu}{1-\mu}. \tag{4-7.111}$$

Leaving the ambiguous terms explicit, we have

$$M < 2m: \quad \sigma(M) = \frac{(4m^2 - M^2)^{1/2}}{4m_\mathrm{D}^2 - M^2} \left[\frac{2\pi}{\mu} \mp i\left(\frac{1}{\mu} \log \frac{1+\mu}{1-\mu} - 2 \right) \right]. \tag{4-7.112}$$

Since there is no physical basis for choosing either of the signs, we use the average of the two forms as in the computation of principal values. That gives the real expression

$$M < 2m: \quad \sigma(M) = 2\pi \frac{2m_\mathrm{D}^2 - M^2}{(4m_\mathrm{D}^2 - M^2)^{3/2}}, \tag{4-7.113}$$

which joins continuously with (4-7.108).

Where does this extrapolation procedure stop? There is another singularity, at $\mu = 1$, which is explicit in (4-7.112). If we notice that the value of μ for $M = 0$, namely $2m/m_\mathrm{D}$, is quite close to unity, it is evident that $\mu = 1$ will occur for such small values of M that we can use the expansion

$$M \ll m: \quad \mu = 1 + \frac{\varepsilon}{2m} - \frac{M^2}{32m^2}. \tag{4-7.114}$$

Thus the singularity occurs for $M = M_0$, where

$$M_0 = 4\gamma. \tag{4-7.115}$$

This time, however, as we proceed below the singularity, where $\mu > 1$, the ambiguous imaginary part of the logarithm implies ambiguous real terms in $\sigma(M)$. The undefined discontinuity at $M = M_0$ tells us to call a halt.

For nonrelativistic values of k, the form factor (4-7.97) is dominated by small values of M, where

$$M_0 < M \ll m: \quad \sigma(M) = \frac{\pi}{4m}. \tag{4-7.116}$$

Thus, the deuteron form factor is

$$F(k) = 1 - k^2 \frac{\pi}{2} \frac{f^2}{m} \int_{4\gamma}^{\infty} \frac{dM}{M^2} \frac{1}{k^2 + M^2 - i\varepsilon}, \tag{4-7.117}$$

identical with (4-7.50). It is reassuring to have this independent evidence for the correctness of the mass extrapolation procedure.

4-8 SCATTERING OF LIGHT BY LIGHT I. LOW FREQUENCIES

We have been discussing multiparticle exchange effects that modify skeletal interactions. Now we turn to examine how multiparticle exchange introduces new classes of interactions. The simplest example, which has great conceptual importance despite its lack of immediate experimental contact, is the scattering of light by light.

Processes involving two spin $\frac{1}{2}$ particle sources and various numbers of photon sources are contained in the coupling term

$$W_{2\ldots} = \tfrac{1}{2} \int (dx)(dx')\eta(x)\gamma^0 G_+^A(x,\,x')\eta(x'),$$

$$G_+^A = G_+ + G_+ eq\gamma A G_+ + G_+ eq\gamma A G_+ eq\gamma A G_+ + \cdots. \qquad (4\text{-}8.1)$$

This is illustrated by

$$W_{22} = \tfrac{1}{2} \int (dx)(dx')\psi(x)\gamma^0 eq\gamma A(x)G_+(x-x')eq\gamma A(x')\psi(x'), \qquad (4\text{-}8.2)$$

which describes a combination of two photon fields as an effective electron-positron source:

$$i\eta(x)\eta(x')|_{\text{eff.}} = eq\gamma A(x)G_+(x-x')eq\gamma A(x')\gamma^0. \qquad (4\text{-}8.3)$$

The corresponding physical process is the collision of two photons to produce an electron-positron pair, or its inverse, which have been discussed in Section 3–13. Now let us consider a causal arrangement in which an electron-positron pair is created by photon fields, labeled A_2, and the subsequent annihilation into two photons is detected by sources, which have associated fields that are designated as A_1. The vacuum amplitude describing the two-particle exchange coupling between effective sources is conveniently written, as in Eq. (4–3.7), by using a trace notation:

$$-\tfrac{1}{2} \int (dx) \cdots (dx''') \,\text{tr}[i\eta_1(x)\eta_1(x')\gamma^0|_{\text{eff.}} G_+(x'-x'')i\eta_2(x'')\eta_2(x''')\gamma^0|_{\text{eff.}}$$

$$\times\, G_+(x'''-x)]. \qquad (4\text{-}8.4)$$

The insertion of (4–8.3) gives

$$-\tfrac{1}{2} \int (dx) \cdots (dx''') \,\text{tr}[eq\gamma A_1(x)G_+(x-x')eq\gamma A_1(x')G_+(x'-x'')$$

$$\times\, eq\gamma A_2(x'')G_+(x''-x''')eq\gamma A_2(x''')G_+(x'''-x)], \qquad (4\text{-}8.5)$$

or, using the more compact notation in which the space-time coordinates join spin and charge indices as matrix labels,

$$- \tfrac{1}{2} \, \mathrm{Tr}[eq\gamma A_1 G_+ eq\gamma A_1 G_+ eq\gamma A_2 G_+ eq\gamma A_2 G_+]. \tag{4-8.6}$$

This vacuum amplitude can be presented as a unified action term,

$$W_{04} = i\tfrac{1}{8} \, \mathrm{Tr}[eq\gamma A G_+ eq\gamma A G_+ eq\gamma A G_+ eq\gamma A G_+]$$

$$= i\tfrac{1}{8} \, \mathrm{Tr}(eq\gamma A G_+)^4, \tag{4-8.7}$$

in which the additional factor of $\tfrac{1}{4}$ records the four equivalent places where one can begin the particular sequence $A_1 A_1 A_2 A_2$, all giving equal contributions in virtue of the cyclic symmetry of the trace.

Here is a process that has no counterpart in the interaction skeleton of Section 3–12, since no charged particles are in evidence. It is stated in a generally applicable space-time form. The validity of that unqualified assertion must, however, be confirmed by applying physical tests. There are two of these, gauge invariance and existence. Granting the existence of the structure, the property of gauge invariance can be verified by a formal matrix calculation, in which it suffices to use an infinitesimal gauge transformation,

$$\delta A = \partial \delta\lambda = i[p, \delta\lambda]. \tag{4-8.8}$$

This induces the change

$$\delta W_{04} = - \tfrac{1}{2} \, \mathrm{Tr}[(eq\gamma A G_+)^3[\gamma p, eq\delta\lambda]G_+], \tag{4-8.9}$$

and the rearrangement

$$eq\gamma A G_+[\gamma p, eq\delta\lambda]G_+ = eq\gamma A G_+[G_+^{-1}, eq\delta\lambda]G_+$$

$$= eq\gamma A[eq\delta\lambda, G_+] = [eq\delta\lambda, eq\gamma A G_+] \tag{4-8.10}$$

implies that

$$\delta W_{04} = - \tfrac{1}{8} \, \mathrm{Tr}[eq\delta\lambda, (eq\gamma A G_+)^3] = 0. \tag{4-8.11}$$

The existence question refers to the situation of complete overlap of the four fields, which is not contemplated in the initial causal arrangement. As the most extreme possibility, we consider such a small space-time region that the vector potentials are sensibly constant over it. Then the multiple space-time integrals of (4–8.7) become, apart from numerical factors,

$$\int (dx) \cdots (dx''') \, \mathrm{tr}[eq\gamma A(x)G_+(x - x')eq\gamma A(x)G_+(x' - x'')$$

$$\times \, eq\gamma A(x)G_+(x'' - x''')eq\gamma A(x)G_+(x''' - x)]$$

$$= \int \frac{(dx)(dp)}{(2\pi)^4} \, \mathrm{tr}\left(eq\gamma A(x)\frac{1}{\gamma p + m - i\varepsilon}\right)^4 \sim \int \frac{(dx)(dp)}{(2\pi)^4} \frac{\mathrm{tr}(eq\gamma A(x)\gamma p)^4}{(p^2)^4}, \tag{4-8.12}$$

where the last form conveys the complementary restriction to very large momenta. Certainly this expression should vanish since it is not gauge invariant. The momentum integral does not seem to exist, however. Yet, the correct value to assign it is zero, provided the extension to unlimited momenta is performed last, in accordance with the picture of extrapolation from initially nonoverlapping fields. This is a consequence of the Lorentz invariance of the integration process, as expressed by the covariant form

$$p_\kappa p_\lambda p_\mu p_\nu \rightarrow \frac{1}{24} (p^2)^2 (g_{\kappa\lambda}g_{\mu\nu} + g_{\kappa\mu}g_{\lambda\nu} + g_{\kappa\nu}g_{\lambda\mu}); \qquad (4\text{--}8.13)$$

a straightforward reduction shows that

$$\text{tr}(\gamma A \gamma p)^4 \rightarrow 0. \qquad (4\text{--}8.14)$$

When the derivatives of the vector potential needed to produce field strengths are included, corresponding inverse powers of momenta appear and the integrals are absolutely convergent at high momenta.

One can exhibit a generalization of (4–8.7) to any number of field products, $\nu \geqslant 4$. Consider the exchange of a particle pair between the two effective sources associated with a very weak field δA, and with an arbitrary field distribution A. The first effective source is

$$i\eta(x)\eta(x')|_{\text{eff.}} = eq\gamma\delta A(x)\gamma^0\delta(x - x'), \qquad (4\text{--}8.15)$$

while the other, obtained from the comparison of $iW_2\ldots$ with

$$\tfrac{1}{2}\left[i \int (dx)\eta(x)\gamma^0\psi(x)\right]^2, \qquad (4\text{--}8.16)$$

is represented by the effective field product

$$i\psi(x)\psi(x')|_{\text{eff.}} = G_+^A(x, x')\gamma^0. \qquad (4\text{--}8.17)$$

The coupling between the two sources, expressed by the vacuum amplitude

$$i\delta W(A) = -\tfrac{1}{2}\int (dx)(dx')\ \text{tr}[i\eta(x')\eta(x)\gamma^0|_{\text{eff.}}i\psi(x)\psi(x')\gamma^0|_{\text{eff.}}], \qquad (4\text{--}8.18)$$

is

$$\delta W(A) = \tfrac{1}{2}i\ \text{Tr}(eq\gamma\delta A G_+^A). \qquad (4\text{--}8.19)$$

Also included here is pair exchange between the sources symbolized by δA, and A acting once, which we do not want to consider again. It is simplest, however, merely to strike out this term in the final result. As the notation $\delta W(A)$ indicates, when we add the infinitesimal field δA to a preexisting field A we generate a differential expression for an action term $W(A)$, which contains all $W_{0\nu}$, $\nu \geqslant 4$. A formal integration is produced by using the integral equation [Eq. (3–12.21)]

$$G_+^A = G_+ + G_+ eq\gamma A G_+^A \qquad (4\text{–}8.20)$$

and its formal solution [Eq. (3–12.22)]

$$G_+^A = (1 - G_+ eq\gamma A)^{-1} G_+ = G_+(1 - eq\gamma A G_+)^{-1}. \qquad (4\text{–}8.21)$$

This gives

$$\delta W(A) = \tfrac{1}{2}i \,\mathrm{Tr}[(1 - eq\gamma A G_+)^{-1} eq\gamma \delta A G_+]$$

$$= - \tfrac{1}{2}i\delta \,\mathrm{Tr}\log(1 - eq\gamma A G_+)$$

$$= - \tfrac{1}{2}i\delta \log \det(1 - eq\gamma A G_+), \qquad (4\text{–}8.22)$$

where the last version refers to the differential property of determinants,

$$\delta \log \det X = \mathrm{Tr}(X^{-1}\delta X). \qquad (4\text{–}8.23)$$

The integrated statement is

$$W(A) = - \tfrac{1}{2}i \log \det(1 - eq\gamma A G_+) = - \tfrac{1}{2}i \,\mathrm{Tr}\log(1 - eq\gamma A G_+), \qquad (4\text{–}8.24)$$

or, in expanded form with $\nu = 2$ omitted,

$$W(A) = i \sum_{\nu = 4,6,\dots} \frac{1}{2\nu} \,\mathrm{Tr}(eq\gamma A G_+)^\nu, \qquad (4\text{–}8.25)$$

where odd powers are missing since q has a vanishing trace. The W_{04} term has been reproduced, which is a reminder that different causal arrangements of a given number of sources can be used to infer the same general space-time coupling of the sources.

This last discussion is easily repeated for spin 0 particles. The particle coupling is

$$W_{2\dots} = \tfrac{1}{2} \int (dx)(dx') K(x) \Delta_+^A(x, x') K(x'), \qquad (4\text{–}8.26)$$

which describes the photon sources in terms of an effective two-particle field,

$$i\phi(x)\phi(x')|_{\text{eff.}} = \Delta_+^A(x, x'). \qquad (4\text{–}8.27)$$

The effective two-particle source description of a weak electromagnetic field is [cf. Eq. (4–3.31)]

$$iK(x)K(x')|_{\text{eff.}} = eq(p\delta A + \delta A p)(x)\delta(x - x'). \qquad (4\text{–}8.28)$$

We can then express the causal coupling between the two photon sources, symbolized by δA and A, by means of the vacuum amplitude

$$i\delta W(A) = \tfrac{1}{2} \int (dx)(dx')\ \mathrm{tr}[iK(x')K(x)|_{\mathrm{eff.}} i\phi(x)\phi(x')|_{\mathrm{eff.}}]$$

$$= \tfrac{1}{2}\ \mathrm{Tr}[eq(p\delta A + \delta A p)\Delta_+^A]. \tag{4-8.29}$$

There is one subtlety here, however, which will be brought out by the formal solution of the equation for Δ_+^A [Eqs. (3–12.27, 28)];

$$\Delta_+^A = [1 - \Delta_+(eq(pA + Ap) - e^2 A^2)]^{-1}\Delta_+$$

$$= \Delta_+[1 - (eq(pA + Ap) - e^2 A^2)\Delta_+]^{-1}. \tag{4-8.30}$$

If the expression for $\delta W(A)$ is to be integrable, in the manner illustrated for spin $\tfrac{1}{2}$, one must change (4–8.29) into

$$i\delta W(A) = \tfrac{1}{2}\ \mathrm{Tr}[(eq(p\delta A + \delta A p) - 2e^2 \delta A A)\Delta_+^A]. \tag{4-8.31}$$

There is no objection to this since δA and A are disjoint and their product vanishes in the causal arrangement for which (4–8.29) was derived. We can now state that

$$W(A) = \tfrac{1}{2}i\ \log \det[1 - (eq(pA + Ap) - e^2 A^2)\Delta_+]$$

$$= \tfrac{1}{2}i\ \mathrm{Tr}\ \log[1 - (eq(pA + Ap) - e^2 A^2)\Delta_+]. \tag{4-8.32}$$

In particular,

$$W_{04} = -\ i\tfrac{1}{8}\ \mathrm{Tr}[eq(pA + Ap)\Delta_+]^4 + i\tfrac{1}{2}\ \mathrm{Tr}[(eq(pA + Ap)\Delta_+)^2 e^2 A^2 \Delta_+]$$

$$-\ i\tfrac{1}{4}\ \mathrm{Tr}[e^2 A^2 \Delta_+]^2, \tag{4-8.33}$$

which can also be derived by considering particle exchange between two pairs of photon sources. The existence and gauge invariance of this expression can be verified, much as in the spin $\tfrac{1}{2}$ discussion.

There are other presentations of $W(A)$ that are particularly useful under special circumstances. Let us begin with the spin 0 situation and write $(\Pi = p - eqA)$

$$\Delta_+^A = \frac{1}{\Pi^2 + m^2 - i\varepsilon} = i\int_0^\infty ds\ \exp[-\ is(\Pi^2 + m^2)], \tag{4-8.34}$$

where $\varepsilon \to +0$ is implicit in the integral as a convergence factor, $\exp(-\varepsilon s)$. Now the differential expression (4–8.31) reads

$$\delta W(A) = -\ \tfrac{1}{2}\int_0^\infty ds\ \mathrm{Tr}[\delta(\Pi^2)\exp(-\ is(\Pi^2 + m^2))] \tag{4-8.35}$$

and

$$W(A) = -\ i\tfrac{1}{2}\int_0^\infty \frac{ds}{s}\ \mathrm{Tr}\ \exp[-\ is(\Pi^2 + m^2)], \tag{4-8.36}$$

although only terms containing at least four field factors are to be retained. A simple proof of gauge invariance becomes available since $A \rightarrow A + \partial\lambda$ implies

$$\Pi \rightarrow \exp(ieq\lambda) \, \Pi \exp(- \, ieq\lambda) \qquad (4\text{–}8.37)$$

and

$$\exp(- \, is\Pi^2) \rightarrow \exp(ieq\lambda) \exp(- \, is\Pi^2) \exp(- \, ieq\lambda), \qquad (4\text{–}8.38)$$

which leaves the trace unaltered.

The spin $\frac{1}{2}$ analogue of (4–8.36) is based on the construction [Eqs. (4–2.1, 3, 4)]

$$G_+^A = \frac{1}{\gamma\Pi + m - i\varepsilon} = (m - \gamma\Pi) \frac{1}{\Pi^2 - eq\sigma F + m^2 - i\varepsilon}$$

$$= \frac{1}{\Pi^2 - eq\sigma F + m^2 - i\varepsilon} (m - \gamma\Pi), \qquad (4\text{–}8.39)$$

where we now write

$$\frac{1}{\Pi^2 - eq\sigma F + m^2 - i\varepsilon} = i \int_0^\infty ds \exp\{- \, is[\Pi^2 - eq\sigma F + m^2]\}. \qquad (4\text{–}8.40)$$

Since the trace of a product containing an odd number of γ matrices is zero (γ^μ and $- \, \gamma^\mu = \gamma_5^{-1}\gamma^\mu\gamma_5$ are equivalent matrices), and

$$- \, eq\gamma\delta A\gamma\Pi - \gamma\Pi eq\gamma\delta A = \delta[(\gamma\Pi)^2]$$

$$= - \, \delta[\Pi^2 - eq\sigma F], \qquad (4\text{–}8.41)$$

the differential form (4–8.19) becomes

$$\delta W(A) = \tfrac{1}{4} \int_0^\infty ds \, \text{Tr}\{\delta(\Pi^2 - eq\sigma F) \exp[- \, is(\Pi^2 - eq\sigma F + m^2)]\}, \qquad (4\text{–}8.42)$$

which gives

$$W(A) = i\tfrac{1}{4} \int_0^\infty \frac{ds}{s} \, \text{Tr} \exp[- \, is(\Pi^2 - eq\sigma F + m^2)]. \qquad (4\text{–}8.43)$$

The utility of these expressions is confined to slowly varying fields, which are effectively constant over appropriate space-time regions. In such situations, the formal similarity of $\Pi^2 + m^2$ or $\Pi^2 - eq\sigma F + m^2$ to a particle Hamiltonian, and of s to a time variable, can be exploited. While this is always possible, it is the constancy of the commutator

$$[\Pi_\mu, \Pi_\nu] = ieqF_{\mu\nu} \qquad (4\text{–}8.44)$$

that produces simple results. We note that

$$[\Pi_\mu, \Pi^2] = 2ieqF_{\mu\nu}\Pi^\nu, \tag{4-8.45}$$

and therefore

$$\Pi_\mu(s) = \exp(is\Pi^2)\,\Pi_\mu\,\exp(-\,is\Pi^2) \tag{4-8.46}$$

obeys the equation of motion

$$\frac{d\Pi_\mu(s)}{ds} = 2eqF_{\mu\nu}\Pi^\nu(s). \tag{4-8.47}$$

The solution is, in matrix notation,

$$\Pi(s) = \exp(2eqFs)\Pi = \Pi\,\exp(-\,2eqFs), \tag{4-8.48}$$

which recalls the antisymmetry of $F_{\mu\nu}$. Now consider the following tensor, which is defined by a trace that does not refer to charge space,

$$\begin{aligned}
T_{\mu\nu} &= \mathrm{Tr}'[\Pi_\mu\Pi_\nu\,\exp(-\,is\Pi^2)] = \mathrm{Tr}'[\Pi_\mu\,\exp(-\,is\Pi^2)\,\Pi_\nu(s)] \\
&= \mathrm{Tr}'[\Pi_\nu(s)\Pi_\mu\,\exp(-\,is\Pi^2)]. \tag{4-8.49}
\end{aligned}$$

An equivalent form is

$$T_{\mu\nu} = \mathrm{Tr}'[\Pi_\mu\Pi_\nu(s)\,\exp(-\,is\Pi^2)] - \mathrm{Tr}'[[\Pi_\mu, \Pi_\nu(s)]\,\exp(-\,is\Pi^2)]. \tag{4-8.50}$$

The commutator that appears here is evaluated as

$$[\Pi_\mu, \Pi_\nu(s)] = [\Pi_\mu, \Pi^\lambda[\exp(-\,2eqFs)]_{\lambda\nu}] = ieq[F\exp(-\,2eqFs)]_{\mu\nu}, \tag{4-8.51}$$

and a return to matrix notation gives

$$T = T\exp(-\,2eqFs) - ieqF\exp(-\,2eqFs)\,\mathrm{Tr}'[\exp(-\,is\Pi^2)], \tag{4-8.52}$$

or

$$\mathrm{Tr}'[\Pi\Pi\,\exp(-\,is\Pi^2)] = -\,ieq\,\frac{F}{\exp(2eqFs) - 1}\,\mathrm{Tr}'[\exp(-\,is\Pi^2)]. \tag{4-8.53}$$

We use this result to evaluate

$$\begin{aligned}
i\frac{d}{ds}\mathrm{Tr}'[\exp(-\,is\Pi^2)] &= \mathrm{Tr}'[\Pi^2\exp(-\,is\Pi^2)] \\
&= -\,ieq\,\mathrm{tr}'\left(\frac{F}{\exp(2eqFs) - 1}\right)\mathrm{Tr}'[\exp(-\,is\Pi^2)].
\end{aligned}$$
$$\tag{4-8.54}$$

The solution of the ensuing differential equation is

$$\text{Tr}'[\exp(-is\Pi^2)] = C \exp\left[-\tfrac{1}{2}\text{tr}'\log\left(\frac{\sinh eqFs}{eqF}\right)\right]$$

$$= \frac{C}{s^2}\left[\left(\det\frac{eFs}{\sinh eFs}\right)\right]^{1/2}, \qquad (4\text{--}8.55)$$

where the latter form employs the dimensionality of space-time, and notes that the sign of q is immaterial.

The constant C can be determined by considering the limit of small s. This situation is dominated by large Π values and the noncommutativity of different Π components ceases to be significant. Using four-dimensional forms of conventional quantum relations, we get

$$s \to 0: \quad \text{Tr}'\exp(-is\Pi^2) = \int (dx)\langle x|\exp(-isp^2)|x\rangle$$

$$= \int \frac{(dx)(dp)}{(2\pi)^4}\exp(-isp^2). \qquad (4\text{--}8.56)$$

The four-dimensional integral over the $(3+1)$-dimensional momentum space is computed as

$$\int \frac{(dp)}{(2\pi)^4}\exp(-isp^2) = \left(\int_{-\infty}^{\infty}\frac{dp_1}{2\pi}\exp(-isp_1^2)\right)^3 \int_{-\infty}^{\infty}\frac{dp_0}{2\pi}\exp(isp_0^2)$$

$$= \left(\frac{1}{2\pi}\left(\frac{\pi}{is}\right)^{1/2}\right)^3 \frac{1}{2\pi}\left(\frac{\pi i}{s}\right)^{1/2} = \frac{1}{(4\pi)^2}\frac{1}{is^2}, \qquad (4\text{--}8.57)$$

and therefore

$$C = -\frac{1}{(4\pi)^2}i\int (dx). \qquad (4\text{--}8.58)$$

To complete the evaluation of $W(A)$ for spin 0 we have only to supply the additional factor of 2, associated with the charge space in the trace, thus obtaining

$$W_{\text{spin 0}}(A) = \int (dx)\mathscr{L}_{\text{spin 0}}(F),$$

$$\mathscr{L}_{\text{spin 0}}(F) = -\frac{1}{(4\pi)^2}\int_0^{\infty}\frac{ds}{s^3}\exp(-im^2s)\left[\det\left(\frac{eFs}{\sinh eFs}\right)\right]^{1/2}. \qquad (4\text{--}8.59)$$

The reality of the Lagrange function is made apparent by deforming the integration path of is to the positive real axis (but, see a later remark):

$$\mathscr{L}_{\text{spin 0}}(F) = \frac{1}{(4\pi)^2}\int_0^{\infty}\frac{ds}{s^3}\exp(-m^2s)\left\{\left[\det\left(\frac{eFs}{\sin eFs}\right)\right]^{1/2} - 1 - \tfrac{1}{3}(es)^2\mathscr{F}\right\}, \qquad (4\text{--}8.60)$$

in which we have now explicitly removed the unwanted terms. The notation used here is

$$\mathscr{F} = -\tfrac{1}{4}F^{\mu\nu}F_{\mu\nu} = \tfrac{1}{2}(\mathbf{E}^2 - \mathbf{H}^2), \tag{4-8.61}$$

to which we add

$$\mathscr{G} = -\tfrac{1}{4}\,{}^*F^{\mu\nu}F_{\mu\nu} = \mathbf{E} \cdot \mathbf{H} \tag{4-8.62}$$

and

$$\mathscr{H}_{\pm} = 2(\mathscr{F} \pm i\mathscr{G}) = (\mathbf{E} \pm i\mathbf{H})^2. \tag{4-8.63}$$

The general evaluation of the determinant can be given by finding the eigenvalues of the tensor F. It is convenient to use the self-dual tensors

$$F_{\pm} = F \pm i\,{}^*F, \qquad {}^*F_{\pm} = \mp iF_{\pm}. \tag{4-8.64}$$

Considered as matrices, the two tensors commute, and the square of each is a multiple of the unit matrix. That can be checked by explicit use of the small number of independent components. The squares are

$$(F_{\pm}^2)_{\mu\nu} = g_{\mu\nu}\mathscr{H}_{\pm}, \tag{4-8.65}$$

where the coefficients $\mathscr{H}_{\pm}$ are found by forming the trace. Equivalent statements are

$$\tfrac{1}{2}(F^2 - {}^*F^2)_{\mu\nu} = g_{\mu\nu}\mathscr{F}, \qquad (F\,{}^*F)_{\mu\nu} = ({}^*FF)_{\mu\nu} = g_{\mu\nu}\mathscr{G}, \tag{4-8.66}$$

from which we deduce

$$(F^4)_{\mu\nu} - 2\mathscr{F}(F^2)_{\mu\nu} - \mathscr{G}^2 g_{\mu\nu} = 0, \tag{4-8.67}$$

the minimum equation of the tensor F. The eigenvalues appear in oppositely signed pairs, $\pm F'$, $\pm F''$, where

$$F', F'' = \tfrac{1}{2}[\mathscr{H}_{+}^{1/2} \pm \mathscr{H}_{-}^{1/2}]. \tag{4-8.68}$$

Accordingly,

$$\left[\det\left(\frac{eFs}{\sin eFs}\right)\right]^{1/2} = \frac{eF's}{\sin eF's}\frac{eF''s}{\sin eF''s} = \frac{2(es)^2 i\mathscr{G}}{\cos(es\mathscr{H}_{-}^{1/2}) - \cos(es\mathscr{H}_{+}^{1/2})}$$

$$= \frac{(es)^2\mathscr{G}}{\operatorname{Im}\cos(es\mathscr{H}^{1/2})}, \tag{4-8.69}$$

where we have finally written just $\mathscr{H}$ in place of $\mathscr{H}_{-}$.

Here, then, is the spin 0 result:

$$\mathscr{L}_{\text{spin}\,0}(F) = \frac{1}{(4\pi)^2}\int_0^{\infty}\frac{ds}{s^3}\exp(-m^2 s)\left[\frac{(es)^2\mathscr{G}}{\operatorname{Im}\cos(es\mathscr{H}^{1/2})} - 1 - \tfrac{1}{3}(es)^2\mathscr{F}\right], \tag{4-8.70}$$

although all that we really want is the term quartic in the fields. For that it is possibly simpler to return to (4–8.60) and use the determinantal expansion

$$\det(1 + Y) = 1 + \operatorname{tr} Y + \tfrac{1}{2}((\operatorname{tr} Y)^2 - \operatorname{tr} Y^2) + \cdots . \tag{4–8.71}$$

This gives

$$\left[\det\left(\frac{eFs}{\sin eFs} \right) \right]^{1/2} = 1 + \tfrac{1}{3}(es)^2 \mathscr{F} + \frac{1}{90}(es)^4(7\mathscr{F}^2 + \mathscr{G}^2) + \cdots \tag{4–8.72}$$

and

$$\text{spin 0:} \quad \mathscr{L}_{04} = \frac{\alpha^2}{90} \frac{1}{m^4}(7\mathscr{F}^2 + \mathscr{G}^2)$$

$$= \frac{\alpha^2}{90} \frac{1}{m^4} \left[\frac{7}{4}(\mathbf{E}^2 - \mathbf{H}^2)^2 + (\mathbf{E} \cdot \mathbf{H})^2 \right]. \tag{4–8.73}$$

The corresponding spin $\tfrac{1}{2}$ result is produced by inserting in the integrand of $\mathscr{L}$ the following trace over the 2×4 dimensional charge-spin space:

$$- \tfrac{1}{4} \operatorname{tr}_{(8)} \exp(eq\sigma Fs) = - \tfrac{1}{2} \operatorname{tr}_{(4)} \cosh(es\sigma F), \tag{4–8.74}$$

where the substitution $is \to s$ has already been made. The algebraic properties of the spin matrices are such that

$$\tfrac{1}{2}\{\sigma_{\kappa\lambda}, \sigma_{\mu\nu}\} = g_{\kappa\mu}g_{\lambda\nu} - g_{\kappa\nu}g_{\lambda\mu} - \varepsilon_{\kappa\lambda\mu\nu}\gamma_5. \tag{4–8.75}$$

Therefore,

$$(\sigma F)^2 = 2(-\mathscr{F} + \gamma_5\mathscr{G}) \tag{4–8.76}$$

and the eigenvalues of σF are

$$(\sigma F)' = \pm i\mathscr{H}_+^{1/2}, \pm i\mathscr{H}_-^{1/2}. \tag{4–8.77}$$

That gives

$$- \tfrac{1}{2} \operatorname{tr}_{(4)} \cosh(es\sigma F) = - 2 \operatorname{Re} \cos(es\mathscr{H}^{1/2}) \tag{4–8.78}$$

and

$$\mathscr{L}_{\text{spin}\frac{1}{2}}(F) = - \frac{2}{(4\pi)^2} \int_0^\infty \frac{ds}{s^3} \exp(-m^2 s) \left[(es)^2 \mathscr{G} \frac{\operatorname{Re} \cos(es\mathscr{H}^{1/2})}{\operatorname{Im} \cos(es\mathscr{H}^{1/2})} - 1 + \tfrac{2}{3}(es)^2 \mathscr{F} \right]. \tag{4–8.79}$$

We use the expansion

$$\operatorname{Re} \cos(es\mathscr{H}^{1/2}) = 1 - (es)^2 \mathscr{F} + \tfrac{1}{6}(es)^4(\mathscr{F}^2 - \mathscr{G}^2) + \cdots \tag{4–8.80}$$

to derive the term of principal interest:

$$\text{spin } \tfrac{1}{2}: \quad \mathscr{L}_{04} = \frac{2\alpha^2}{45} \frac{1}{m^4} (4\mathscr{F}^2 + 7\mathscr{G}^2)$$

$$= \frac{2\alpha^2}{45} \frac{1}{m^4} [(\mathbf{E}^2 - \mathbf{H}^2)^2 + 7(\mathbf{E} \cdot \mathbf{H})^2]. \qquad (4\text{–}8.81)$$

Several applications can be made of these low-energy Lagrange functions. Inasmuch as comparisons with experiment are not at immediate issue we shall be content to omit numerical factors (but not π) and infer general orders of magnitude. The T matrix element for photon-photon scattering is derived from iW_{04} as the coefficient of $iJ^*_{k_1\lambda_1} iJ^*_{k_1'\lambda_1'} iJ_{k_2\lambda_2} iJ_{k_2'\lambda_2'}$. In the center of mass frame where all photon energies equal $\tfrac{1}{2}M$, the presence of four field strengths introduces the factor $(\tfrac{1}{2}M)^4$ and

$$\langle 1_{k_1\lambda_1} 1_{k_1'\lambda_1'} | T | 1_{k_2\lambda_2} 1_{k_2'\lambda_2'} \rangle \sim (d\omega_{k_1} \cdots d\omega_{k_2'})^{1/2} \alpha^2 \left(\frac{M}{2m}\right)^4. \qquad (4\text{–}8.82)$$

Since this is an elastic collision, the ratio of the kinematical factors involved in the differential cross section definition is $\sim (1/\pi^2)(1/M^2)$ and the total cross section emerges as

$$M \ll 2m: \quad \sigma \sim \frac{\alpha^4}{\pi} \left(\frac{M}{2m}\right)^6 \frac{1}{m^2}. \qquad (4\text{–}8.83)$$

As a variant of photon-photon scattering, we note that a region containing a macroscopic electromagnetic field is an anisotropic medium for photon propagation. In the example of a magnetic field of strength H, the deviation from unity of the propagation parameters is of the order of $\alpha^2 H^2 / m^4$. The quartic coupling of photon fields also states that an extended photon source can emit or absorb three real photons. This is of interest because the process exists, if weakly, below the mass threshold for particle pair exchange at $M = 2m$. The weight factor $a(M^2)$ in the structure of the modified photon propagation function [Eq. (4–3.82)] is inferred as

$$M \ll 2m: \quad a(M^2) \sim \frac{\alpha^4}{m^8} \frac{1}{M^2} \int (k^{0'})^2 \, d\omega_{k'} (k^{0''})^2 \, d\omega_{k''} (k^{0'''})^2 \, d\omega_{k'''}$$

$$\times (2\pi)^3 \delta(k' + k'' + k''' - k)$$

$$\sim \left(\frac{\alpha}{\pi}\right)^4 \left(\frac{M^2}{4m^2}\right)^3 \frac{1}{m^2}. \qquad (4\text{–}8.84)$$

The existence of this effect, with a threshold at $M = 0$, implies that the initial long range deviation from the Coulomb interaction of static charges has an algebraic rather than an exponential dependence on distance:

$$2m|\mathbf{x}| \gg 1: \quad \mathscr{D}(\mathbf{x}) - \frac{1}{4\pi|\mathbf{x}|} \sim \frac{1}{4\pi|\mathbf{x}|} \left(\frac{\alpha}{\pi}\right)^4 \left(\frac{1}{2m|\mathbf{x}|}\right)^8. \qquad (4\text{–}8.85)$$

Finally, let us note one use of the general Lagrange function, say (4–8.79) referring to spin $\frac{1}{2}$ particles, which concerns a region occupied by a strong electric field E. In the limit of vanishing magnetic field (invariantly characterized by $\mathscr{G} = 0$, $\mathscr{F} > 0$), we find that

$$\mathscr{L}_{\text{spin }\frac{1}{2}}(E) = -\frac{1}{8\pi^2} \int_0^\infty \frac{ds}{s^3} \exp(-m^2 s) \left[eEs \cot(eEs) - 1 + \tfrac{1}{3}(eEs)^2\right]. \qquad (4\text{–}8.86)$$

One remark is needed, however. The variable now called s originally ranged along the positive imaginary axis, and then its path was deformed to the positive real axis. We must recall that sense of approach to the real axis, from above, since the integrand has singularities on the real axis. They occur at

$$eEs_n = n\pi, \qquad n = 1, 2, \ldots. \qquad (4\text{–}8.87)$$

The necessary deformation of the contour into the upper half-plane near these singularities provides $\mathscr{L}$ with an imaginary part,

$$\text{Im } \mathscr{L}_{\text{spin }\frac{1}{2}}(E) = \frac{1}{8\pi} \sum_{n=1}^\infty \frac{1}{s_n^2} \exp(-m^2 s_n), \qquad (4\text{–}8.88)$$

or

$$2 \text{ Im } \mathscr{L}_{\text{spin }\frac{1}{2}}(E) = \frac{1}{4\pi} \left(\frac{eE}{\pi}\right)^2 \sum_{n=1}^\infty \frac{1}{n^2} \exp\left(-n\frac{\pi m^2}{eE}\right)$$

$$\gtrsim \frac{1}{4\pi} \left(\frac{eE}{\pi}\right)^2 \exp\left(-\frac{\pi m^2}{eE}\right). \qquad (4\text{–}8.89)$$

Since the vacuum persistence probability is

$$|\exp(iW)|^2 = \exp(-2 \text{ Im } W), \qquad (4\text{–}8.90)$$

we recognize in $2 \text{ Im } \mathscr{L}$ a measure of the probability, per unit time and unit spatial volume, that an electron-positron pair has been created. This process is interesting conceptually, for no finite number of encounters with the static electric field, in the sense of a scattering description, can produce the energy needed to create the particles.

4-9 SCATTERING OF LIGHT BY LIGHT II. FORWARD SCATTERING

In the preceding section we exhibited the space-time form of couplings that involve only the electromagnetic field, and we also used these forms directly for calculations, in the special circumstance of slowly varying fields. With more general situations, however, it is usually preferable to consider an appropriate causal arrangement and then perform the space-time extrapolation. We are recognizing now that source theory is flexible; it is not committed to any special calculational method and is free to choose the most convenient one. Indeed, it is the interplay and synthesis of various calculational devices, each adapted to specific circumstances, that constitutes the general source theory computational method.

Let us consider the arrangement in which two photons collide to create a charged particle pair, and then the two photons emitted in the subsequent annihilation of the particles are detected. For spin 0 particles, we can use the coupling (4–8.33), where

$$A = A_1 + A_2, \tag{4–9.1}$$

and retain just the terms, symbolized by $A_1A_1A_2A_2$, that describe particle creation or annihilation through the combined efforts of two simple photon sources rather than by individual extended photon sources. The corresponding vacuum amplitude is

$$\tfrac{1}{2}\,\mathrm{Tr}[(2eq\not{p}A_1{}_+2eq\not{p}A_1 - (eqA_1)^2)\varDelta_+(2eq\not{p}A_2{}_+2eq\not{p}A_2 - (eqA_2)^2)\varDelta_+], \tag{4–9.2}$$

where we have simplified the writing by replacing $\not{p}A + A\not{p}$ with $2\not{p}A$, as is appropriate to a Lorentz gauge. We can see here the individual factors that describe the effective two-particle sources which are associated with the twofold action of the electromagnetic field. This is the causal structure with which one could have begun in order to derive Eq. (4–8.33). The causal form of the fields is given by

$$A_1^\mu(x) = \sum_{k_1\lambda_1} iJ^*_{k_1\lambda_1}(d\omega_{k_1})^{1/2}e^{\mu\,*}_{k_1\lambda_1}\exp(-ik_1x),$$

$$A_2^\mu(x) = \sum_{k_2\lambda_2} (d\omega_{k_2})^{1/2}e^\mu_{k_2\lambda_2}\exp(ik_2x)\,iJ_{k_2\lambda_2}, \tag{4–9.3}$$

and the introduction of the causal forms of the propagation function $\varDelta_+$ produces the vacuum amplitude

$$-\sum_{k_1\lambda_1\dots} iJ^*_{k_1\lambda_1}iJ^*_{k_1'\lambda_1'}iJ_{k_2\lambda_2}iJ_{k_2'\lambda_2'}(d\omega_{k_1}\cdots d\omega_{k_2'})^{1/2}(2\pi)^4\,\delta(k_1 + k_1' - k_2 - k_2')$$

$$\times\, e^4 I_{k_1\lambda_1\dots k_2'\lambda_2'}, \tag{4–9.4}$$

where, in a simplified notation with real polarization vectors,

$$I = \int d\omega_p\, d\omega_{p'}\,(2\pi)^4\delta(p + p' - k_2 - k_2')\left[-2\frac{e_1'\not{p}'e_1\not{p}}{(p - k_1)^2 + m^2} - 2\frac{e_1\not{p}'e_1'\not{p}}{(p - k_1')^2 + m^2}\right.$$

$$\left. - e_1e_1'\right]\left[-2\frac{e_2'\not{p}e_2\not{p}'}{(p' - k_2)^2 + m^2} - 2\frac{e_2\not{p}e_2'\not{p}'}{(p' - k_2')^2 + m^2} - e_2e_2'\right]. \tag{4–9.5}$$

The kinematical properties of the momenta reduce this to

$$I = \int d\dot{\omega}_p \, d\omega_{p'} (2\pi)^4 \delta(p + p' - k_2 - k_2') \left[\frac{e_1' p' e_1 p}{p k_1} + \frac{e_1 p' e_1' p}{p k_1'} - e_1 e_1' \right]$$

$$\times \left[\frac{e_2' p e_2 p'}{p' k_2} + \frac{e_2 p e_2' p'}{p' k_2'} - e_2 e_2' \right]. \tag{4-9.6}$$

This structure will be used only to produce the coupling that describes forward (and backward) scattering of the photons. For that situation, considered in the rest frame of the total momentum, there is one preferred spatial direction with photons moving in either sense along this direction, as expressed by the momentum relations

$$\mathbf{k}_1 = -\mathbf{k}_1' = \mathbf{k}_2 = -\mathbf{k}_2', \qquad k_1^0 = k_1^{0\prime} = k_2^0 = k_2^{0\prime} = \tfrac{1}{2}M. \tag{4-9.7}$$

In the gauge where polarization vectors are purely spatial, all are perpendicular to this common direction of motion. The integrations of (4-9.6) involve the variable z, the cosine of the angle between $\mathbf{p} = -\mathbf{p}'$ and the preferred direction, and an angle in the plane perpendicular to that direction. Averages for the latter integration process are given by

$$\langle \mathbf{e} \cdot \mathbf{p} \mathbf{e}' \cdot \mathbf{p} \rangle = \tfrac{1}{2} \mathbf{e} \cdot \mathbf{e}'(1 - z^2) \left[\left(\frac{M}{2} \right)^2 - m^2 \right] \tag{4-9.8}$$

and

$$\langle \mathbf{e}_1 \cdot \mathbf{p} \mathbf{e}_1' \cdot \mathbf{p} \mathbf{e}_2 \cdot \mathbf{p} \mathbf{e}_2' \cdot \mathbf{p} \rangle = \tfrac{1}{8} (\mathbf{e}_1 \cdot \mathbf{e}_1' \mathbf{e}_2 \cdot \mathbf{e}_2' + \mathbf{e}_1 \cdot \mathbf{e}_2 \mathbf{e}_1' \cdot \mathbf{e}_2' + \mathbf{e}_1 \cdot \mathbf{e}_2' \mathbf{e}_1' \cdot \mathbf{e}_2)$$

$$\times (1 - z^2)^2 \left[\left(\frac{M}{2} \right)^2 - m^2 \right]^2. \tag{4-9.9}$$

The use of these averages produces

$$I = \frac{1}{8\pi} \left(1 - \frac{4m^2}{M^2} \right)^{1/2} [a \mathbf{e}_1 \cdot \mathbf{e}_1' \mathbf{e}_2 \cdot \mathbf{e}_2' + \tfrac{1}{2} b (\mathbf{e}_1 \cdot \mathbf{e}_1' \mathbf{e}_2 \cdot \mathbf{e}_2'$$

$$+ \mathbf{e}_1 \cdot \mathbf{e}_2 \mathbf{e}_1' \cdot \mathbf{e}_2' + \mathbf{e}_1 \cdot \mathbf{e}_2' \mathbf{e}_1' \cdot \mathbf{e}_2)], \tag{4-9.10}$$

where

$$a(M^2) = 1 - \int_{-1}^{1} dz \, \frac{[1 - (4m^2/M^2)](1 - z^2)}{1 - [1 - (4m^2/M^2)]z^2} = -1 + \frac{1 - v^2}{v} \log \frac{1 + v}{1 - v}, \tag{4-9.11}$$

and

$$b(M^2) = \tfrac{1}{2} \int_{-1}^{1} dz \left[\frac{[1 - (4m^2/M^2)](1 - z^2)}{1 - [1 - (4m^2/M^2)]z^2} \right]^2$$

$$= -a(M^2) + \frac{1-v^2}{2}\left[1 + \frac{1-v^2}{2v}\log\frac{1+v}{1-v}\right]. \tag{4-9.12}$$

The results of the integrations have been stated in terms of the variable

$$v = \left(1 - \frac{4m^2}{M^2}\right)^{1/2}. \tag{4-9.13}$$

The causal picture is emphasized by writing

$$(2\pi)^4\delta(k_1 + k_1' - k_2 - k_2')$$

$$= \int (2\pi)^4\delta(k_1 + k_1' - k)\, d\omega_k \frac{dM^2}{2\pi}(2\pi)^4\delta(k - k_2 - k_2')$$

$$= -i\int\frac{dM^2}{2\pi}\int (dx)(dx')\exp[-i(k_1 + k_1')x]\left[i\int d\omega_k \exp[ik(x - x')]\right]$$

$$\times \exp[i(k_2 + k_2')x'], \tag{4-9.14}$$

which makes explicit the individual coordinate dependences of the field strengths, and the propagation function that causally connects the two regions. In order to give a covariant space-time form to the vacuum amplitude, we must replace the polarization vector structure of (4–9.10) by equivalent field strength combinations. Consider first the situation in which all polarization vectors are parallel, thereby reducing (4–9.10) to

$$I = \frac{1}{8\pi}\left(1 - \frac{4m^2}{M^2}\right)^{1/2}(a + \tfrac{3}{2}b). \tag{4-9.15}$$

We observe that, generally,

$$\mathcal{F}_1(x) = \tfrac{1}{2}(\mathbf{E}_1^2 - \mathbf{H}_1^2)(x)$$

$$= -\sum iJ^*_{k_1\lambda_1}J^*_{k_1'\lambda_1'}(d\omega_{k_1}\, d\omega_{k_1'})^{1/2}$$

$$\times \exp[-i(k_1 + k_1')x]\,\tfrac{1}{2}[k_1^0 k_1^{0'}\mathbf{e}_{k_1\lambda_1}\cdot\mathbf{e}_{k_1'\lambda_1'} - \mathbf{k}_1\times\mathbf{e}_{k_1\lambda_1}\cdot\mathbf{k}_1'\times\mathbf{e}_{k_1'\lambda_1'}]$$

$$\tag{4-9.16}$$

and that the polarization vector factor reduces, under the circumstances now being considered, to

$$-\tfrac{1}{2}k_1 k_1' = -\tfrac{1}{4}(k_1 + k_1')^2 = \tfrac{1}{4}M^2. \tag{4-9.17}$$

A similar remark applies to $\mathcal{F}_2(x')$. Accordingly, for the special situation of parallel polarizations, the vacuum amplitude (4–9.4) can be presented as

$$i\alpha^2 \int dM^2 \left(1 - \frac{4m^2}{M^2}\right)^{1/2} \frac{16}{(M^2)^2} (a + \tfrac{3}{2}b)(M^2) \int (dx)(dx') \mathscr{F}_1(x) \Delta_+(x - x', M^2) \mathscr{F}_2(x').$$

$$(4\text{-}9.18)$$

The corresponding action expression is

$$8\alpha^2 \int_{(2m)^2}^{\infty} \frac{dM^2}{(M^2)^2} \left(1 - \frac{4m^2}{M^2}\right)^{1/2} (a + \tfrac{3}{2}b)(M^2) \int (dx)(dx') \mathscr{F}(x) \Delta_+(x - x', M^2) \mathscr{F}(x').$$

$$(4\text{-}9.19)$$

If we consider the limit of slowly varying fields, where $1/(2m)$ sets the scale of length, one can replace $\mathscr{F}(x')$ by $\mathscr{F}(x)$ in (4–9.19), which introduces

$$\int (dx') \Delta_+(x - x', M^2) = \frac{1}{M^2}. \qquad (4\text{-}9.20)$$

Then the action term can be represented by a Lagrange function, which is

$$8\alpha^2 \int_{(2m)^2}^{\infty} \frac{dM^2}{(M^2)^3} \left(1 - \frac{4m^2}{M^2}\right)^{1/2} (a + \tfrac{3}{2}b)(M^2)(\mathscr{F}(x))^2$$

$$= \int_0^1 dv \, v^2(1 - v^2)(a + \tfrac{3}{2}b) \frac{\alpha^2}{m^4} (\mathscr{F}(x))^2. \qquad (4\text{-}9.21)$$

The integrals that appear here are

$$\int_0^1 dv \, v^2(1 - v^2)a = \frac{2}{45}, \qquad \int_0^1 dv \, v^2(1 - v^2)b = \frac{1}{45}, \qquad (4\text{-}9.22)$$

and the coefficient of $(\alpha^2/m^4)\mathscr{F}^2$ becomes 7/90, in agreement with that part of the Lagrange function in (4–8.73). Note that the initial limitation to forward scattering which is actually a restriction on momentum transfer in the collision, ceases to be a constraint on the scattering angle in this limit of small momenta.

The choice of parallel polarizations is one example in which the polarization vectors of the photons do not change in scattering. The other example refers to perpendicular polarizations of the initial photons and of the final photons:

$$e_1 \cdot e_1' = e_2 \cdot e_2' = e_1 \cdot e_2' = e_2 \cdot e_1' = 0,$$

$$e_1 \cdot e_2 = e_1' \cdot e_2' = 1, \qquad (4\text{-}9.23)$$

when (4–9.10) becomes

$$I = \frac{1}{8\pi} \left(1 - \frac{4m^2}{M^2}\right)^{1/2} \tfrac{1}{2}b. \qquad (4\text{-}9.24)$$

A related field structure is

$$\mathscr{G}_1(x) = \mathbf{E}_1(x) \cdot \mathbf{H}_1(x)$$

$$= - \sum i J^*_{k_1 \lambda_1} i J^*_{k_1' \lambda_1'} (d\omega_{k_1} \, d\omega_{k_1'})^{1/2} \exp[- i(k_1 + k_1')x]$$

$$\times \tfrac{1}{2}[k_1^0 e_{k_1 \lambda_1} \cdot \mathbf{k}_1' \times e_{k_1' \lambda_1'} + k_1^{0'} e_{k_1' \lambda_1'} \cdot \mathbf{k}_1 \times e_{k_1 \lambda_1}]; \qquad (4\text{–}9.25)$$

the general polarization vector factor here reduces to

$$\tfrac{1}{2} M \mathbf{k}_1 \cdot \mathbf{e}_1 \times \mathbf{e}_1' = \pm \tfrac{1}{4} M^2, \qquad (4\text{–}9.26)$$

where the $\pm$ sign depends upon the particular orientation of the perpendicular vectors $\mathbf{e}_1$ and $\mathbf{e}_1'$. But the same $\pm$ sign appears in $\mathscr{G}_2(x')$ since the polarization vectors are unaltered in the scattering act. Therefore the vacuum amplitude (4–9.4) becomes, in this situation,

$$i\alpha^2 \int dM^2 \left(1 - \frac{4m^2}{M^2}\right)^{1/2} \frac{16}{(M^2)^2} \, \tfrac{1}{2}b(M^2) \int (dx)(dx') \mathscr{G}_1(x) \varDelta_+(x - x', M^2) \mathscr{G}_2(x'),$$

$$(4\text{–}9.27)$$

and the corresponding action expression is

$$4\alpha^2 \int_{(2m)^2}^{\infty} \frac{dM^2}{(M^2)^2} \left(1 - \frac{4m^2}{M^2}\right)^{1/2} b(M^2) \int (dx)(dx') \mathscr{G}(x) \varDelta_+(x - x', M^2) \mathscr{G}(x'). \quad (4\text{–}9.28)$$

In the limit of slowly varying fields there is a Lagrange function term:

$$4\alpha^2 \int_{(2m)^2}^{\infty} \frac{dM^2}{(M^2)^3} \left(1 - \frac{4m^2}{M^2}\right)^{1/2} b(M^2) (\mathscr{G}(x))^2$$

$$= \tfrac{1}{2} \int_0^1 dv \, v^2(1 - v^2) b \, \frac{\alpha^2}{m^4} (\mathscr{G}(x))^2 = \frac{1}{90} \frac{\alpha^2}{m^4} (\mathscr{G}(x))^2, \qquad (4\text{–}9.29)$$

which also agrees with the corresponding term of (4–8.73).

Now that we have established contact with the low frequency results by considering special polarization assignments, what can we say about the general polarization vector combination in (4–9.10)? The term with the coefficient a is easily given a gauge invariant representation since the polarization vector factor in $\mathscr{F}_1(x)\mathscr{F}_2(x')$ is

$$(\tfrac{1}{4}M^2)^2 \mathbf{e}_1 \cdot \mathbf{e}_1' \mathbf{e}_2 \cdot \mathbf{e}_2'. \qquad (4\text{–}9.30)$$

But the more elaborate structure with coefficient b cannot, in general, be expressed by products of the two scalars, $\mathscr{F}_1(x)\mathscr{F}_2(x')$ and $\mathscr{G}_1(x)\mathscr{G}_2(x')$. It is necessary to use the tensor

$$F^2(x)^{\mu\nu} = F^{\mu\alpha}(x)F_\alpha{}^\nu(x). \qquad (4\text{–}9.31)$$

Now,

$$F_1^2(x)^{\mu\nu}F_2^2(x')_{\mu\nu} = \sum iJ_{k_1\lambda_1}^* iJ_{k_1'\lambda_1'}^* iJ_{k_2\lambda_2} iJ_{k_2'\lambda_2'}(d\omega_{k_1}\cdots d\omega_{k_2'})^{1/2}$$

$$\times (f_{k_1\lambda_1})^{\mu\alpha}(f_{k_1'\lambda_1'})_\alpha{}^\nu(f_{k_2\lambda_2})_{\mu\beta}(f_{k_2'\lambda_2'})^\beta{}_\nu, \qquad (4\text{-}9.32)$$

where, omitting the compensating $\pm i$ factors, we have

$$(f)^{\mu\nu} = k^\mu e^\nu - k^\nu e^\mu. \qquad (4\text{-}9.33)$$

Evaluated in the center of mass frame, and restricted to forward scattering conditions, the polarization factor in (4–9.32) is, indeed,

$$\tfrac{1}{8}M^4(\mathbf{e}_1\cdot\mathbf{e}_1'\mathbf{e}_2\cdot\mathbf{e}_2' + \mathbf{e}_1\cdot\mathbf{e}_2\mathbf{e}_1'\cdot\mathbf{e}_2' + \mathbf{e}_1\cdot\mathbf{e}_2'\mathbf{e}_1'\cdot\mathbf{e}_2). \qquad (4\text{-}9.34)$$

As we have seen, the tensor combination can be replaced by scalars in the two situations where no change of polarization occurs:

$$F_1^2(x)^{\mu\nu}F_2^2(x')_{\mu\nu} \rightarrow \begin{cases}\text{parallel:} & 6\mathscr{F}_1(x)\mathscr{F}_2(x') \\ \text{perpendicular:} & 2\mathscr{G}_1(x)\mathscr{G}_2(x').\end{cases} \qquad (4\text{-}9.35)$$

Thus the additional term that is required is given by the difference,

$$F_1^2(x)^{\mu\nu}F_2^2(x')_{\mu\nu} - 6\mathscr{F}_1(x)\mathscr{F}_2(x') - 2\mathscr{G}_1(x)\mathscr{G}_2(x'). \qquad (4\text{-}9.36)$$

It would seem that we should add the space-time extrapolation of this coupling to the ones already known, as comprised in Eqs. (4–9.19) and (4–9.28). But, if we do so, the static interaction will be changed. That follows from the nonzero value of (4–9.36) in its unified form, when the distinction between x and x' is removed,

$$\text{tr } F^4 - 6\mathscr{F}^2 - 2\mathscr{G}^2 = 2(\mathscr{F}^2 + \mathscr{G}^2) \neq 0. \qquad (4\text{-}9.37)$$

The proper procedure should be clear. The static interaction provides a normalization condition for the more general calculation. To avoid altering the already correct result, the space-time extrapolation of the contribution containing (4–9.36) must be performed with an additional contact term, designed to remove this contribution at low frequencies:

$$\Delta_+(x - x', M^2) - \frac{1}{M^2}\delta(x - x') = \frac{1}{M^2}\partial^2\Delta_+(x - x', M^2). \qquad (4\text{-}9.38)$$

The complete action expression obtained in this way, presented in momentum space, is

$$W_{04} = 4\alpha^2\int_{(2m)^2}^\infty \frac{dM^2}{(M^2)^2}\left(1 - \frac{4m^2}{M^2}\right)^{1/2}\int\frac{(dk)}{(2\pi)^4}\frac{1}{k^2 + M^2 - i\varepsilon}$$

$$\times \left[(2a + 3b)(M^2)\mathscr{F}(-k)\mathscr{F}(k) + b(M^2)\mathscr{G}(-k)\mathscr{G}(k) - \tfrac{1}{2}b(M^2)\right.$$

$$\left.\times \frac{k^2}{M^2}(F^2(-k)^{\mu\nu}F^2(k)_{\mu\nu} - 6\mathscr{F}(-k)\mathscr{F}(k) - 2\mathscr{G}(-k)\mathscr{G}(k))\right] \quad (4\text{-}9.39)$$

or, alternatively,

$$W_{04} = \frac{1}{90} \frac{\alpha^2}{m^4} \int \frac{(dk)}{(2\pi)^4} [7\mathscr{F}(-k)\mathscr{F}(k) + \mathscr{G}(-k)\mathscr{G}(k)]$$

$$- 4\alpha^2 \int_{(2m)^2}^{\infty} \frac{dM^2}{(M^2)^3} \left(1 - \frac{4m^2}{M^2}\right)^{1/2} \int \frac{(dk)}{(2\pi)^4} \frac{k^2}{k^2 + M^2 - i\varepsilon}$$

$$\times [2a(M^2)\mathscr{F}(-k)\mathscr{F}(k) + \tfrac{1}{2}b(M^2)F^2(-k)^{\mu\nu}F^2(k)_{\mu\nu}]. \qquad (4\text{-}9.40)$$

One can also rewrite the tensor combination in the form

$$F^2(-k)^{\mu\nu}F^2(k)_{\mu\nu} = 4\mathscr{F}(-k)\mathscr{F}(k) + \tfrac{1}{4}(F^{\kappa\lambda}F^{\mu\nu})(-k)(F_{\kappa\lambda}F_{\mu\nu})(k)$$

$$+ \tfrac{1}{4}(*F^{\kappa\lambda}F^{\mu\nu})(-k)(F_{\kappa\lambda}*F_{\mu\nu})(k), \qquad (4\text{-}9.41)$$

so that the relevant term of (4-9.39) reads

$$F^2(-k)^{\mu\nu}F^2(k)_{\mu\nu} - 6\mathscr{F}(-k)\mathscr{F}(k) - 2\mathscr{G}(-k)\mathscr{G}(k)$$

$$= \tfrac{1}{4}(F^{\kappa\lambda}F^{\mu\nu})(-k)(F_{\kappa\lambda}F_{\mu\nu})(k) - 2\mathscr{F}(-k)\mathscr{F}(k)$$

$$+ \tfrac{1}{4}(*F^{\kappa\lambda}F^{\mu\nu})(-k)(F_{\kappa\lambda}*F_{\mu\nu})(k) - 2\mathscr{G}(-k)\mathscr{G}(k). \qquad (4\text{-}9.42)$$

Perhaps it should be emphasized that the additional factor of $-k^2/M^2$, which appears in (4-9.39), is not needed to bring about the existence of the spectral integral. Only the normalization requirement imposed at low frequencies demands its presence.

When spin $\tfrac{1}{2}$ charged particles are involved, one can study these photon-photon scattering phenomena with either of two calculational methods. The first one starts from the vacuum amplitude expression (4-8.6) while the other exploits the similarity between spin 0 and spin $\tfrac{1}{2}$ couplings. On comparing (4-8.36) and (4-8.43), we see that the substitution $\Pi^2 \to \Pi^2 - eq\sigma F$ and an additional factor of $-\tfrac{1}{2}$ converts $W(A)_{\text{spin }0}$ into $W(A)_{\text{spin }\frac{1}{2}}$, provided one extends the trace operation to include the spinor indices. The second method is more economical since much of the calculation is common to both spin values. The vacuum amplitude derived in this way from the spin 0 structure (4-9.2) is

$$- \tfrac{1}{4} \text{Tr}[(eq(2pA_1 + \sigma F_1)\Delta_+ eq(2pA_1 + \sigma F_1) - (eqA_1)^2)\Delta_+$$

$$\times (eq(2pA_2 + \sigma F_2)\Delta_+ eq(2pA_2 + \sigma F_2) - (eqA_2)^2)\Delta_+]. \qquad (4\text{-}9.43)$$

One facilitates the spin part of this calculation by writing

$$\sigma F = \boldsymbol{\sigma} \cdot \mathbf{H} + \gamma_5 \boldsymbol{\sigma} \cdot \mathbf{E} = -(\gamma_5 \partial_0 + i\boldsymbol{\sigma} \cdot \boldsymbol{\nabla})\boldsymbol{\sigma} \cdot \mathbf{A}, \qquad (4\text{-}9.44)$$

which uses the field strength construction of the radiation gauge, far from sources. If we retain the vacuum amplitude representation (4-9.4), the appropriate structure of $I_{\text{spin}\,\frac{1}{2}}$ is derived from (4–9.5) by introducing the spinor operation $(-\frac{1}{2})$ tr, while performing the modifications indicated by

$$\mathbf{e}_1 \cdot \mathbf{p} \to \mathbf{e}_1 \cdot \mathbf{p} - \tfrac{1}{4}M(i\gamma_5 + \boldsymbol{\sigma} \cdot \mathbf{n})\boldsymbol{\sigma} \cdot \mathbf{e}_1,$$

$$-\mathbf{e}_1' \cdot \mathbf{p}' \to -\mathbf{e}_1' \cdot \mathbf{p}' - \tfrac{1}{4}M(i\gamma_5 - \boldsymbol{\sigma} \cdot \mathbf{n})\boldsymbol{\sigma} \cdot \mathbf{e}_1',$$

$$\mathbf{e}_2 \cdot \mathbf{p} \to \mathbf{e}_2 \cdot \mathbf{p} + \tfrac{1}{4}M(i\gamma_5 + \boldsymbol{\sigma} \cdot \mathbf{n})\boldsymbol{\sigma} \cdot \mathbf{e}_2,$$

$$-\mathbf{e}_2' \cdot \mathbf{p}' \to -\mathbf{e}_2' \cdot \mathbf{p}' + \tfrac{1}{4}M(i\gamma_5 - \boldsymbol{\sigma} \cdot \mathbf{n})\boldsymbol{\sigma} \cdot \mathbf{e}_2', \qquad (4\text{--}9.45)$$

where $\mathbf{n}$ is the unit vector associated with the preferred direction of (4–9.7). Note also that

$$(i\gamma_5 \pm \boldsymbol{\sigma} \cdot \mathbf{n})\boldsymbol{\sigma} \cdot \mathbf{e} = \boldsymbol{\sigma} \cdot \mathbf{e}(i\gamma_5 \mp \boldsymbol{\sigma} \cdot \mathbf{n}) \qquad (4\text{--}9.46)$$

and

$$(i\gamma_5 \pm \boldsymbol{\sigma} \cdot \mathbf{n})^2 = 2(1 \pm i\gamma_5\boldsymbol{\sigma} \cdot \mathbf{n}). \qquad (4\text{--}9.47)$$

The values of the spin traces encountered here are given by

$$\tfrac{1}{4}\operatorname{tr} \boldsymbol{\sigma} \cdot \mathbf{e}\,\boldsymbol{\sigma} \cdot \mathbf{e}' = \mathbf{e} \cdot \mathbf{e}' \qquad (4\text{--}9.48)$$

and

$$\tfrac{1}{4}\operatorname{tr} \boldsymbol{\sigma} \cdot \mathbf{e}\boldsymbol{\sigma} \cdot \mathbf{e}'\boldsymbol{\sigma} \cdot \mathbf{e}''\boldsymbol{\sigma} \cdot \mathbf{e}''' = \mathbf{e} \cdot \mathbf{e}'\mathbf{e}'' \cdot \mathbf{e}''' - \mathbf{e} \cdot \mathbf{e}''\mathbf{e}' \cdot \mathbf{e}''' + \mathbf{e} \cdot \mathbf{e}'''\mathbf{e}' \cdot \mathbf{e}''; \qquad (4\text{--}9.49)$$

the latter can be derived by reducing the $\boldsymbol{\sigma}$ products or by systematically commuting one factor from left to right in the trace. The outcome of the calculation is expressed by

$$I_{\text{spin}\,\frac{1}{2}} = -2I_{\text{spin}\,0} + \frac{1}{8\pi}\left(1 - \frac{4m^2}{M^2}\right)^{1/2}[c(\mathbf{e}_1 \cdot \mathbf{e}_1'\mathbf{e}_2 \cdot \mathbf{e}_2' - \mathbf{e}_1 \cdot \mathbf{e}_2'\mathbf{e}_1' \cdot \mathbf{e}_2$$
$$+ \mathbf{e}_1 \cdot \mathbf{e}_2\mathbf{e}_1' \cdot \mathbf{e}_2') + 2d(\mathbf{e}_1 \cdot \mathbf{e}_2'\mathbf{e}_1' \cdot \mathbf{e}_2 - \mathbf{e}_1 \cdot \mathbf{e}_1'\mathbf{e}_2 \cdot \mathbf{e}_2')], \qquad (4\text{--}9.50)$$

where .

$$c(M^2) = \int_{-1}^{1} dz \frac{1}{1 - v^2z^2} = \frac{1}{v}\log\frac{1+v}{1-v},$$

$$d(M^2) = \int_{-1}^{1} dz \frac{v^2(1 - z^2)}{(1 - v^2z^2)^2} = \tfrac{1}{2}(1 + v^2)c(M^2) - 1. \qquad (4\text{--}9.51)$$

We also observe that

$$\int_{0}^{1} dv \, v^2(1 - v^2)c = \tfrac{1}{3}. \qquad (4\text{--}9.52)$$

This is the only integral needed to discuss the low frequency limits, for parallel and perpendicular polarizations that do not alter in scattering. The changes in the factors contained in (4–9.21) and (4–9.29) are indicated by

$$\int_0^1 dv\, v^2(1 - v^2)(a + \tfrac{3}{2}b) \rightarrow -2\int_0^1 dv\, v^2(1 - v^2)(a + \tfrac{3}{2}b) + \int_0^1 dv\, v^2(1 - v^2)c,$$

$$\tfrac{1}{2}\int_0^1 dv\, v^2(1 - v^2)b \rightarrow -\int_0^1 dv\, v^2(1 - v^2)b + \int_0^1 dv\, v^2(1 - v^2)c, \quad (4\text{–}9.53)$$

and the corresponding numerical values are

$$\mathscr{F}^2: \quad \frac{7}{90} \rightarrow -\frac{7}{45} + \frac{1}{3} = 2\frac{4}{45},$$

$$\mathscr{G}^2: \quad \frac{1}{90} \rightarrow -\frac{1}{45} + \frac{1}{3} = 2\frac{7}{45}. \quad (4\text{–}9.54)$$

These are indeed the coefficients exhibited in (4–8.81).

To produce a space-time extrapolation for arbitrary polarization vector assignments, we observe that the two vector combinations of (4–9.50) can be written as follows,

$$\mathbf{e}_1 \cdot \mathbf{e}_1'\mathbf{e}_2 \cdot \mathbf{e}_2' - \mathbf{e}_1 \cdot \mathbf{e}_2'\mathbf{e}_1' \cdot \mathbf{e}_2 + \mathbf{e}_1 \cdot \mathbf{e}_2\mathbf{e}_1' \cdot \mathbf{e}_2' = \mathbf{e}_1 \cdot \mathbf{e}_1'\mathbf{e}_2 \cdot \mathbf{e}_2' + (\mathbf{e}_1 \times \mathbf{e}_1') \cdot (\mathbf{e}_2 \times \mathbf{e}_2'),$$

$$2(\mathbf{e}_1 \cdot \mathbf{e}_2'\mathbf{e}_1' \cdot \mathbf{e}_2 - \mathbf{e}_1 \cdot \mathbf{e}_1'\mathbf{e}_2 \cdot \mathbf{e}_2') = (\mathbf{e}_1 \cdot \mathbf{e}_1'\mathbf{e}_2 \cdot \mathbf{e}_2' + \mathbf{e}_1 \cdot \mathbf{e}_2\mathbf{e}_1' \cdot \mathbf{e}_2'$$

$$+ \mathbf{e}_1 \cdot \mathbf{e}_2'\mathbf{e}_1' \cdot \mathbf{e}_2) - 3\mathbf{e}_1 \cdot \mathbf{e}_1'\mathbf{e}_2 \cdot \mathbf{e}_2'$$

$$- (\mathbf{e}_1 \times \mathbf{e}_1') \cdot (\mathbf{e}_2 \times \mathbf{e}_2').$$

$$(4\text{–}9.55)$$

The first of these will be recognized as the polarization vector structure of $\mathscr{F}_1(x)\mathscr{F}_2(x') + \mathscr{G}_1(x)\mathscr{G}_2(x')$, while the second refers to the combination of (4–9.36). The latter does not contribute to either of the elastic polarization vector assignments, and its space-time extrapolation requires a contact term that avoids a low frequency contribution from this coupling. The result is

$$W_{04,\,\text{spin}\,\tfrac{1}{2}} = -2W_{04,\,\text{spin}\,0} + 4\alpha^2 \int_{(2m)^2}^{\infty} \frac{dM^2}{(M^2)^2}\left(1 - \frac{4m^2}{M^2}\right)^{1/2} \int \frac{(dk)}{(2\pi)^4}\frac{1}{k^2 + M^2 - i\varepsilon}$$

$$\times \left[2c(M^2)(\mathscr{F}(-k)\mathscr{F}(k) + \mathscr{G}(-k)\mathscr{G}(k)) - d(M^2)\frac{k^2}{M^2}(F^2(-k)^{\mu\nu}F^2(k)_{\mu\nu}\right.$$

$$\left. - 6\mathscr{F}(-k)\mathscr{F}(k) - 2\mathscr{G}(-k)\mathscr{G}(k))\right], \quad (4\text{–}9.56)$$

which has the equivalent form

$$
\begin{aligned}
W_{04,\,\text{spin}\frac{1}{2}} = {}& -2W_{04,\,\text{spin}\,0} + \frac{1}{3}\frac{\alpha^2}{m^4}\int\frac{(dk)}{(2\pi)^4}\,(\mathscr{F}(-k)\mathscr{F}(k) + \mathscr{G}(-k)\mathscr{G}(k)) \\
& -4\alpha^2\int_{(2m)^2}^{\infty}\frac{dM^2}{(M^2)^3}\left(1 - \frac{4m^2}{M^2}\right)^{1/2}\int\frac{(dk)}{(2\pi)^4}\frac{k^2}{k^2 + M^2 - i\varepsilon} \\
& \times [2c(M^2)(\mathscr{F}(-k)\mathscr{F}(k) + \mathscr{G}(-k)\mathscr{G}(k)) + d(M^2)(F^2(-k)^{\mu\nu}F^2(k)_{\mu\nu} \\
& -6\mathscr{F}(-k)\mathscr{F}(k) - 2\mathscr{G}(-k)\mathscr{G}(k))].
\end{aligned}
\tag{4-9.57}
$$

4-10 SCATTERING OF LIGHT BY LIGHT III.
DOUBLE SPECTRAL FORMS

We come at last to the general situation of photon-photon scattering. To produce the required coupling the following causal situation is considered. An extended photon source, J_2, emits a pair of charged particles. Each of these particles is individually deflected by extended photon sources operating in a space-like sense, J_a, J_b, and finally the two particles are detected by an extended photon source, J_1. The fields associated with the four sources do not overlap, and $A_a + A_b$ is causally intermediate between A_1 and A_2. Accordingly, the spin 0 vacuum amplitude deduced from (4-8.33) has the form

$$
\tfrac{1}{2}\,\mathrm{Tr}[2eqpA_1\varDelta_+2eqpA_a\varDelta_+2eqpA_2\varDelta_+2eqpA_b\varDelta_+] + (a\leftrightarrow b),
\tag{4-10.1}
$$

where account has been taken of the four equivalent positions in which A_1, for example, can be placed. The required sequence is such that A_a and A_b occur between A_1 and A_2, corresponding to the causal order. The special feature of this causal arrangement is that all four propagation functions describe real particles. We shall write the fields of the four extended sources as

$$
A(x) = \int\frac{(dk)}{(2\pi)^4}\exp(ikx)\,A(k),
\tag{4-10.2}
$$

with the associated vacuum amplitude taking the form

$$
\int\frac{(dk_1)}{(2\pi)^4}\frac{(dk_a)}{(2\pi)^4}\frac{(dk_b)}{(2\pi)^4}\frac{(dk_2)}{(2\pi)^4}(2\pi)^4\delta(k_1 + k_a + k_b + k_2)e^4 I.
\tag{4-10.3}
$$

The structure of the scalar integral I is indicated by

$$
I = \int\left(\prod_1^4 d\omega\right)\left(\prod_1^4 2pA\right)\left(\prod_1^3 (2\pi)^4\delta\right),
\tag{4-10.4}
$$

since there are four invariant momentum space measures for that number of real particles, four vector potentials representing the action of the related sources, and three delta function factors that establish momentum conservation at the corresponding interactions. The fourth such delta function is already exhibited in (4–10.3). It is more convenient to replace the three-dimensional momentum measures by restricted four-dimensional ones,

$$d\omega \to \frac{(dp)}{(2\pi)^3} \delta(p^2 + m^2), \qquad p^0 > 0, \tag{4–10.5}$$

and then use the delta functions to eliminate all but one of the particle momenta. The choice of the latter is arbitrary, and the possibilities are illustrated by

$$I = \int (dp)\, \delta((p + \tfrac{1}{2}k_2)^2 + m^2)\, \delta((-p + \tfrac{1}{2}k_2)^2 + m^2)\, \delta((p + \tfrac{1}{2}k_2 + k_a)^2 + m^2)$$

$$\times\, \delta((-p + \tfrac{1}{2}k_2 + k_b)^2 + m^2)(2p + k_a - k_b)A_1(k_1)(2p + k_2 + k_a)A_a(k_a)$$

$$\times\, (2p - k_2 - k_b)A_b(k_b)2pA_2(k_2) + (a \leftrightarrow b). \tag{4–10.6}$$

Here the variable p is not a particle momentum; the actual particle momenta are displayed in the various delta functions. The momentum factors that accompany the vector potentials are produced by adding the two relevant particle momenta, each multiplied by the appropriate charge sign factor. This procedure removes the gauge restriction that is implicit in the combination $2pA$.

The four delta functions in (4–10.6) can also be written as

$$\delta(2pk_2)\, \delta(p^2 + \tfrac{1}{4}k_2^2 + m^2)\, \delta(2pk_a + k_2k_a + k_a^2)\, \delta(-2pk_b + k_2k_b + k_b^2). \tag{4–10.7}$$

They supply four conditions on the components of p, which determine this vector (almost) completely. In the rest frame of the time-like vector k_2, where $k_2^0 = M_2$, we have $p^0 = 0$, according to the first factor in (4–10.7). The second one fixes the magnitude of the momentum $\mathbf{p}$,

$$|\mathbf{p}|^2 = \tfrac{1}{4}M_2^2 - m^2. \tag{4–10.8}$$

The other two delta functions determine components of $\mathbf{p}$ along the directions of $\mathbf{k}_a$ and $\mathbf{k}_b$, which vectors define a plane. Then the magnitude, but not the sign, of the p-component perpendicular to this plane is supplied by (4–10.8). If a real value does not appear, the integral vanishes. Since p is uniquely determined, apart from the sign ambiguity we have noted, the essential integration process in (4–10.6) refers entirely to the delta functions. One can easily state the value of this invariant integral in the special coordinate system where $\mathbf{k}_a$, $\mathbf{k}_b$ occupy the xy-plane and $\mathbf{k}_a$ coincides with the x-axis,

$$I_0 = \int (dp)\left(\prod_1^4 \delta\right) = \frac{1}{8}\frac{1}{|M_2 k_{ax}k_{by}p_z|}. \tag{4–10.9}$$

The reference to a special coordinate system is removed in

$$I_0 = \frac{1}{8} \frac{1}{(-\Delta)^{1/2}},$$

(4-10.10)

where

$$(-\Delta)^{1/2} = |\varepsilon_{\kappa\lambda\mu\nu} k_2^{\kappa} k_a^{\lambda} k_b^{\mu} p^{\nu}|.$$

(4-10.11)

The symbol $(-\Delta)^{1/2}$ anticipates the possibility of giving the determinant that appears in (4–10.11) an explicitly invariant form by squaring and using the determinantal multiplication property (Gram determinant). There is, however, one pitfall associated with the indefinite Minkowski metric which we can avoid by expressing all vectors in Euclidean form ($V_4 = iV^0$). Since this introduces an explicit factor of i in the determinant of (4–10.11), we see that

$$\Delta = \det \begin{pmatrix} k_2^2, & k_2 k_a, & k_2 k_b, & k_2 p \\ \vdots & \vdots & \vdots & \vdots \\ p k_2, & p k_a, & p k_b, & p^2 \end{pmatrix}.$$

(4-10.12)

This formula does not necessarily provide the easiest way to construct Δ, however.

A general expression for the vector p, which satisfies explicitly the constraint $k_2 p = 0$, is stated in

$$p^{\mu} = a\left(k_a - k_2 \frac{k_2 k_a}{k_2^2}\right)^{\mu} + b\left(k_b - k_2 \frac{k_2 k_b}{k_2^2}\right)^{\mu} + c\varepsilon^{\mu\nu\kappa\lambda} k_{a\nu} k_{b\kappa} k_{2\lambda}.$$

(4-10.13)

The last term gives a covariant form to the p_z component. Multiplication by the vectors k_a and k_b supplies the equations

$$-\tfrac{1}{2}(k_2 k_a + k_a^2) = a\left(k_a^2 - \frac{(k_2 k_a)^2}{k_2^2}\right) + b\left(k_a k_b - \frac{k_2 k_a k_2 k_b}{k_2^2}\right),$$

$$\tfrac{1}{2}(k_2 k_b + k_b^2) = a\left(k_a k_b - \frac{k_2 k_a k_2 k_b}{k_2^2}\right) + b\left(k_b^2 - \frac{(k_2 k_b)^2}{k_2^2}\right),$$

(4-10.14)

which determine the coefficients a and b. To find c, we square the combination (4–10.13). In doing so, one encounters

$$a^2\left(k_a^2 - \frac{(k_2 k_a)^2}{k_2^2}\right) + 2ab\left(k_a k_b - \frac{k_2 k_a k_2 k_b}{k_2^2}\right) + b^2\left(k_b^2 - \frac{(k_2 k_b)^2}{k_2^2}\right)$$

$$= -\tfrac{1}{2}a(k_2 k_a + k_a^2) + \tfrac{1}{2}b(k_2 k_b + k_b^2),$$

(4-10.15)

and also

$$(\varepsilon^{\mu\nu\kappa\lambda} k_{a\nu} k_{b\kappa} k_{2\lambda})(\varepsilon_{\mu\nu'\kappa'\lambda'} k_a^{\nu'} k_b^{\kappa'} k_2^{\lambda'}) = -k_2^2 D,$$

(4-10.16)

where D is the determinant of the coefficient array in (4–10.14):

$$D = \left(k_a^2 - \frac{(k_2 k_a)^2}{k_2^2}\right)\left(k_b^2 - \frac{(k_2 k_b)^2}{k_2^2}\right) - \left(k_a k_b - \frac{k_2 k_a k_2 k_b}{k_2^2}\right)^2$$

$$= k_a^2 k_b^2 - (k_a k_b)^2 - \frac{1}{k_2^2}[k_a^2 (k_2 k_b)^2 + k_b^2 (k_2 k_a)^2 - 2k_a k_b k_2 k_a k_2 k_b]. \quad (4\text{–}10.17)$$

The condition for the nonvanishing of the integral $I_0(p_z^2 > 0)$ reads

$$- k_2^2 D c^2 = \tfrac{1}{2}a(k_2 k_a + k_a^2) - \tfrac{1}{2}b(k_2 k_b + k_b^2) - m^2 - \tfrac{1}{4}k_2^2 > 0; \quad (4\text{–}10.18)$$

it has been left in implicit form for compactness. Another useful result is produced by multiplying (4–10.13) with p_μ. This reproduces the structure realized by squaring (4–10.13), with the exception that the c^2 term is replaced by one linear in c. The comparison of the two supplies the relation [which is also contained in (4–10.9)]

$$- \varDelta = (k_2^2 D c)^2 = (-k_2^2 D)(-k_2^2 D c^2). \quad (4\text{–}10.19)$$

The latter form combines (4–10.16) and (4–10.18) to give an alternative evaluation of $\varDelta$.

The causal arrangement under consideration refers entirely to extended photon sources, with k_1, k_2 time-like, and k_a, k_b space-like vectors. After the corresponding space-time coupling has been established, an extrapolation will be made to the situation of interest, where $k_1^2 = k_a^2 = k_b^2 = k_2^2 = 0$. We shall illustrate the algebraic relations just discussed by utilizing these kinematical simplifications appropriate to real photons. The two variables needed to discuss the photon-photon collision are conveniently chosen as

$$M_a^2 = - (k_2 + k_a)^2 \rightarrow - 2k_2 k_a,$$

$$M_b^2 = - (k_2 + k_b)^2 \rightarrow - 2k_2 k_b, \quad (4\text{–}10.20)$$

where the second version refers to the real photon situation. Some quantities derived for this circumstance are

$$- k_2^2 D \rightarrow - \tfrac{1}{4}M_a^2 M_b^2 (M_a^2 + M_b^2),$$

$$a \rightarrow - \tfrac{1}{2}\frac{M_b^2}{M_a^2 + M_b^2}, \qquad b \rightarrow \tfrac{1}{2}\frac{M_a^2}{M_a^2 + M_b^2}, \quad (4\text{–}10.21)$$

where we have used

$$k_1^2 = - (k_a + k_b + k_2)^2 \rightarrow 0 \quad (4\text{–}10.22)$$

to evaluate

$$2k_a k_b \rightarrow M_a^2 + M_b^2. \quad (4\text{–}10.23)$$

The positiveness criterion (4–10.18) now becomes

$$\frac{1}{4}\frac{M_a^2 M_b^2}{M_a^2 + M_b^2} > m^2, \tag{4–10.24}$$

and, with the aid of (4–10.19), one obtains

$$\Delta = \tfrac{1}{4}M_a^2 M_b^2 (M_a^2 + M_b^2)\left(\frac{1}{4}\frac{M_a^2 M_b^2}{M_a^2 + M_b^2} - m^2\right). \tag{4–10.25}$$

Of course, the same result is derived from the determinant of (4–10.12),

$$\cdot\,\Delta = \det\begin{pmatrix} 0 & -\tfrac{1}{2}M_a^2 & -\tfrac{1}{2}M_b^2 & 0 \\ -\tfrac{1}{2}M_a^2 & 0 & \tfrac{1}{2}(M_a^2 + M_b^2) & \tfrac{1}{4}M_a^2 \\ -\tfrac{1}{2}M_b^2 & \tfrac{1}{2}(M_a^2 + M_b^2) & 0 & -\tfrac{1}{4}M_b^2 \\ 0 & \tfrac{1}{4}M_a^2 & -\tfrac{1}{4}M_b^2 & -m^2 \end{pmatrix}, \tag{4–10.26}$$

but the work seems more ponderous. The introduction of the variables

$$u_{a,b} = 1 - \frac{4m^2}{M_{a,b}^2}, \tag{4–10.27}$$

which range between 0 and 1, converts the inequality (4–10.24) into

$$u_a + u_b > 1, \tag{4–10.28}$$

and the expression for Δ becomes

$$\Delta = (4m^4)^2\,\frac{u_a + u_b - 1}{(1 - u_a)^2 (1 - u_b)^2}. \tag{4–10.29}$$

The momentum combinations appearing in (4–10.20),

$$K_a = k_2 + k_a = -(k_1 + k_b),$$
$$K_b = k_2 + k_b = -(k_1 + k_a), \tag{4–10.30}$$

are associated with two different ways of considering the causal coupling among the sources in terms of two-particle exchanges ($M_{a,b} > 2m$). Sources $J_2 + J_a$ exchange a pair of real particles with sources $J_1 + J_b$, and $J_2 + J_b$ exchange a pair of particles with $J_1 + J_a$. We shall make these momenta K_a and K_b explicit by introducing the factor

$$1 = \int (dK_a)(dK_b)\delta(K_a - k_2 - k_a)\delta(K_b - k_2 - k_b)$$

$$= \int (d\xi_a)(d\xi_b)\frac{(dK_a)}{(2\pi)^4}\frac{(dK_b)}{(2\pi)^4}\exp[i(K_a - k_2 - k_a)\xi_a]\exp[i(K_b - k_2 - k_b)\xi_b]. \tag{4–10.31}$$

It is combined with the total momentum delta function,

$$(2\pi)^4 \, \delta(k_1 + k_a + k_b + k_2) = \int (dx) \exp[i(k_1 + k_a + k_b + k_2)x], \quad \text{(4–10.32)}$$

to give

$$(2\pi)^4\delta(k_1 + k_a + k_b + k_2)$$

$$= \int (dx)(d\xi_a)(d\xi_b) \frac{dM_a^2}{2\pi} \frac{dM_b^2}{2\pi} \, d\omega_a \, d\omega_b \exp(iK_a\xi_a) \exp(iK_b\xi_b)$$

$$\times \exp(ik_1 x) \exp[ik_a(x - \xi_a)] \exp[ik_b(x - \xi_b)] \exp[ik_2(x - \xi_a - \xi_b)]. \quad \text{(4–10.33)}$$

The causal significance of this form becomes apparent on considering the structure

$$\int \frac{(dk_1)}{(2\pi)^4} \cdots \frac{(dk_2)}{(2\pi)^4} \, (2\pi)^4\delta(k_1 + \cdots + k_2)A_1(k_1) \cdots A_2(k_2)$$

$$= -\int \frac{dM_a^2}{2\pi} \frac{dM_b^2}{2\pi} \int (dx)(d\xi_a)(d\xi_b)A_1(x)A_a(x - \xi_a)A_b(x - \xi_b)A_2(x - \xi_a - \xi_b)$$

$$\times \left[i \int d\omega_a \exp(iK_a\xi_a) \right] \left[i \int d\omega_b \exp(iK_b\xi_b) \right]. \quad \text{(4–10.34)}$$

According to the causal arrangement of the fields, the vectors ξ_a and ξ_b are time-like, with positive time components. We recognize the appropriate causal forms of the propagation functions

$$\Delta_+(\xi_{a,b}, M_{a,b}^2) = \int \frac{(dk)}{(2\pi)^4} \frac{\exp(ik\xi_{a,b})}{k^2 + M_{a,b}^2 - i\varepsilon}$$

$$= i \int d\omega_{a,b} \exp(iK_{a,b}\xi_{a,b}), \qquad \xi_{a,b}^0 > 0. \quad \text{(4–10.35)}$$

Although other details must be added, involving questions of gauge invariance and contact terms, this is the essence of the space-time extrapolation process. As a convenient way of expressing the result, we return to four-dimensional momentum space and state the space-time extrapolation of (4–10.34) in the double spectral form

$$-\int \frac{(dk_1)}{(2\pi)^4} \cdots \frac{(dk_2)}{(2\pi)^4} \, (2\pi)^4\delta(k_1 + \cdots + k_2)A_1(k_1) \cdots A_2(k_2)$$

$$\times \int \frac{dM_a^2}{2\pi} \frac{dM_b^2}{2\pi} \frac{1}{(k_2 + k_a)^2 + M_a^2 - i\varepsilon} \frac{1}{(k_2 + k_b)^2 + M_b^2 - i\varepsilon}. \quad \text{(4–10.36)}$$

It is instructive to apply what we have learned to a simplified scalar field problem, in which we compare a noncausal calculation of the coupling, analogous to (4–10.1), with the double spectral form evaluation. This will be done only in the limit as all photon momenta approach zero. The phrase "noncausal calculation" refers to the direct use of the propagation function in its four-dimensional form, as distinguished from the causal calculation leading to the double spectral form. The alternative computations are displayed on opposite sides of the equation

$$\int \frac{(dp)}{(2\pi)^4} \frac{1}{(p^2 + m^2 - i\varepsilon)^4} = -\int \frac{dM_a^2 \, dM_b^2}{2\pi} \frac{1}{2\pi} \frac{1}{M_a^2} \frac{1}{M_b^2} \frac{1}{8} \frac{1}{(-\Delta)^{1/2}} \cdot \quad (4\text{–}10.37)$$

The left side can be evaluated in several ways. One can transform to a Euclidean metric ($p_0 = ip_4$), and then perform the single radial momentum integral, utilizing the surface area of a unit sphere in four dimensions, $2\pi^2$,

$$i \frac{2\pi^2}{(2\pi)^4} \int_0^\infty \frac{\tfrac{1}{2} p^2 \, dp^2}{(p^2 + m^2)^4} = i \frac{1}{(4\pi)^2} \frac{1}{6m^4} \cdot \quad (4\text{–}10.38)$$

Or, we can exploit the technique introduced in (4–8.34), which is used here in the form

$$\frac{1}{(p^2 + m^2 - i\varepsilon)^4} = \frac{1}{3!} \int_0^\infty ds \, s^3 \exp[- is(p^2 + m^2)], \quad (4\text{–}10.39)$$

and then apply the integral (4–8.57) to get

$$\int \frac{(dp)}{(2\pi)^4} \frac{1}{(p^2 + m^2 - i\varepsilon)^4} = - i \frac{1}{(4\pi)^2} \frac{1}{6} \int_0^\infty ds \, s \exp(- ism^2) = i \frac{1}{(4\pi)^2} \frac{1}{6m^4} \cdot (4\text{–}10.40)$$

When we turn to the other side of Eq. (4–10.37), an important aspect of the extrapolation procedure is brought to our attention. In the causal arrangement that produced the integral (4–10.9, 10), the quantity Δ is necessarily negative. But after the extrapolation to real photons has been performed, Δ has become positive, as exhibited in (4–10.25). This poses the question of which square root of -1 to adopt, in evaluating $(-\Delta)^{1/2}$. The comparison of the two sides in (4–10.37) shows that

$$(-\Delta)^{1/2} = i\Delta^{1/2}. \quad (4\text{–}10.41)$$

We complete this test by verifying the numerical equality of the two sides. In the notation of (4–10.27), this is stated as

$$\frac{1}{8} \int \frac{du_a \, du_b}{(u_a + u_b - 1)^{1/2}} = \frac{1}{6}, \quad (4\text{–}10.42)$$

where u_a and u_b range over the interval between 0 and 1, subject to the positiveness restriction that is evident in the denominator of the integral. Performing the integrations successively, we get

$$\frac{1}{4}\int du_a \, d(u_a + u_b - 1)^{1/2} = \frac{1}{4}\int_0^1 du_a \, u_a^{1/2} = \frac{1}{6}, \qquad (4\text{-}10.43)$$

as required.

We must now turn to the vector potential factors in (4–10.6) and the problem of making explicit the gauge invariance that does hold in the causal arrangement, in order to preserve it after space-time extrapolation. Since our interest in photon-photon scattering is primarily didactic, we shall avoid the relatively unrewarding complexities of arbitrary polarizations by considering only the simplest polarization assignment. It is the one in which all polarization vectors are parallel to each other, and perpendicular to the plane of scattering. In that circumstance the vector potential factors of (4–10.6) become

$$(2p_z)^4 \prod_1^4 A = \left(\frac{M_a^2 M_b^2}{M_a^2 + M_b^2} - 4m^2\right)^2 A_1 A_a A_b A_2, \qquad (4\text{-}10.44)$$

where the various functions are the single nonvanishing vector components, those perpendicular to the scattering plane. To introduce field strengths, let us consider the product

$$F^{\mu\lambda}(k)F_\lambda{}^\nu(k') = (ik^\mu A^\lambda(k) - ik^\lambda A^\mu(k))(ik'_\lambda A^\nu(k') - ik'^\nu A_\lambda(k'))$$

$$= kk' A^\mu(k)A^\nu(k') + k^\mu k'^\nu A(k)A(k'), \qquad (4\text{-}10.45)$$

where $k'A(k)$ and $kA(k')$ have been set equal to zero, as is appropriate to the situation where the spatial polarization vectors are perpendicular to the plane of scattering. One way to use this relation, in order to give gauge invariant expression to the product of four vector potentials, is indicated by

$$(F(k)F(k'))(F(k'')F(k''')) = (kk')(k''k''')A(k)A(k')A(k'')A(k'''), \quad (4\text{-}10.46)$$

which employs the notation

$$(F(k)F(k')) = \tfrac{1}{2}F^{\mu\nu}(k)F_{\mu\nu}(k'). \qquad (4\text{-}10.47)$$

The same product of four single-component potentials can be provided with different gauge invariant interpretations depending upon the pairing of field strengths. Here are examples:

$$A_1 A_a A_b A_2 = \frac{(F_1 F_a)(F_b F_2)}{(\tfrac{1}{2}M_b^2)^2} = \frac{(F_1 F_b)(F_a F_2)}{(\tfrac{1}{2}M_a^2)^2}$$

$$= \frac{(F_1 F_2)(F_a F_b) - (F_1 F_a)(F_b F_2) - (F_1 F_b)(F_a F_2)}{\tfrac{1}{2}M_a^2 M_b^2}. \qquad (4\text{-}10.48)$$

All agree in the initial causal situation, but differ after the space-time extrapolation has been performed. The three field structures in (4–10.48) can be expressed generally as the product $A_1 \cdots A_2$, multiplied by momentum factors which are, respectively,

$$\left[-\frac{(k_2 + k_b)^2}{M_b{}^2} \right]^2, \quad \left[-\frac{(k_2 + k_a)^2}{M_a{}^2} \right]^2, \quad \left[-\frac{(k_2 + k_a)^2}{M_a{}^2} \right]\left[-\frac{(k_2 + k_b)^2}{M_b{}^2} \right]. \quad (4\text{–}10.49)$$

When used in the context of the double spectral form (4–10.36), the various possibilities differ by single spectral forms, as illustrated in

$$\frac{(k^2/M^2)^2}{(k^2 + M^2)(k'^2 + M'^2)} - \frac{(k^2/M^2)(k'^2/M'^2)}{(k^2 + M^2)(k'^2 + M'^2)}$$

$$= \frac{k^2}{M^2}\left[\frac{1}{M^2}\frac{1}{k'^2 + M'^2} - \frac{1}{M'^2}\frac{1}{k^2 + M^2} \right]. \quad (4\text{–}10.50)$$

This element of arbitrariness in enforcing gauge invariance indicates that added single spectral forms can be present, thus requiring additional physical information. That information is forthcoming in the single spectral form appropriate to forward scattering, which was derived in the preceding section.

The results obtained thus far are united in the following action expression:

$$W_{04} = \alpha^2 \int \frac{(dk)}{(2\pi)^4} \cdots \frac{(dk''')}{(2\pi)^4}\ (2\pi)^4 \delta(k + \cdots + k''')A(k) \cdots A(k''')$$

$$\times \left[\int \frac{dM_a{}^2 dM_b{}^2}{M_a{}^2 M_b{}^2} \phi_2(M_a{}^2, M_b{}^2)\frac{(k' + k''')^2}{(k' + k''')^2 + M_a{}^2 - i\varepsilon}\frac{(k'' + k''')^2}{(k'' + k''')^2 + M_b{}^2 - i\varepsilon} \right.$$

$$\left. + \int \frac{dM^2}{(M^2)^2} \phi_1(M^2)\frac{[(k' + k''')^2]^2}{(k' + k''')^2 + M^2 - i\varepsilon} \right], \quad (4\text{–}10.51)$$

where

$$\text{spin 0:} \quad \phi_2(M_a{}^2, M_b{}^2) = \frac{1}{8\Delta^{1/2}}\left(\frac{M_a{}^2 M_b{}^2}{M_a{}^2 + M_b{}^2} - 4m^2 \right)^2$$

$$= \frac{1}{2}\frac{(1 - u_a)(1 - u_b)(u_a + u_b - 1)^{3/2}}{(2 - u_a - u_b)^2}. \quad (4\text{–}10.52)$$

We have elected to use the most symmetrical of the gauge invariant representations detailed in (4–10.48, 49). It has also been recognized that, owing to the symmetry in k', k'', and those arising from the equalities

$$(k' + k''')^2 = (k + k'')^2, \qquad (k'' + k''')^2 = (k + k')^2, \qquad (4\text{–}10.53)$$

the initial causal process is contained in eight equivalent terms of (4–10.51). The single spectral integral, with its as yet unknown weight factor $\phi_1(M^2)$, is the momentum version of the form seen in (4–9.19), with field strengths replaced by vector potentials in accordance with

$$(F(k)F(k')) = -\tfrac{1}{2}(k+k')^2 A(k)A(k').$$ (4-10.54)

The four photon momenta obey the relation

$$(k+k''')^2 + (k'+k''')^2 + (k''+k''')^2 = 0.$$ (4-10.55)

In the situation of forward scattering, one of the three combinations vanishes, and the other two are equal in magnitude but opposite in sign. This leads to the following effective substitution,

$$\frac{(k'+k''')^2}{(k'+k''')^2 + M_a{}^2 - i\varepsilon} \frac{(k''+k''')^2}{(k''+k''')^2 + M_b{}^2 - i\varepsilon}$$

$$\rightarrow -\frac{1}{M_a{}^2 + M_b{}^2} \frac{[(k'+k''')^2]^2}{(k'+k''')^2 + M_a{}^2 - i\varepsilon},$$ (4-10.56)

which is verified by comparing the three terms produced on each side by symmetrization in k, k', k'', while utilizing the symmetry of ϕ_2 in $M_a{}^2$, $M_b{}^2$. The identification of the resulting single spectral form with (4–9.19) then gives

$$\phi_1(M^2) - M^2 \int \frac{dM'^2}{M'^2} \frac{\phi_2(M^2, M'^2)}{M^2 + M'^2}$$

$$= \frac{1}{2}\left(1 - \frac{4m^2}{M^2}\right)^{1/2}\left(a + \frac{3}{2}b\right)(M^2)$$

$$= \tfrac{1}{4}v + \tfrac{3}{8}v(1-v^2) + \tfrac{1}{4}(1-v^2)(\tfrac{3}{4}(1-v^2)-1)\log\frac{1+v}{1-v}.$$ (4-10.57)

Performing the necessary integration, we deduce that $(u = v^2)$

$$\text{spin 0:}\quad \phi_1(M^2) = \tfrac{1}{2}v + (1-v^2)\left(2v - \log\frac{1+v}{1-v}\right),$$ (4-10.58)

which completes the specification of the spectral forms in (4–10.51).

To produce the spin $\tfrac{1}{2}$ counterpart of these considerations, we have only to make the replacement

$$\prod_1^4 2pA \rightarrow -\tfrac{1}{2}\operatorname{tr}[(2pA_1 + \sigma F_1)(2pA_a + \sigma F_a)(2pA_2 + \sigma F_2)(2pA_b + \sigma F_b)],$$ (4-10.59)

where the multiplication order [Eq. (4–10.1)] reflects the causal arrangement. The right side of (4–10.59) is calculated as

$$- 2[(2p_z)^4 A_1 \cdots A_2 + (2p_z)^2 T_1 + T_2], \qquad (4\text{-}10.60)$$

in which

$$T_1 = \sum A A \tfrac{1}{4} \operatorname{tr} \sigma F \sigma F = A_1 A_a(F_2 F_b) + A_2 A_b(F_1 F_a) + A_1 A_b(F_2 F_a)$$
$$+ A_2 A_a(F_1 F_b) + A_1 A_2(F_a F_b) + A_a A_b(F_1 F_2) \qquad (4\text{-}10.61)$$

and

$$T_2 = \tfrac{1}{4} \operatorname{tr}(\sigma F_1 \sigma F_a \sigma F_2 \sigma F_b) = (F_1 F_a)(F_2 F_b) + (F_1 F_b)(F_2 F_a) - (F_1 F_2)(F_a F_b).$$
$$(4\text{-}10.62)$$

The latter assertion, which is analogous to (4–9.49), refers specifically to the situation under consideration where all electric fields are perpendicular to the magnetic fields. In the causal arrangement the value of T_1 is

$$T_1 = - [(k_2 + k_a)^2 + (k_2 + k_b)^2 + (k_a + k_b)^2] A_1 \cdots A_2 = 0, \quad (4\text{-}10.63)$$

while

$$T_2 = \tfrac{1}{4}[((k_2 + k_a)^2)^2 + ((k_2 + k_b)^2)^2 - ((k_a + k_b)^2)^2] A_1 \cdots A_2$$
$$= - \tfrac{1}{2} M_a{}^2 M_b{}^2 A_1 \cdots A_2. \qquad (4\text{-}10.64)$$

Accordingly, the weight factor of the double spectral form is derived from that of spin 0 by the substitution

$$(2p_z)^4 \rightarrow - 2(2p_z)^4 + M_a{}^2 M_b{}^2, \qquad (4\text{-}10.65)$$

and

$$\text{spin } \tfrac{1}{2}: \quad \phi_2(M_a{}^2, M_b{}^2) = \tfrac{1}{2}(u_a + u_b - 1)^{-1/2} - \frac{(1 - u_a)(1 - u_b)(u_a + u_b - 1)^{3/2}}{(2 - u_a - u_b)^2}.$$
$$(4\text{-}10.66)$$

The discussion is completed by using forward scattering information in the manner of Eq. (4–10.57), but applied to the convenient combinations

$$\phi'_1(M^2) = \phi_1(M^2)_{\text{spin } \frac{1}{2}} + 2\phi_1(M^2)_{\text{spin } 0},$$

$$\phi'_2(M_a{}^2, M_b{}^2) = \phi_2(M_a{}^2, M_b{}^2)_{\text{spin } \frac{1}{2}} + 2\phi_2(M_a{}^2, M_b{}^2)_{\text{spin } 0} = \tfrac{1}{2}(u_a + u_b - 1)^{-1/2},$$
$$(4\text{-}10.67)$$

and the analogous forward scattering amplitude combination [cf. Eq. (4–9.53)]

$$[- 2(a + \tfrac{3}{2}b) + c] + 2(a + \tfrac{3}{2}b) = c(M^2) = \frac{1}{v} \log \frac{1 + v}{1 - v}. \qquad (4\text{-}10.68)$$

Performing the integration in

$$\phi'_1(M^2) - M^2 \int \frac{dM'^2}{M'^2} \frac{\phi'_2(M^2, M'^2)}{M^2 + M'^2} = \frac{1}{2} \log \frac{1 + v}{1 - v}, \tag{4-10.69}$$

we get

$$\phi'_1(M^2) = \log \frac{1 + v}{1 - v} \tag{4-10.70}$$

and therefore

$$\text{spin } \tfrac{1}{2}: \quad \phi_1(M^2) = v + (3 - 2v^2)\left(\log \frac{1 + v}{1 - v} - 2v\right). \tag{4-10.71}$$

For the actual application to photon-photon scattering in the center of mass frame, with total energy M and scattering angle θ, the three momentum combinations that appear in (4–10.55) are, individually,

$$- M^2, \quad M^2 \sin^2\tfrac{1}{2}\theta, \quad M^2 \cos^2\tfrac{1}{2}\theta. \tag{4-10.72}$$

The T matrix element is

$$\langle 1_{k_1} 1_{k_1'} | T | 1_{k_2} 1_{k_2'} \rangle = (d\omega_{k_1} \cdots d\omega_{k_2'})^{1/2} 8\alpha^2 t, \tag{4-10.73}$$

with

$$t = \int \frac{dM_a^2}{M_a^2} \frac{dM_b^2}{M_b^2} \phi_2(M_a^2, M_b^2) \left[- \frac{M^2}{M_a^2 - M^2 - i\varepsilon} \left(\frac{M^2 \sin^2\tfrac{1}{2}\theta}{M^2 \sin^2\tfrac{1}{2}\theta + M_b^2} \right. \right.$$
$$\left. + \frac{M^2 \cos^2\tfrac{1}{2}\theta}{M^2 \cos^2\tfrac{1}{2}\theta + M_b^2} \right) + \frac{M^2 \sin^2\tfrac{1}{2}\theta}{M^2 \sin^2\tfrac{1}{2}\theta + M_a^2} \frac{M^2 \cos^2\tfrac{1}{2}\theta}{M^2 \cos^2\tfrac{1}{2}\theta + M_b^2} \right]$$
$$+ \int \frac{dM'^2}{(M'^2)^2} \phi_1(M'^2) \left[\frac{M^4}{M'^2 - M^2 - i\varepsilon} + \frac{M^4 \sin^4\tfrac{1}{2}\theta}{M^2 \sin^2\tfrac{1}{2}\theta + M'^2} \right.$$
$$\left. + \frac{M^4 \cos^4\tfrac{1}{2}\theta}{M^2 \cos^2\tfrac{1}{2}\theta + M'^2} \right], \tag{4-10.74}$$

and the differential cross section is computed as

$$\frac{d\sigma}{d\Omega} = \frac{\alpha^4}{\pi^2} \frac{1}{M^2} |t|^2. \tag{4-10.75}$$

Concerning the details of the angular distribution we shall only remark that at low energies, $M \ll m$, the scattering amplitude t is proportional to

$$1 + \sin^4\tfrac{1}{2}\theta + \cos^4\tfrac{1}{2}\theta = \tfrac{1}{2}(3 + \cos^2\theta), \tag{4-10.76}$$

while at high energies, $M \gg m$, there is a logarithmic variation with scattering angle.

4–11 H-PARTICLE ENERGY DISPLACEMENTS. NONRELATIVISTIC DISCUSSION

Occasional reference has been made to the energy level displacements of bound systems. There is even an explicit formula stated [Eq. (4–4.84)] in which, however, the precise meaning of an atomic excitation energy is left open. Unlike the scattering discussed in Section 4–4, which is of interest primarily at the high energies that are attainable experimentally, the problem of energy displacements is dominated by the characteristic low energies of bound systems. Accordingly, we shall initiate the discussion of H-particle energy displacements with an instructive nonrelativistic treatment.

There has been frequent consideration of nonrelativistic limits to relativistic dynamical derivations. This time we shall work directly with nonrelativistic dynamics, although, in doing so, we utilize without extended explanation the procedures that have already received their general space-time description. In the nonrelativistic circumstances conveyed by the energy expression

$$p^0 \cong m + T(\mathbf{p}), \qquad T(\mathbf{p}) = \mathbf{p}^2/2m, \tag{4–11.1}$$

the propagation function $\Delta_+(x - x')$ has the causal forms $(x^0 = t)$

$$x^0 > x^{0'}: \quad \Delta_+(x - x') \cong i\frac{\exp[-im(x^0 - x^{0'})]}{2m}\int \frac{(d\mathbf{p})}{(2\pi)^3}$$
$$\times \exp\{i[\mathbf{p}\cdot(\mathbf{r} - \mathbf{r}') - T(\mathbf{p})(t - t')]\},$$

$$x^0 < x^{0'}: \quad \Delta_+(x - x') \cong i\frac{\exp[im(x^0 - x^{0'})]}{2m}\int \frac{(d\mathbf{p})}{(2\pi)^3}$$
$$\times \exp\{-i[\mathbf{p}\cdot(\mathbf{r} - \mathbf{r}') - T(\mathbf{p})(t - t')]\}. \tag{4–11.2}$$

The nonrelativistic origin of energy is introduced by multiplying $\Delta_+(x - x')$ with $\exp[im(x^0 - x^{0'})]$. (The alternative of multiplication by $\exp[-im(x^0 - x^{0'})]$ is equivalent. It interchanges the roles of particle and antiparticle, which are given separate descriptions nonrelativistically.) In the limit that m is regarded as an arbitrarily large energy, we reach the nonrelativistic form of the propagation function:

$$G(\mathbf{r} - \mathbf{r}', t - t') = \text{Lim}(-2m)\exp[im(x^0 - x^{0'})]\Delta_+(x - x')$$
$$= \begin{cases} t > t': & -i\int \frac{(d\mathbf{p})}{(2\pi)^3}\exp\{i[\mathbf{p}\cdot(\mathbf{r} - \mathbf{r}') - T(\mathbf{p})(t - t')]\}, \\ t < t': & 0. \end{cases}$$

$$\tag{4–11.3}$$

This retarded function is a Green's function of the inhomogeneous Schrödinger equation,

$$\left[i \frac{\partial}{\partial t} - \frac{1}{2m}\left(\frac{1}{i}\boldsymbol{\nabla}\right)^2\right] G(\mathbf{r} - \mathbf{r}', t - t') = \delta(\mathbf{r} - \mathbf{r}')\delta(t - t'). \tag{4-11.4}$$

The momentum space version of this equation is

$$\left[i \frac{\partial}{\partial t} - T(\mathbf{p})\right] G(\mathbf{p}, t - t') = \delta(t - t'), \tag{4-11.5}$$

where

$$G(\mathbf{p}, t - t') = \int (d\mathbf{r}) \exp[-i\mathbf{p}\cdot(\mathbf{r} - \mathbf{r}')]\, G(\mathbf{r} - \mathbf{r}', t - t'). \tag{4-11.6}$$

The general time transform function,

$$G(\ , E) = \int_{-\infty}^{\infty} dt \exp[iE(t - t')]\, G(\ , t - t'), \tag{4-11.7}$$

obeys equations appropriate to the choice of spatial variables:

$$\left[E - \frac{1}{2m}\left(\frac{1}{i}\boldsymbol{\nabla}\right)^2\right] G(\mathbf{r} - \mathbf{r}', E) = \delta(\mathbf{r} - \mathbf{r}'), \tag{4-11.8}$$

or

$$[E - T(\mathbf{p})]G(\mathbf{p}, E) = 1. \tag{4-11.9}$$

Since the Green's function is retarded, in time, the transform function (4–11.7) exists for complex values of E that are confined to the upper half-plane, Im $E > 0$. Accordingly, the appropriate solution of (4–11.9) for real E is produced by approaching the real axis from the half-plane of regularity:

$$G(\mathbf{p}, E) = \frac{1}{E + i\varepsilon - T(\mathbf{p})}, \qquad \varepsilon \to +0. \tag{4-11.10}$$

One verifies directly that the implied time behavior is that of a retarded function,

$$G(\mathbf{p}, t - t') = \int_{-\infty}^{\infty} \frac{dE}{2\pi} \exp[-iE(t - t')] \frac{1}{E + i\varepsilon - T(\mathbf{p})}$$

$$= \eta(t - t')(1/i) \exp[-iT(\mathbf{p})(t - t')]. \tag{4-11.11}$$

An action expression that incorporates the inhomogeneous Schrödinger equation is given by

$$W(\eta^*, \eta) = \int (d\mathbf{r})\, dt \left[-\eta^*(\mathbf{r}, t)\psi(\mathbf{r}, t) - \psi^*(\mathbf{r}, t)\eta(\mathbf{r}, t)\right.$$

$$+ \psi^*(\mathbf{r}, t)\left(i\,\frac{\partial}{\partial t} - \frac{1}{2m}\left(\frac{1}{i}\boldsymbol{\nabla}\right)^2\right)\psi(\mathbf{r}, t)\bigg].$$ (4–11.12)

The implied field equations are

$$\left[i\,\frac{\partial}{\partial t} - \frac{1}{2m}\left(\frac{1}{i}\boldsymbol{\nabla}\right)^2\right]\psi(\mathbf{r}, t) = \eta(\mathbf{r}, t),$$

$$\left[-i\,\frac{\partial}{\partial t} - \frac{1}{2m}\left(-\frac{1}{i}\boldsymbol{\nabla}\right)^2\right]\psi^*(\mathbf{r}, t) = \eta^*(\mathbf{r}, t).$$ (4–11.13)

As the notation indicates, these equations are in complex conjugate relationship. The field equation for ψ is solved by

$$\psi(\mathbf{r}, t) = \int (d\mathbf{r}')\, dt'\, G(\mathbf{r} - \mathbf{r}', t - t')\eta(\mathbf{r}', t').$$ (4–11.14)

One uses this solution to find the explicit source dependence of W,

$$W(\eta^*, \eta) = -\int (d\mathbf{r})\, dt\, \eta^*(\mathbf{r}, t)\psi(\mathbf{r}, t)$$

$$= -\int (d\mathbf{r})\, dt\, (d\mathbf{r}')\, dt'\, \eta^*(\mathbf{r}, t)G(\mathbf{r} - \mathbf{r}', t - t')\eta(\mathbf{r}', t').$$ (4–11.15)

The minus sign that appears here, in contrast to the relativistic forms, reflects the sign factor contained in the relation of (4–11.3); it has been introduced in order to conform with nonrelativistic conventions concerning Green's functions. The corresponding action principle definition of fields is

$$\delta W(\eta^*, \eta) = -\int (d\mathbf{r})\, dt\, [\delta\eta^*(\mathbf{r}, t)\psi(\mathbf{r}, t) + \psi^*(\mathbf{r}, t)\,\delta\eta(\mathbf{r}, t)].$$ (4–11.16)

It implies the additional field expression

$$\psi^*(\mathbf{r}', t') = \int (d\mathbf{r})\, dt\, \eta^*(\mathbf{r}, t)G(\mathbf{r} - \mathbf{r}', t - t').$$ (4–11.17)

Notice that this is not the complex conjugate of $\psi(\mathbf{r}', t')$. The field ψ is related to its source η at earlier times, while ψ^* is linked to values of η^* at later times. Alternative choices of variables enable the explicit expression of W, for example, to be given such forms as

$$W(\eta^*, \eta) = -\int \frac{(d\mathbf{p})}{(2\pi)^3}\, dt\, dt'\, \eta^*(\mathbf{p}, t)G(\mathbf{p}, t - t')\eta(\mathbf{p}, t'),$$ (4–11.18)

with

$$\eta(\mathbf{p}, t) = \int (d\mathbf{r}) \exp(-i\mathbf{p} \cdot \mathbf{r}) \, \eta(\mathbf{r}, t), \qquad \eta^*(\mathbf{p}, t) = \int (d\mathbf{r}) \exp(i\mathbf{p} \cdot \mathbf{r}) \, \eta^*(\mathbf{r}, t),$$

$$(4\text{–}11.19)$$

and

$$W(\eta^*, \eta) = -\int \frac{(d\mathbf{p}) \, dE}{(2\pi)^4} \, \eta^*(\mathbf{p}, E) G(\mathbf{p}, E) \eta(\mathbf{p}, E), \qquad (4\text{–}11.20)$$

in which

$$\eta(\ , E) = \int dt \, \exp(iEt) \, \eta(\ , t), \qquad \eta^*(\ , E) = \int dt \, \exp(-iEt) \, \eta^*(\ , t). \quad (4\text{–}11.21)$$

Electromagnetic interactions are introduced by the radiation gauge substitutions

$$\frac{1}{i} \boldsymbol{\nabla} \rightarrow \frac{1}{i} \boldsymbol{\nabla} + e\mathbf{A}(\mathbf{r}, t), \qquad \boldsymbol{\nabla} \cdot \mathbf{A}(\mathbf{r}, t) = 0$$

$$i \frac{\partial}{\partial t} \rightarrow i \frac{\partial}{\partial t} - V(\mathbf{r}, t), \qquad V(\mathbf{r}, t) = -eA^0(\mathbf{r}, t), \qquad (4\text{–}11.22)$$

which are written for a particle of charge $-e$. In the radiation gauge attention focuses on the vector photon source $\mathbf{J}$. After removing the instantaneous Coulomb interaction of the charge density J^0 [cf. Eq. (3–15.51)], we can eliminate this time component with the aid of the conservation condition

$$\boldsymbol{\nabla} \cdot \mathbf{J}(\mathbf{r}, t) + \frac{\partial}{\partial t} J^0(\mathbf{r}, t) = 0. \qquad (4\text{–}11.23)$$

Thus, the photon contribution to W can be exhibited in the dyadic form [cf. Eqs. (3–15.52, 53)]

$$W(\mathbf{J}) = \tfrac{1}{2} \int (d\mathbf{r}) \, dt \, (d\mathbf{r}') \, dt' \, \mathbf{J}(\mathbf{r}, t) \cdot \mathbf{D}(\mathbf{r} - \mathbf{r}', t - t') \cdot \mathbf{J}(\mathbf{r}', t'), \quad (4\text{–}11.24)$$

where

$$t > t': \quad \mathbf{D}(\mathbf{r} - \mathbf{r}', t - t') = i \int d\omega_k \exp\{i[\mathbf{k} \cdot (\mathbf{r} - \mathbf{r}')$$

$$- k^0(t - t')]\} \left(1 - \frac{\mathbf{k}\mathbf{k}}{(k^0)^2}\right). \qquad (4\text{–}11.25)$$

The nonrelativistic situations we shall consider are such that the momentum carried by the photon is relatively negligible, which is to say that the typical photon wavelength is large in comparison with the spatial dimensions of the system, $|\mathbf{k} \cdot (\mathbf{r} - \mathbf{r}')| \ll 1$. Then the photon propagation function can be simplified to (a factor of e^2 is also supplied)

$$t > t': \quad e^2 \mathbf{D}(t - t') \cong i \frac{2\alpha}{3\pi} \int dk^0 \, k^0 \exp[- ik^0(t - t')] \, \mathbf{1}. \tag{4-11.26}$$

In arriving at this expression we have introduced spherical coordinates for the $\mathbf{k}$ space,

$$d\omega_k = \frac{|\mathbf{k}|^2 \, d|\mathbf{k}| \, d\Omega}{(2\pi)^3 2k^0} = \frac{1}{4\pi^2} \frac{d\Omega}{4\pi} |\mathbf{k}| \, dk^0, \tag{4-11.27}$$

and performed the angular integrations,

$$\mathbf{k}\mathbf{k} \to \tfrac{1}{3} |\mathbf{k}|^2 \, \mathbf{1}, \tag{4-11.28}$$

after which the photon momentum property,

$$|\mathbf{k}| = k^0, \tag{4-11.29}$$

was used.

First let us derive some known nonrelativistic results concerning modified propagation functions and form factors. The action term

$$- \int (d\mathbf{r}) \, dt \, \psi^*(\mathbf{r}, t) \frac{e}{m} \mathbf{p} \cdot \mathbf{A}(\mathbf{r}, t) \psi(\mathbf{r}, t), \qquad \mathbf{p} = - i\boldsymbol{\nabla}, \tag{4-11.30}$$

characterizes an extended particle source $\eta_2(\mathbf{r}, t)$ as the effective two-particle emission source

$$i\mathbf{J}_2(\mathbf{r}', t')\eta_2(\mathbf{r}, t)|_{\text{eff.}} = \delta(\mathbf{r} - \mathbf{r}') \, \delta(t - t') \frac{e}{m} \mathbf{p}\psi_2(\mathbf{r}, t), \tag{4-11.31}$$

and similarly for an extended absorption source $\eta^*_1(\mathbf{r}, t)$,

$$i\eta^*_1(\mathbf{r}, t)\mathbf{J}_1(\mathbf{r}', t')|_{\text{eff.}} = \psi^*_1(\mathbf{r}, t) \frac{e}{m} \mathbf{p} \, \delta(\mathbf{r} - \mathbf{r}') \, \delta(t - t'). \tag{4-11.32}$$

It is actually only the transverse parts of these vectors that are effective, but these are selected automatically by the photon propagation function (4-11.25). The vacuum amplitude term that represents the exchange of a photon and a free particle is

$$- \int (d\mathbf{r}_1) \, dt_1 \cdots i\eta^*_1(\mathbf{r}_1, t_1)\mathbf{J}_1(\mathbf{r}'_1, t'_1)|_{\text{eff.}} \cdot \mathbf{D}(\mathbf{r}'_1 - \mathbf{r}'_2, t'_1 - t'_2)G(\mathbf{r}_1 - \mathbf{r}_2, t_1 - t_2)$$

$$\cdot i\mathbf{J}_2(\mathbf{r}'_2, t'_2)\eta_2(\mathbf{r}_2, t_2)|_{\text{eff.}}$$

$$= - \frac{e^2}{m^2} \int (d\mathbf{r}) \, dt \, (d\mathbf{r}') \, dt' \, \psi^*_1(\mathbf{r}, t) \mathbf{p} \cdot \mathbf{D}(\mathbf{r} - \mathbf{r}', t - t') G(\mathbf{r} - \mathbf{r}', t - t') \cdot \mathbf{p} \psi_2(\mathbf{r}', t').$$

$$(4\text{-}11.33)$$

The use of the simplified photon propagation function (4–11.26), and of the particle propagation function (4–11.3), converts this vacuum amplitude into

$$- i \, \frac{2\alpha}{3\pi} \frac{1}{m^2} \int dk^0 \, k^0 \int \frac{(d\mathbf{p})}{(2\pi)^3} \, dt \, dt' \, \psi^*_1(\mathbf{p}, t) \frac{1}{i} \exp\{- i[T(\mathbf{p}) + k^0](t - t')\} \, \mathbf{p}^2 \psi_2(\mathbf{p}, t').$$

$$(4\text{-}11.34)$$

We recognize in $(1/i) \exp(- i[T(\mathbf{p}) + k^0](t - t'))$ the $t > t'$ expression for the propagation function of a particle that has the energy $T(\mathbf{p}) + k^0$. Its general form will be given by (4–11.11), modified appropriately by the substitution $T \rightarrow T + k^0$, to which can be added time contact terms, that is, $\delta(t - t')$ and a finite number of its derivatives. Accordingly, the time extrapolation of (4–11.34) is

$$- i \, \frac{2\alpha}{3\pi} \frac{1}{m^2} \int dk^0 \, k^0 \int \frac{(d\mathbf{p}) \, dE}{(2\pi)^4} \, \psi^*(\mathbf{p}, E)$$

$$\times \left[\frac{1}{E + i\varepsilon - T(\mathbf{p}) - k^0} + \text{contact terms} \right] \mathbf{p}^2 \psi(\mathbf{p}, E), \quad (4\text{-}11.35)$$

where the contact terms now appear as a polynomial in the energy parameter E. A sufficient form for them is fixed by the requirement that this additional coupling should refer to sources, rather than fields, to avoid altering the initial description of the free particle. The needed factors, those displayed in

$$(E - T(\mathbf{p}))\psi(\mathbf{p}, E) = \eta(\mathbf{p}, E), \qquad \psi^*(\mathbf{p}, E)(E - T(\mathbf{p})) = \eta^*(\mathbf{p}, E), \quad (4\text{-}11.36)$$

are produced by the combination

$$\frac{1}{E + i\varepsilon - T - k^0} + \frac{1}{k^0} + \frac{E - T}{(k^0)^2} = \left(\frac{E - T}{k^0} \right)^2 \frac{1}{E + i\varepsilon - T - k^0}. \quad (4\text{-}11.37)$$

The modified propagation function is then obtained as

$$\bar{G}(\mathbf{p}, E) = \frac{1}{E + i\varepsilon - T(\mathbf{p})} + \frac{2\alpha}{3\pi} \frac{\mathbf{p}^2}{m^2} \int \frac{dk^0}{k^0} \frac{1}{E + i\varepsilon - T(\mathbf{p}) - k^0}. \quad (4\text{-}11.38)$$

One will recognize here the structure of the soft photon result that is stated in Eq. (4–1.83) for spin 0, and in Eq. (4–1.95) for spin $\frac{1}{2}$.

The motion of the particle in a static potential $V(\mathbf{r})$ is represented, initially, by altering the Green's function differential equation to

$$\left[i \, \frac{\partial}{\partial t} - V(\mathbf{r}) - \frac{1}{2m} \left(\frac{1}{i} \boldsymbol{\nabla} \right)^2 \right] G^V(\mathbf{r}, \mathbf{r}', t - t') = \delta(\mathbf{r} - \mathbf{r}')\delta(t - t'). \quad (4\text{-}11.39)$$

An equivalent integral equation, presented in an abstract notation, is

$$G^V = G + GVG^V, \tag{4-11.40}$$

which has the formal solution

$$G^V = (1 - GV)^{-1}G = G + GVG + \cdots . \tag{4-11.41}$$

A more explicit statement of these first terms is given by

$$G^V(\mathbf{p}, t; \mathbf{p}', t') = \delta(\mathbf{p} - \mathbf{p}')G(\mathbf{p}, t - t') + \int dt_1\, G(\mathbf{p}, t - t_1) V(\mathbf{p} - \mathbf{p}')$$
$$\times G(\mathbf{p}', t_1 - t') + \cdots , \tag{4-11.42}$$

where

$$V(\mathbf{p} - \mathbf{p}') = \int (d\mathbf{r})\, \exp[- i(\mathbf{p} - \mathbf{p}') \cdot \mathbf{r}]\, V(\mathbf{r}). \tag{4-11.43}$$

The modified description of motion in the potential V is produced by using the propagation function G^V in Eq. (4–11.33).

Let us consider first the linear V term in the expansion of (4–11.41). It represents the effect of single scattering by the potential in a causal arrangement involving two extended particle sources that exchange a particle and a photon. The observation that

$$\exp[- ik^0(t - t')] \int dt_1 \exp[- iT(\mathbf{p})(t - t_1)]\, V(\mathbf{p} - \mathbf{p}')\, \exp[- iT(\mathbf{p}')(t_1 - t')]$$
$$= \int dt_1 \exp[- i(T(\mathbf{p}) + k^0)(t - t_1)]\, V(\mathbf{p} - \mathbf{p}')\, \exp[- i(T(\mathbf{p}') + k^0)(t_1 - t')] \tag{4-11.44}$$

leads directly to the extrapolated form of the vacuum amplitude:

$$- i\, \frac{2\alpha}{3\pi}\, \frac{1}{m^2} \int dk^0\, k^0\, \frac{(d\mathbf{p})}{(2\pi)^3}\, \frac{(d\mathbf{p}')}{(2\pi)^3}\, \frac{dE}{2\pi}\, \psi^*(\mathbf{p}, E) \left[\mathbf{p} \cdot \frac{1}{E + i\varepsilon - T(\mathbf{p}) - k^0}\, V(\mathbf{p} - \mathbf{p}') \right.$$
$$\left. \times \frac{1}{E + i\varepsilon - T(\mathbf{p}') - k^0}\, \mathbf{p}' - \left(\frac{1}{k^0}\right)^2 \frac{1}{2}\, (\mathbf{p}^2 V(\mathbf{p} - \mathbf{p}') + V(\mathbf{p} - \mathbf{p}')\mathbf{p}'^2) \right] \psi(\mathbf{p}', E). \tag{4-11.45}$$

A contact term, which is independent of E, has also been added. Its existence is implied by the requirement that a constant potential V produce only a displacement of the energy origin, $E \to E - V$. When this substitution is introduced in (4–11.35, 37), and the terms linear in the constant V are selected, we recognize the counterparts of the two contributions in (4–11.45) where a symmetrization has been used to give general meaning to the product $\mathbf{p}^2 V$.

In the application of the vacuum amplitude (4–11.45) to single scattering by the potential, the fields obey

$$\psi^*(\mathbf{p}, E)(E - T(\mathbf{p})) = 0, \qquad (E - T(\mathbf{p}'))\psi(\mathbf{p}', E) = 0 \qquad (4\text{–}11.46)$$

and the vacuum amplitude reduces to

$$i\, \frac{\alpha}{3\pi}\, \frac{1}{m^2} \int \frac{dk^0}{k^0}\, \frac{(d\mathbf{p})}{(2\pi)^3}\, \frac{(d\mathbf{p}')}{(2\pi)^3}\, \frac{dE}{2\pi}\, \psi^*(\mathbf{p}, E)(\mathbf{p} - \mathbf{p}')^2 V(\mathbf{p} - \mathbf{p}')\psi(\mathbf{p}', E). \qquad (4\text{–}11.47)$$

The potential effective in scattering the particle from momentum $\mathbf{p}'$ to momentum $\mathbf{p}$ is thus given by

$$\left[1 - \frac{\alpha}{3\pi}\, \frac{(\mathbf{p} - \mathbf{p}')^2}{m^2} \int \frac{dk^0}{k^0} \right] V(\mathbf{p} - \mathbf{p}'). \qquad (4\text{–}11.48)$$

We recognize the essential structure of the charge form factor (4–4.81). The comparison is sharpened by introducing into the spectral integral $\int dk^0/k^0$ an upper limit K, which represents the boundary of applicability of the nonrelativistic treatment, and a lower limit associated with a finite photon mass, μ. The effect of the latter, as contained in the altered momentum

$$|\mathbf{k}| = ((k^0)^2 - \mu^2)^{1/2}, \qquad (4\text{–}11.49)$$

appears in Eqs. (4–11.25, 27), and brings about the modification

$$\int \frac{dk^0}{k^0} \to \int \frac{|\mathbf{k}|\, dk^0}{(k^0)^2}\, \frac{3}{2} \left(1 - \frac{1}{3}\, \frac{|\mathbf{k}|^2}{(k^0)^2} \right)$$

$$= \int_\mu^K \frac{dk^0}{k^0} \left(1 - \frac{\mu^2}{(k^0)^2} \right)^{1/2} \left(1 + \frac{1}{2}\, \frac{\mu^2}{(k^0)^2} \right)$$

$$\cong \log \frac{2K}{\mu} - \frac{5}{6}, \qquad \frac{K}{\mu} \gg 1; \qquad (4\text{–}11.50)$$

the integral is performed conveniently by using the transformation

$$k^0 = \mu \cosh \theta. \qquad (4\text{–}11.51)$$

Now the comparison with the spin $\tfrac{1}{2}$ charge form factor (4–4.81) produces a precise limit to the nonrelativistic discussion,

$$\text{spin } \tfrac{1}{2}: \quad \log \frac{2K}{m} = \frac{5}{6} - \frac{3}{8} = \frac{11}{24}. \qquad (4\text{–}11.52)$$

For spin 0 particles, according to the relation (4–4.132) and its consequence

$$\int_{(2m)^2}^\infty \frac{dM^2}{(M^2)^2} (f_1(M^2) - f(M^2)) = \frac{\alpha}{8\pi}\, \frac{1}{m^2} \int_0^1 dv = \frac{\alpha}{8\pi}\, \frac{1}{m^2}, \qquad (4\text{–}11.53)$$

this is replaced by

$$\text{spin } 0: \quad \log \frac{2K}{m} = \frac{5}{6} - \frac{3}{4} = \frac{1}{12}. \tag{4-11.54}$$

The situation in which an unlimited number of interactions with the potential V can occur is represented by the use of the complete Green's function G^V in (4–11.33). Alternative integral equations obeyed by the function are given by

$$G^V = G + GVG^V = G + G^VVG, \tag{4-11.55}$$

according to the algebraic equivalence stated in

$$G^V = (1 - GV)^{-1}G = G(1 - VG)^{-1}. \tag{4-11.56}$$

The two equations are combined in

$$G^V = G + GVG + GVG^VVG, \tag{4-11.57}$$

which gives an exact form to the remainder after two terms of the expansion in powers of V. The formal treatment of this remainder term is straightforward. Multiplication of G^V by the photon propagation function effectively adds the photon energy to the kinetic or total energy of the particle, and there are no further physical normalization conditions to require the presence of contact terms. We present the complete result as an additional action term,

$$-\int (d\mathbf{r}) \, dt \, (d\mathbf{r}') \, dt' \, \psi^*(\mathbf{r}, t) \, \delta V(\mathbf{r}, t; \mathbf{r}', t') \psi(\mathbf{r}', t'), \tag{4-11.58}$$

where, written in a matrix notation,

$$
\begin{aligned}
\delta V = \frac{2\alpha}{3\pi} \frac{1}{m^2} \int dk^0 \, k^0 \left[\mathbf{p}^2 \left(\frac{1}{E + i\varepsilon - T - k^0} + \frac{1}{k^0} + \frac{E - T}{(k^0)^2} \right) \right. \\
+ \mathbf{p} \cdot \frac{1}{E + i\varepsilon - T - k^0} \, V \, \frac{1}{E + i\varepsilon - T - k^0} \, \mathbf{p} - \left(\frac{1}{k^0} \right)^2 \frac{1}{2} \left(\mathbf{p}^2 V + V\mathbf{p}^2 \right) \\
\left. + \mathbf{p} \cdot \frac{1}{E + i\varepsilon - T - k^0} \, V \, \frac{1}{E + i\varepsilon - H - k^0} \, V \, \frac{1}{E + i\varepsilon - T - k^0} \mathbf{p} \right],
\end{aligned}
\tag{4-11.59}
$$

and

$$H = T + V. \tag{4-11.60}$$

This additional interaction will be applied to bound states which obey, initially, the eigenvalue equations

$$\psi^*(E - H) = 0, \quad (E - H)\psi = 0. \tag{4-11.61}$$

The use of these equations in (4–11.58, 59) produces the cancellation of the $\mathbf{p}^2(E - T)$ and $-\frac{1}{2}(\mathbf{p}^2 V + V\mathbf{p}^2)$ terms. A further simplification results from the algebraic identity [essentially (4–11.55)],

$$\frac{1}{E + i\varepsilon - T - k^0} = \left[1 - \frac{1}{E + i\varepsilon - T - k^0}V\right]\frac{1}{E + i\varepsilon - H - k^0}$$

$$= \frac{1}{E + i\varepsilon - H - k^0}\left[1 - V\frac{1}{E + i\varepsilon - T - k^0}\right], \quad (4\text{–}11.62)$$

which, applied to the first term of (4–11.59), makes explicit the linear dependence upon the potential V. We write this version as

$$\delta V = \delta V^{(1)} + \delta V^{(2)}, \quad (4\text{–}11.63)$$

where

$$\delta V^{(1)} = \frac{\alpha}{3\pi}\frac{1}{m^2}\int dk^0 \left[\mathbf{p}^2 \frac{1}{E + i\varepsilon - T - k^0}V + V\frac{1}{E + i\varepsilon - T - k^0}\mathbf{p}^2\right.$$

$$\left. + 2k^0\mathbf{p}\cdot\frac{1}{E + i\varepsilon - T - k^0}V\frac{1}{E + i\varepsilon - T - k^0}\mathbf{p}\right] \quad (4\text{–}11.64)$$

and

$$\delta V^{(2)} = \frac{2\alpha}{3\pi}\frac{1}{m^2}\int dk^0\, k^0\mathbf{p}\cdot\frac{1}{E + i\varepsilon - T - k^0}V\frac{1}{E + i\varepsilon - H - k^0}V\frac{1}{E + i\varepsilon - T - k^0}\mathbf{p}. \quad (4\text{–}11.65)$$

Note that, in the approximation of single scattering, where the field equations (4–11.61) are replaced by (4–11.46), the additional scattering potential of (4–11.48) is regained from $\delta V^{(1)}$. The structure of $\delta V^{(2)}$ is well-defined nonrelativistically. Owing to the three powers of k^0 in the denominator the spectral integral converges at high energies. And, there is no infrared difficulty since $T - E$ never vanishes for bound states, where $E < 0$. Not even the existence of a zero eigenvalue for $H - E$, yielding the singular factor $-1/k^0$, is significant since the multiplicative matrix elements vanish in the limit $k^0 \to 0$:

$$\left\langle \cdots E\left|\frac{1}{m}\,\mathbf{p}\,\frac{1}{E - T}V\right|\cdots E\right\rangle = \left\langle \cdots E\left|\frac{1}{m}\,\mathbf{p}\right|\cdots E\right\rangle$$

$$= \left\langle \cdots E\left|\frac{1}{i}\,[\mathbf{r}, H]\right|\cdots E\right\rangle = 0. \quad (4\text{–}11.66)$$

All this suggests, correctly, that the $\delta V^{(2)}$ contribution is a relatively minor one. It is $\delta V^{(1)}$ that produces the major part of the energy level displacement, quite apart from the logarithmic dependence upon the large energy value K.

To facilitate the calculation of this dominant contribution we rewrite $\delta V^{(1)}$ as ($i\varepsilon$ is without effect and is omitted)

$$
\delta V^{(1)} = \frac{\alpha}{3\pi} \frac{1}{m^2} \int \frac{dk^0}{k^0} \left[- \frac{k^0}{T - E + k^0} [\mathbf{p}, [\cdot \mathbf{p}, V]] \frac{k^0}{T - E + k^0} \right.
$$

$$
- \mathbf{p}^2 \frac{k^0}{T - E + k^0} V \left(1 - \frac{k^0}{T - E + k^0} \right)
$$

$$
\left. - \left(1 - \frac{k^0}{T - E + k^0} \right) V \frac{k^0}{T - E + k^0} \mathbf{p}^2 \right]. \qquad (4\text{--}11.67)
$$

The Coulomb potential

$$
V(\mathbf{r}) = - Z\alpha/r \qquad (4\text{--}11.68)
$$

is such that

$$
- [\mathbf{p}, [\cdot \mathbf{p}, V]] = \nabla^2 V = 4\pi Z\alpha \, \delta(\mathbf{r}). \qquad (4\text{--}11.69)
$$

For a given state of interest, one with wave function $\psi(\mathbf{r})$, the delta function selects the value at the origin of an auxiliary wave function which is defined by

$$
\chi(\mathbf{r}, k^0) = \frac{k^0}{T - E + k^0} \, \psi(\mathbf{r}) \qquad (4\text{--}11.70)
$$

or

$$
\left(- \frac{1}{2m} \nabla^2 - E + k^0 \right) \chi(\mathbf{r}, k^0) = k^0 \psi(\mathbf{r}). \qquad (4\text{--}11.71)
$$

The energy displacement is given by

$$
\langle \delta V^{(1)} \rangle = \frac{\alpha}{3\pi} \frac{1}{m^2} \int \frac{dk^0}{k^0} \left[4\pi Z\alpha |\chi(0, k^0)|^2 + 4m(k^0 + |E|) \int (d\mathbf{r}) \frac{Z\alpha}{r} |\psi(\mathbf{r}) - \chi(\mathbf{r}, k^0)|^2 \right.
$$

$$
\left. - 4m|E| \, \mathrm{Re} \int (d\mathbf{r}) \psi(\mathbf{r})^* \frac{Z\alpha}{r} (\psi(\mathbf{r}) - \chi(\mathbf{r}, k^0)) \right]. \qquad (4\text{--}11.72)
$$

It is convenient to present it in terms of dimensionless variables, which are constructed from the Bohr radius

$$
a_0 = 1/(mZ\alpha) \qquad (4\text{--}11.73)
$$

and the Bohr energy values

$$
|E_n| = m \frac{Z^2\alpha^2}{2n^2} = \frac{1}{2m(na_0)^2} = \frac{Z^2}{n^2} \, \mathrm{Ry}. \qquad (4\text{--}11.74)
$$

(We have also introduced the so-called Rydberg energy unit.) The variables are

$$x = \frac{r}{na_0}, \qquad \frac{4s}{(1-s)^2} = \frac{k^0}{|E_n|}, \tag{4–11.75}$$

and the wavefunctions are correspondingly redefined by

$$\psi(x) = [\pi(na_0)^3]^{1/2}\psi(r),$$

$$\chi(x, s) = [\pi(na_0)^3]^{1/2}\chi(r, k^0), \tag{4–11.76}$$

which gives

$$\langle \delta V^{(1)} \rangle_n = \frac{8}{3\pi} \frac{Z^4 \alpha^3}{n^3} \text{Ry} \int_0^{\to 1} \frac{ds}{s} \frac{1+s}{1-s} \left[|\chi(0, s)|^2 \right.$$

$$+ \left(\frac{1+s}{1-s} \right)^2 \int \frac{(dx)}{2\pi x} |\psi(x) - \chi(x, s)|^2 - \text{Re} \int \frac{(dx)}{2\pi x} \psi(x)^* (\psi(x) - \chi(x, s)) \bigg]. \tag{4–11.77}$$

The simplest example of this calculation is provided by the $n = 1$, $l = 0$ ground state where

$$\psi_{10}(x) = \exp(-x),$$

$$\chi_{10}(x, s) = \exp(-x) - \frac{(1-s)^2 \exp(-x) - \exp\{-[(1+s)/(1-s)]x\}}{2s} \tag{4–11.78}$$

and

$$\chi_{10}(0, s) = s. \tag{4–11.79}$$

The successive integrations are exhibited in

$$\langle \delta V^{(1)} \rangle_{10} = \frac{8}{3\pi} Z^4 \alpha^3 \text{Ry} \int_0^{\to 1} \frac{ds}{s} \frac{1+s}{1-s} \left[s^2 + \frac{(1-s^2)^2}{2s^2} \log \frac{1}{1-s^2} - \frac{1}{2}(1-s)^2 \right]$$

$$= \frac{8}{3\pi} Z^4 \alpha^3 \text{Ry} \left[\log \frac{K}{Z^2 \text{Ry}} - \left(\frac{17}{4} - 2 \log 2 \right) \right], \tag{4–11.80}$$

where the result involves the number

$$\frac{17}{4} - 2 \log 2 = 2.8637. \tag{4–11.81}$$

Concerning the $\delta V^{(2)}$ contribution we note that, on introducing the complete set of wavefunctions $\psi_{Ea}(r)$, we have generally

$$\langle \delta V^{(2)} \rangle_{nl} = -\frac{2\alpha}{3\pi} \frac{1}{m^2} \int \frac{dk^0}{k^0} \sum_{Ea} \frac{|\int(dr)\chi_{nlm}(r, k^0)^* p V \psi_{Ea}(r)|^2}{E - E_n + k^0 - i\varepsilon}. \tag{4–11.82}$$

For the ground state, in particular, all the matrix elements that appear here are such that $E - E_1 > 0$, and this additional energy displacement is negative. Its numerical value[1] is indicated by the following alteration in the additive constant (4–11.81),

$$2.8637 + 0.1105 = 2.9742. \tag{4–11.83}$$

Thus, quite apart from the dominant logarithm, $\log m/|E_1| \sim 10.5$, the additive constant is fixed to within four percent by this elementary calculation.

It is advantageous, for a similar discussion of other s states, to begin with the following generating function (essentially that of the Laguerre polynomials)

$$\frac{1}{(1-t)^2} \exp\{-[(1+t)/(1-t)]x\} = \sum_{n=1}^{\infty} t^{n-1} n \psi_{n0}(x), \tag{4–11.84}$$

which has the same exponential form as the wave function $\psi_{10}(x)$, and reduces to it on placing $t = 0$. Accordingly, the generating function for the $\chi_{n0}(x, s)$,

$$\chi(x, s, t) = \sum_{n=1}^{\infty} t^{n-1} n \chi_{n0}(x, s), \tag{4–11.85}$$

also resembles $\chi_{10}(x, s)$. Produced by solving the differential equation

$$\left[-\nabla^2 + \left(\frac{1+s}{1-s}\right)^2 \right] \chi(x, s, t) = \frac{4s}{(1-s)^2} \frac{1}{(1-t)^2} \exp\{-[(1+t)/(1-t)]x\}, \tag{4–11.86}$$

this generating function is

$$\chi(x, s, t) = \frac{4s}{(1-s)^2(1-t)^2} \frac{1}{\left(\frac{1+s}{1-s}\right)^2 - \left(\frac{1+t}{1-t}\right)^2} \left[\exp\{-[(1+t)/(1-t)]x\} \right.$$

$$\left. -2 \frac{\frac{1+t}{1-t}}{\left(\frac{1+s}{1-s}\right)^2 - \left(\frac{1+t}{1-t}\right)^2} \frac{\exp\{-[(1+t)/(1-t)]x\} - \exp\{-[(1+s)/(1-s)]x\}}{x} \right]. \tag{4–11.87}$$

On evaluating it at the origin we get

---

[1] The necessary calculations have been performed most effectively with the aid of a momentum space construction of the H-particle Green's function, which is described in *Quantum Kinematics and Dynamics*, W. A. Benjamin, Inc., Menlo Park, 1970. The details can be found in the Harvard, 1967, thesis of Michael Lieber. See also *Phys. Rev.* **174**, 2037 (1968).

$$\sum_{n=1}^{\infty} t^{n-1} n \chi_{n0}(0, s) = \frac{s}{(1 - st)^2} = \frac{d}{dt} \sum_{n=1}^{\infty} (st)^n \qquad (4\text{-}11.88)$$

and therefore

$$\chi_{n0}(0, s) = s^n. \qquad (4\text{-}11.89)$$

Although it would be possible to construct generating functions for the integrals that appear in (4–11.77), we shall be content to consider only the $n = 2$ s-state. The required wave functions are

$$\psi_{20}(\mathbf{x}) = (1 - x) \exp(- x),$$

$$\chi_{20}(\mathbf{x}) - \psi_{20}(\mathbf{x}) = \frac{(1 - s)^2}{s} \left[\exp(- x) - \frac{(1 + s)^2}{2s} \right.$$

$$\left. \times \frac{\exp(- x) - \exp\{- [(1 + s)/(1 - s)]x\}}{x} \right], \quad (4\text{-}11.90)$$

and

$$\langle \delta V^{(1)} \rangle_{20} = \frac{1}{3\pi} Z^4 \alpha^3 \, \mathrm{Ry} \int_0^{\rightarrow 1} \frac{ds}{s} \frac{1 + s}{1 - s}$$

$$\times \left[s^4 + \frac{(1 - s^2)^2}{s^2} \left(\frac{(1 + s^2)^2}{2s^2} \log \frac{1}{1 - s^2} - s^2 - \frac{1}{2} \right) - \frac{1}{4}(1 - s)^2(1 + s^2) \right]$$

$$= \frac{1}{3\pi} Z^4 \alpha^3 \, \mathrm{Ry} \left[\log \frac{K}{Z^2 \, \mathrm{Ry}} - \left(\frac{899}{144} - \frac{16}{3} \log 2 \right) \right]. \qquad (4\text{-}11.91)$$

The value of the numerical constant that appears here, and of the additional effect associated with $\delta V^{(2)}$, which again involves only negative terms, is indicated by

$$2.5463 + 0.2655 = 2.8118. \qquad (4\text{-}11.92)$$

While the contribution of $\delta V^{(2)}$ has increased, relative to the $n = 1$ situation, it is still only a few percent of the complete effect.

The 1s and 2s states we have been discussing are rather special ones since they are stable. (Of course, the 2s level is not completely stable but this refinement does not appear in our physically limited treatment.) The instability of other levels is manifested in the structure of $\delta V^{(2)}$ [Eq. (4–11.82)] by the appearance of an imaginary part,

$$\mathrm{Im} \langle \delta V^{(2)} \rangle_{nl} = - \frac{2\alpha}{3} \frac{1}{m^2} \int dk^0 \sum_{Ea} \delta(E - E_n + k^0) k^0 |\langle nlm| \mathbf{p} |Ea \rangle|^2, \quad (4\text{-}11.93)$$

where the integral that occurs in (4–11.82) has been simplified with the aid of the energy restriction imposed by the delta function,

$$- \int (d\mathbf{r}) \psi_{nlm}^* \frac{k^0}{T - E_n + k^0} \mathbf{p}(T - E)\psi_{Ea} = -k^0 \langle nlm|\mathbf{p}|Ea\rangle. \quad (4\text{-}11.94)$$

We recognize the structure of the spontaneous emission probabilities per unit time, Eq. (3–15.69), which are added to give the total decay rate [cf. Eq. (3–16.41)],

$$\mathrm{Im}\langle \delta V^{(2)}\rangle_{nl} = -\tfrac{1}{2} \sum_{E < E_n, a} \gamma_{Ea \leftarrow nl}$$

$$= -\tfrac{1}{2}\gamma_{nl}. \quad (4\text{-}11.95)$$

This modification of the energy, appearing in the time propagation phase factor

$$\exp[-i(E_{nl} - \tfrac{1}{2}i\gamma_{nl})t], \quad (4\text{-}11.96)$$

correctly represents the instability of the nl state through photon emission transitions to H-particles of lesser energy. (Perhaps we should recall that the earlier discussion of H-particle instability was completely phenomenological concerning energy values. Now we are in the process of deepening our understanding of the energy level structure.)

For states of nonzero angular momentum both wave functions, $\psi(\mathbf{r})$ and $\chi(\mathbf{r}, k^0)$, vanish at the origin. That removes the major term of $\delta V^{(1)}$, Eq. (4–11.77). The residual contribution and that of $\delta V^{(2)}$ combine to give a quite small displacement. It is illustrated for the $2p$ level by

$$\mathrm{Re}\langle \delta V^{(1)} + \delta V^{(2)}\rangle_{21} = \frac{1}{3\pi} Z^4\alpha^3 \, \mathrm{Ry}(0.0300). \quad (4\text{-}11.97)$$

Relativistic effects, in the experimentally interesting example of spin $\tfrac{1}{2}$, are partly included by the relation (4–11.52), which converts (4–11.91, 92) into

$$\langle \delta V^{(1)} + \delta V^{(2)}\rangle_{20} = \frac{1}{3\pi} Z^4\alpha^3 \, \mathrm{Ry}\left[\log\frac{1}{(Z\alpha)^2} - 2.8118 + \frac{11}{24}\right]. \quad (4\text{-}11.98)$$

There is also the magnetic moment effect displayed in (4–4.124), for example. It produces the following energy displacement,

$$\frac{\alpha}{2\pi} \frac{1}{2m} \int (d\mathbf{r}) \psi^* \gamma^0 \gamma_5 \boldsymbol{\sigma} \cdot \boldsymbol{\nabla} V \psi. \quad (4\text{-}11.99)$$

The wave functions that appear here obey the Dirac equation, which can be written as

$$(m + E - V)\psi = (\gamma_5 \boldsymbol{\sigma} \cdot \boldsymbol{\nabla} + m\gamma^0)\psi, \quad (4\text{-}11.100)$$

and

$$\psi^*(m + E - V) = \psi^*(\gamma_5 \boldsymbol{\sigma} \cdot \boldsymbol{\nabla}^T + m\gamma^0),$$

$$\psi^* \boldsymbol{\nabla}^T = -\boldsymbol{\nabla}\psi^*. \quad (4\text{-}11.101)$$

These equations are combined in

$$2(m + E - V)\psi^*\gamma^0\gamma_5\boldsymbol{\sigma} \cdot \nabla V\psi$$

$$= \psi^*\gamma^0(\boldsymbol{\sigma} \cdot \nabla^T\boldsymbol{\sigma} \cdot \nabla V - \boldsymbol{\sigma} \cdot \nabla V\boldsymbol{\sigma} \cdot \nabla)\psi$$

$$= \psi^*\gamma^0(\nabla^T \cdot \nabla V - \nabla V \cdot \nabla + i\boldsymbol{\sigma} \cdot (\nabla^T \times \nabla V - \nabla V \times \nabla))\psi \quad (4\text{-}11.102)$$

and, using a nonrelativistic approximation that neglects $E - V$ relative to m, we express the energy displacement (4–11.99) as

$$\frac{\alpha}{2\pi}\frac{1}{4m^2} \langle \nabla^2 V + 2\boldsymbol{\sigma} \cdot \nabla V \times \mathbf{p} \rangle. \quad (4\text{-}11.103)$$

Comparison with (4–11.67, 69) and its multiplicative factor of $(\alpha/3\pi)(1/m^2)$ shows that the first term of (4–11.103), which is effective only in s-states, adds the constant $\frac{3}{8}$ to the dominant logarithm.

In states of nonzero orbital angular momentum and a given total angular momentum quantum number $j = l \pm \frac{1}{2}$, (4–11.103) becomes

$$\frac{Z\alpha^2}{4\pi m^2}\left\langle \frac{\boldsymbol{\sigma} \cdot \mathbf{L}}{r^3} \right\rangle_{nlj} = \frac{Z\alpha^2}{4\pi m^2}\left\langle \frac{1}{r^3} \right\rangle_{nl} \times \begin{cases} j = l + \frac{1}{2}: & l, \\ j = l - \frac{1}{2}: & -l - 1 \end{cases}. \quad (4\text{-}11.104)$$

To review the derivation of the known expectation value that appears here, one notes that the average radial force vanishes in a stationary state,

$$\left\langle \frac{d}{dr}\left(\frac{l(l+1)}{2mr^2} - \frac{Z\alpha}{r} \right) \right\rangle = -\frac{l(l+1)}{m}\left\langle \frac{1}{r^3} \right\rangle + Z\alpha\left\langle \frac{1}{r^2} \right\rangle = 0, \quad (4\text{-}11.105)$$

while the familiar linear dependence of the principal quantum number n upon the orbital quantum number l, $n = n_r + l + 1$, shows that

$$\frac{d}{dl}E_n = -\frac{2E_n}{n} = \left\langle \frac{d}{dl}\frac{l(l+1)}{2mr^2} \right\rangle = \frac{l + \frac{1}{2}}{m}\left\langle \frac{1}{r^2} \right\rangle. \quad (4\text{-}11.106)$$

This gives

$$\left\langle \frac{1}{r^3} \right\rangle_{nl} = \frac{1}{l(l+1)}\frac{1}{a_0}\left\langle \frac{1}{r^2} \right\rangle_{nl} = \frac{1}{(na_0)^3}\frac{1}{l(l+\frac{1}{2})(l+1)}, \quad (4\text{-}11.107)$$

and the energy displacement induced by the additional magnetic moment in states with $l > 0$ appears as

$$\frac{1}{\pi}Z^4\alpha^3 \text{ Ry } \frac{1}{n^3}\frac{1}{2l+1} \times \begin{cases} j = l + \frac{1}{2}: & \dfrac{1}{l+1} \\ j = l - \frac{1}{2}: & -\dfrac{1}{l} \end{cases}. \quad (4\text{-}11.108)$$

Incidentally, the previously stated result for the $s_{1/2}$ levels is also produced by this formula.

The combination of all the effects we have discussed, including that of vacuum polarization, gives the following expression for the relative displacement of the initially degenerate $2s_{1/2}$ and $2p_{1/2}$ hydrogenic levels:

$$E_{2s_{1/2}} - E_{2p_{1/2}} = \frac{1}{3\pi} Z^4 \alpha^3 \, \mathrm{Ry} \left[\log \frac{1}{Z^2 \alpha^2} - 2.8118 + \frac{5}{6} - \frac{1}{5} \right]$$

$$- \frac{1}{3\pi} Z^4 \alpha^3 \, \mathrm{Ry}[0.0300 - \tfrac{1}{8}]$$

$$= \frac{1}{3\pi} Z^4 \alpha^3 \left(\frac{M}{M+m} \right)^3 \mathrm{Ry} \left[\log \left(\frac{1}{Z^2 \alpha^2} \frac{M+m}{M} \right) - 2.8418 + \frac{91}{120} \right].$$

$$(4\text{--}11.109)$$

In writing the last line we have also introduced some of the more obvious mass corrections for a realistic H-particle. These recognize that it is the reduced mass of the electron (m) and nucleus (M) that enters the Bohr radius

$$a_0 = \frac{1}{Z\alpha} \frac{M+m}{mM} \tag{4--11.110}$$

and the Bohr energy values

$$|E_n| = \frac{Z^2 \alpha^2}{2n^2} \frac{mM}{M+m} = \frac{Z^2}{n^2} \frac{M}{M+m} \mathrm{Ry} \tag{4--11.111}$$

(the Rydberg unit for infinite mass is retained). Using the value

$$\alpha = \frac{1}{137.036} \tag{4--11.112}$$

and the energy unit (conventionally stated as a frequency)

$$\frac{1}{3\pi} \alpha^3 \, \mathrm{Ry} = 135.644 \text{ MHz}, \tag{4--11.113}$$

we deduce from (4–11.109) this frequency shift for hydrogen,

$$\mathrm{H}: \quad E_{2s_{1/2}} - E_{2p_{1/2}} = 1050.55 \text{ MHz}. \tag{4--11.114}$$

A recent measurement of the level splitting gave 1057.90 ± 0.10 MHz. The agreement to better than one percent is impressive, particularly since one can expect an improved relativistic treatment to produce additional effects of relative magnitude $\sim \alpha = 7.3 \times 10^{-3}$.

Still preoccupied with ancient history, Harold asks a question.

H.: Since you were first to state the additional magnetic moment, and the radiative correction to scattering of the electron, it would be surprising if you had not been the first to arrive at the energy displacement formula you have just rederived. Were you?

S.: I believe that I was, although at the time (1947) I was not at all convinced of the correctness of the result. Let me review some of that history which can also be traced through the various papers collected in *Selected Papers on Quantum Electrodynamics*, Dover Publications, Inc., New York, 1958. Discussions between V. Weisskopf and myself, prior to and during the famous Shelter Island Conference of June, 1947, produced agreement that relativistically calculated electrodynamic effects should give a finite splitting of hydrogenic energy levels. Shortly after, H. Bethe performed his nonrelativistic calculation, which left unsettled the value of the constant that accompanies the dominant logarithm. That month of June, 1947, held two other significant events for me. I stopped smoking, and I got married. Following an extended honeymoon tour of the country, I returned to the relativistic problem. Using the noncovariant operator field theory then in vogue, I introduced a canonical transformation that isolated the physical effects associated with an externally applied electromagnetic field. When this was chosen to be a homogeneous magnetic field the additional magnetic moment of $\alpha/2\pi$ magnetons was obtained. An inhomogeneous electric field gave just the results that are stated in (4–11.98), but the accompanying spin-orbit coupling had the wrong factor to be identified with the additional magnetic moment—relativistic invariance was violated. Inserting the right factor gives (as we now know) the correct answer, but there was no conviction then in such a procedure. Attention therefore shifted to the development of manifestly covariant calculational methods. Although they greatly reduced the labor of the computations, there was a period of confusion concerning the proper joining of high and low frequency contributions. By this time (1948) other groups, variously using noncovariant and covariant methods, had attacked the problem and arrived at a variety of answers. It was Weisskopf, by the way, who insisted on the particular number with which all eventually agreed; this was the almost forgotten result of my earlier calculation. (I still recall the shock I experienced when I happened to compare them and discovered their identity. The classification of processes in the first method was not a particularly physical one and only the final answer for the additive constant was stated, as a certain fractional number. The covariant technique, on the other hand, gave a clear physical separation of various effects which were therefore left explicit in the form of corresponding fractions. Somehow, with the memory of the noncovariant calculation resolutely suppressed, it had never occurred to me to combine all these fractions into one.) As in the history of the scattering formula, there is a moral here for us. The artificial separation of high and low frequencies, which are handled in different ways, must be avoided. We shall see that the relativistic treatment of energy displacements has this desirable feature.

Before leaving the nonrelativistic arena, however, it is interesting to follow the workings of a method which unites elastic and inelastic processes in scattering, thereby avoiding that artificial separation. The vacuum amplitude representing single particle exchange in a causal arrangement is, for a free particle,

$$- i \int (d\mathbf{r}) \cdots dt' \, \eta^*(\mathbf{r}, t) G(\mathbf{r} - \mathbf{r}', t - t') \eta(\mathbf{r}', t'), \qquad (4\text{-}11.115)$$

where

$$t > t': \quad G(\mathbf{r} - \mathbf{r}', t - t') = - i \sum_{\mathbf{p}} \psi_{\mathbf{p}}(\mathbf{r}, t) \psi_{\mathbf{p}}(\mathbf{r}', t')^*,$$

$$\psi_{\mathbf{p}}(\mathbf{r}, t) = \left[\frac{(d\mathbf{p})}{(2\pi)^3} \right]^{1/2} \exp[i(\mathbf{p} \cdot \mathbf{r} - T(\mathbf{p})t)]. \qquad (4\text{-}11.116)$$

Using the source definitions

$$\eta_{\mathbf{p}} = \int (d\mathbf{r}) \, dt \, \psi_{\mathbf{p}}(\mathbf{r}, t)^* \eta(\mathbf{r}, t), \qquad \eta_{\mathbf{p}}^* = \int (d\mathbf{r}) \, dt \, \eta^*(\mathbf{r}, t) \psi_{\mathbf{p}}(\mathbf{r}, t), \quad (4\text{-}11.117)$$

we find that the vacuum amplitude (4–11.115) is

$$\sum_{\mathbf{p}} (- i\eta_{\mathbf{p}}^*)(- i\eta_{\mathbf{p}}), \qquad (4\text{-}11.118)$$

which represents the fact that a particle, emitted with momentum $\mathbf{p}$, is detected with certainty in the same circumstance. Now let us replace G with G^V and again extract the coefficient of $(- i\eta_{\mathbf{p}}^*)(- i\eta_{\mathbf{p}})$. The absolute square of this probability amplitude states the probability that the particle persists in its initial state despite the action of the potential V. We retain only the first two terms of the expansion (4–11.41), which gives the persistence probability amplitude

$$1 - i \int (d\mathbf{r}) \cdots dt' \, \psi_{\mathbf{p}}(\mathbf{r}, t)^* [V(\mathbf{r}) \, \delta(\mathbf{r} - \mathbf{r}') \, \delta(t - t')$$

$$+ V(\mathbf{r}) G(\mathbf{r} - \mathbf{r}', t - t') V(\mathbf{r}')] \psi_{\mathbf{p}}(\mathbf{r}', t'). \qquad (4\text{-}11.119)$$

The time integrations reduce to one extended over the entire interval, T, between the emission and absorption acts. The probability amplitude then reads

$$1 - iT \frac{(d\mathbf{p})}{(2\pi)^3} \left[V(0) + \int \frac{(d\mathbf{p}')}{(2\pi)^3} \frac{V(\mathbf{p} - \mathbf{p}') V(\mathbf{p}' - \mathbf{p})}{E + i\varepsilon - T(\mathbf{p}')} \right] \qquad (4\text{-}11.120)$$

where $E = T(\mathbf{p})$ and $V(0)$ is $V(\mathbf{p} - \mathbf{p})$. From this the persistence probability is derived as

$$1 - T \frac{(d\mathbf{p})}{(2\pi)^3} (-2) \operatorname{Im} \left[V(0) + \int \frac{(d\mathbf{p}')}{(2\pi)^3} \frac{|V(\mathbf{p} - \mathbf{p}')|^2}{E + i\varepsilon - T(\mathbf{p}')} \right]$$

$$= 1 - T \frac{(d\mathbf{p})}{(2\pi)^3} \int \frac{(d\mathbf{p}')}{(2\pi)^3} 2\pi \, \delta(E - T(\mathbf{p}')) |V(\mathbf{p} - \mathbf{p}')|^2, \qquad (4\text{-}11.121)$$

which also states the total probability of a scattering process. That probability, per unit time, divided by the incident particle flux $[(d\mathbf{p})/(2\pi)^3][|\mathbf{p}|/m]$, gives the total scattering cross section

$$\sigma = \frac{m}{|\mathbf{p}|} \int \frac{(d\mathbf{p}')}{(2\pi)^3} 2\pi \, \delta(E - T(\mathbf{p}')) |V(\mathbf{p} - \mathbf{p}')|^2. \qquad (4\text{-}11.122)$$

Spherical coordinates in momentum space are introduced by

$$(d\mathbf{p}') = d\Omega |\mathbf{p}'|^2 \, d|\mathbf{p}'| = d\Omega |\mathbf{p}'| m \, dT(\mathbf{p}'), \qquad (4\text{-}11.123)$$

where $d\Omega$ is the element of solid angle. The evident implication concerning the differential cross section for elastic scattering is

$$\frac{d\sigma}{d\Omega} = \left(\frac{m}{2\pi} \right)^2 |V(\mathbf{p} - \mathbf{p}')|^2, \qquad (4\text{-}11.124)$$

which is indeed the well-known first Born approximation.

All this leads up to the analogous consideration of the scattering potential $V + \delta V$. The two parts of δV are stated in Eqs. (4–11.64), or (4–11.67) with $i\varepsilon$ reinstated, and (4–11.65) where H is to be replaced by T for our limited purpose. The probability amplitude (4–11.120) continues to apply, with the substitutions

$$V(0) \to V(0) + \frac{2\alpha}{3\pi} \frac{\mathbf{p}^2}{m^2} \int \frac{(d\mathbf{p}') \, dk^0}{(2\pi)^3 \, k^0} \frac{|V(\mathbf{p} - \mathbf{p}')|^2}{E + i\varepsilon - T(\mathbf{p}') - k^0} \qquad (4\text{-}11.125)$$

and

$$V(\mathbf{p} - \mathbf{p}') \to V(\mathbf{p} - \mathbf{p}') \left[1 + \frac{\alpha}{3\pi} \frac{1}{m^2} \int \frac{dk^0}{k^0} \left(k^0 \frac{(\mathbf{p} - \mathbf{p}')^2}{E + i\varepsilon - T(\mathbf{p}') - k^0} \right. \right.$$

$$\left. \left. - \mathbf{p}^2 \frac{E - T(\mathbf{p}')}{E + i\varepsilon - T(\mathbf{p}') - k^0} \right) \right], \qquad (4\text{-}11.126)$$

where the last factor (not its complex conjugate) also applies to $V(\mathbf{p}' - \mathbf{p})$. There is complete cancellation, in the analogue of (4–11.120), between the terms that contain $\mathbf{p}^2$ explicitly. The resulting expression for the total cross section is

$$\sigma = \frac{m}{|\mathbf{p}|} (-2) \operatorname{Im} \int \frac{(d\mathbf{p}')}{(2\pi)^3} \frac{|V(\mathbf{p} - \mathbf{p}')|^2}{E + i\varepsilon - T(\mathbf{p}')}$$

$$\times \left(1 + \frac{2\alpha}{3\pi} \frac{(\mathbf{p} - \mathbf{p}')^2}{m^2} \int_0^K dk^0 \frac{1}{E + i\varepsilon - T(\mathbf{p}') - k^0} \right), \qquad (4\text{-}11.127)$$

which form emphasizes that no singularity at $k^0 = 0$ appears. Then, to facilitate the extraction of the imaginary part, we write

$$\frac{1}{E + i\varepsilon - T(\mathbf{p}')} \frac{1}{E + i\varepsilon - T(\mathbf{p}') - k^0}$$

$$= \frac{1}{k^0} \left[\frac{1}{E + i\varepsilon - T(\mathbf{p}') - k^0} - \frac{1}{E + i\varepsilon - T(\mathbf{p}')} \right], \qquad (4\text{-}11.128)$$

and this gives the modification of the cross section in the form

$$\delta\sigma = -\frac{m}{|\mathbf{p}|} \frac{2\alpha}{3\pi} \int \frac{(d\mathbf{p}')}{(2\pi)^3} \frac{(\mathbf{p} - \mathbf{p}')^2}{m^2} |V(\mathbf{p} - \mathbf{p}')|^2 \int_0^K \frac{dk^0}{k^0} 2\pi [\delta(E - T(\mathbf{p}'))$$

$$- \delta(E - T(\mathbf{p}') - k^0)]. \qquad (4\text{-}11.129)$$

The two terms clearly exhibit the decrease in the cross section for purely elastic processes, and the increase associated with inelastic processes. The practical impossibility of distinguishing the two classes of events, as $k^0 \to 0$, is offset by the precise cancellation of the two kinds of contributions in that limit. We can use (4-11.123) to infer the modification of the differential cross section in angle,

$$\delta\left(\frac{d\sigma}{d\Omega} \right) = -\left(\frac{m}{2\pi} \right)^2 \frac{2\alpha}{3\pi} \int dT(\mathbf{p}') \frac{|\mathbf{p}'|}{|\mathbf{p}|} (\mathbf{p} - \mathbf{p}')^2 |V(\mathbf{p} - \mathbf{p}')|^2 \int_0^K \frac{dk^0}{k^0} [\delta(E - T)$$

$$- \delta(E - T - k^0)], \qquad (4\text{-}11.130)$$

which still leaves free the selection of the range of kinetic energy for the scattered particle.

Let us consider the situation of essentially elastic scattering, as characterized by

$$E - k^0_{\min} < T(\mathbf{p}') < E + 0, \qquad (4\text{-}11.131)$$

where $k^0_{\min}$ represents the accuracy with which scattered particle energies can be measured. Under the assumption that $k^0_{\min} \ll E$ we have $|\mathbf{p}'| \simeq |\mathbf{p}|$, and the integrals of (4-11.130) reduce to

$$\int_{E - k^0_{\min}}^{E + 0} dT \int_0^K \frac{dk^0}{k^0} [\delta(E - T) - \delta(E - T - k^0)]$$

$$= \int_0^K \frac{dk^0}{k^0} [1 - \eta(k^0_{\min} - k^0)] = \int_{k^0_{\min}}^K \frac{dk^0}{k^0} = \log \frac{K}{k^0_{\min}}. \qquad (4\text{-}11.132)$$

The resulting modification in the differential cross section for essentially elastic scattering is

$$\delta\left(\frac{d\sigma}{d\Omega}\right) \bigg/ \frac{d\sigma}{d\Omega} = -\frac{2\alpha}{3\pi}\frac{(\mathbf{p}-\mathbf{p}')^2}{m^2}\log\frac{K}{k^0_{\min}}. \qquad (4\text{–}11.133)$$

In the example of spin 0 particles, relativistic effects are introduced by the substitution of (4–11.54),

$$\log\frac{K}{k^0_{\min}} = \log\frac{m}{2k^0_{\min}} + \frac{1}{12} \qquad (4\text{–}11.134)$$

(we are still considering slow particles, of course), and by adding the contribution of vacuum polarization. The latter, for spin 0 particles, is $\frac{1}{5}$ of the spin $\frac{1}{2}$ contribution which alters the added constant in (4–11.134) to

$$\frac{1}{12} - \frac{1}{40} = \frac{7}{120}. \qquad (4\text{–}11.135)$$

We recognize the spin 0 result stated in Eq. (4–4.99).

4–12 A RELATIVISTIC SCATTERING CALCULATION

It has been remarked that, in processes where photons can be emitted, a decomposition into elastic and inelastic cross sections is artificial owing to the experimental impossibility of making that distinction for sufficiently soft photons. More realistic than the elastic cross section is the cross section for essentially elastic scattering, as defined by the limited experimental ability to perform the necessary energetic control. We have just described, in a nonrelativistic context, a technique for the direct calculation of the essentially elastic cross section. Now let us apply that idea to the relativistic scattering of charged particles in a Coulomb field. Since we wish to illustrate a method, rather than obtain new results, it will suffice to consider spinless particles (the spin $\frac{1}{2}$ results are then easily inferred). It is convenient, however, to modify slightly the objective of the calculation relative to that of Section 4–4. There, processes were selected in which the emitted photon had an energy less than $k^0_{\min} \ll p^0 - m$. We propose to replace this criterion for almost elasticity by a restriction on the mass of the final two-particle state:

$$m^2 \leqslant M^2 < (m + \delta M)^2, \qquad \delta M \ll m. \qquad (4\text{–}12.1)$$

Accordingly, using the method of Section 4–4, we shall first construct the essentially elastic cross section that embodies this criterion.

The probability amplitude (3–14.61) for soft photon emission yields an emission probability with the kinematical factors $d\omega_{p_1}\, d\omega_k$. We transfer attention to the total momentum

$$P = p_1 + k \tag{4-12.2}$$

by writing

$$d\omega_{p_1} = \frac{(dp_1)}{(2\pi)^3} \delta(p_1^2 + m^2) = \frac{(dP)}{(2\pi)^3} \delta((P - k)^2 + m^2)$$

$$= d\omega_P \, dM^2 \, \delta(2kP + M^2 - m^2). \tag{4-12.3}$$

Then, using the total momentum for the comparison, rather than the particle momentum, the relative probability for photon emission, replacing Eq. (4-4.93), becomes

$$e^2 \int_{(m+\mu)^2}^{(m+\delta M)^2} dM^2 \, d\omega_k \, \delta(2Pk + M^2 - m^2) \left(\frac{p_1}{kp_1} - \frac{p_2}{kp_2}\right)^2. \tag{4-12.4}$$

The lower limit of the M^2 integral recalls that the photon is temporarily assigned the mass μ. The basic photon momentum integral, evaluated in the rest frame of P and then simplified in accordance with $M^2 - m^2 < 2m \, \delta M \ll m^2$, is

$$\int d\omega_k \, \delta(2Pk + M^2 - m^2) = \int \frac{dk^0((k^0)^2 - \mu^2)^{1/2}}{(2\pi)^2} \delta(-2Mk^0 + M^2 - m^2)$$

$$= \frac{1}{(2\pi)^2} \frac{1}{2M} \left[\left(\frac{M^2 - m^2}{2M}\right)^2 - \mu^2\right]^{1/2}$$

$$\cong \frac{1}{(2\pi)^2} \frac{1}{2M} ((M - m)^2 - \mu^2)^{1/2}. \tag{4-12.5}$$

We exhibit it by rewriting the integral (4-12.4) as

$$\frac{\alpha}{\pi} \int_\mu^{\delta M} d(M - m)((M - m)^2 - \mu^2)^{1/2} \left\langle -\frac{m^2}{(kp_1)^2} - \frac{m^2}{(kp_2)^2} - \frac{2p_1 p_2}{kp_1 kp_2}\right\rangle, \tag{4-12.6}$$

where the average indicated by the expectation value notation refers to an angle. It is illustrated by

$$\left\langle \frac{1}{(kp_2)^2}\right\rangle = \frac{1}{2} \int_{-1}^{1} dz \, \frac{1}{(k^0 p_2{}^0 - |\mathbf{k}| \, |\mathbf{p}_2|z)^2} = \frac{1}{(k^0)^2((p_2{}^0)^2 - |\mathbf{p}_2|^2) + \mu^2 |\mathbf{p}_2|^2} \tag{4-12.7}$$

or

$$\left\langle \frac{1}{(kp_2)^2}\right\rangle \cong \frac{1}{m^2(M - m)^2 + \mu^2 q^2(1 + (q^2/4m^2))}. \tag{4-12.8}$$

The last evaluation has been performed in the rest frame of $P \cong p_1$, where, in the notation of Section 4-4,

$$q = p_1 - p_2, \tag{4-12.9}$$

we have

$$|\mathbf{p}_2|^2 = \left(\frac{p_1 p_2}{m}\right)^2 - m^2 = q^2\left(1 + \frac{q^2}{4m^2}\right). \tag{4-12.10}$$

To complete the computation of (4–12.4), we note that

$$\frac{1}{(kp_1)^2} \cong \frac{1}{m^2(M-m)^2}, \tag{4-12.11}$$

while

$$\frac{1}{kp_1 kp_2} = \int_{-1}^{1} \frac{1}{2}\, dv\, \frac{1}{\left[k\left(p_1\dfrac{1+v}{2} + p_2\dfrac{1-v}{2}\right)\right]^2} \tag{4-12.12}$$

leads to

$$\left\langle\frac{1}{kp_1 kp_2}\right\rangle \cong \int_{-1}^{1} \frac{1}{2}\, dv\, \frac{1}{(M-m)^2\left(m^2 + \dfrac{1-v^2}{4}q^2\right) + \mu^2\left(\dfrac{1-v}{2}\right)^2 q^2\left(1 + \dfrac{q^2}{4m^2}\right)}, \tag{4-12.13}$$

according to the relations

$$-\left(p_1\frac{1+v}{2} + p_2\frac{1-v}{2}\right)^2 = m^2 + \frac{1-v^2}{4}q^2,$$

$$-p_1\left(p_1\frac{1+v}{2} + p_2\frac{1-v}{2}\right) = m^2 + \frac{1-v}{4}q^2. \tag{4-12.14}$$

The spectral integrals that appear here are of the form [cf. Eq. (4–4.97)]

$$\int_{\mu}^{\delta M} d(M-m)((M-m)^2 - \mu^2)^{1/2}\frac{1}{(M-m)^2 + \lambda^2\mu^2}$$

$$= \log\frac{2\delta M}{\mu} - \frac{1}{2}\left(1 + \frac{1}{\lambda^2}\right)^{1/2}\log\frac{\left(1 + \dfrac{1}{\lambda^2}\right)^{1/2} + 1}{\left(1 + \dfrac{1}{\lambda^2}\right)^{1/2} - 1}, \tag{4-12.15}$$

and the resulting evaluation of (4–12.4) can be presented as

$$\frac{\alpha}{2\pi}\log\frac{2\delta M}{\mu}\frac{q^2}{m^2}\int_{0}^{1} dv\,\frac{1+v^2}{1 + \dfrac{1-v^2}{4}\dfrac{q^2}{m^2}} - \frac{\alpha}{\pi}\Theta\left(\frac{q^2}{m^2}\right). \tag{4-12.16}$$

Here, expressed in the variable θ,

$$\cosh \theta = 1 + q^2/2m^2, \tag{4-12.17}$$

we have

$$\Theta(\theta) = \cosh \theta \int_{-1}^{1} \frac{1}{2} \, dv \, \frac{1}{\xi_+\xi_-} \frac{\xi_+ + \xi_-}{\xi_+ - \xi_-} \log \frac{\xi_+}{\xi_-} - 1 - \theta \coth \theta, \tag{4-12.18}$$

where

$$\xi_\pm = \tfrac{1}{2}(1 + v) + \tfrac{1}{2}(1 - v) \exp(\pm \theta). \tag{4-12.19}$$

This is not the most felicitous presentation of the function Θ, however. We shall do better by observing that

$$\xi_\pm = \exp(\pm \tfrac{1}{2}\theta) \, [\cosh \tfrac{1}{2}\theta \mp v \sinh \tfrac{1}{2}\theta], \tag{4-12.20}$$

which leads to the expression,

$$\frac{1}{\cosh \theta} [\Theta + 1 + \theta \coth \theta]$$

$$= \int_{-1}^{1} \frac{1}{2} \, dv \, \frac{1}{1 + (1 - v^2) \sinh^2\tfrac{1}{2}\theta} \frac{1 + (1 - v) \sinh^2\tfrac{1}{2}\theta}{(1 - v) \sinh\tfrac{1}{2}\theta \cosh\tfrac{1}{2}\theta}$$

$$\times \left[\log \frac{\cosh\tfrac{1}{2}\theta - v \sinh\tfrac{1}{2}\theta}{\cosh\tfrac{1}{2}\theta + v \sinh\tfrac{1}{2}\theta} + \theta \right]. \tag{4-12.21}$$

Now, if we explicitly enforce a symmetrization between v and $-v$ as integration variables, and perform some partial integrations, there emerges as the equivalent of the right side of (4–12.21):

$$\int_{0}^{1} dv \, \frac{\log(1 + (1 - v^2) \sinh^2\tfrac{1}{2}\theta) - \log(1 - v^2)}{1 + (1 - v^2) \sinh^2\tfrac{1}{2}\theta} + \frac{2\theta}{\sinh \theta} \log 2. \tag{4-12.22}$$

The recognition that

$$\theta \coth \theta - 1 = \frac{q^2}{4m^2} \int_{0}^{1} dv \, \frac{1 + v^2}{1 + \dfrac{1 - v^2}{4} \dfrac{q^2}{m^2}}, \tag{4-12.23}$$

combined with the identity (4–4.109), then produces the result

$$\Theta\left(\frac{q^2}{m^2}\right) = \frac{q^2}{4m^2} \int_{0}^{1} dv(1 + v^2) \, \frac{2 \log\left(\dfrac{4v}{1 - v^2}\right) - 1}{1 + \dfrac{1 - v^2}{4} \dfrac{q^2}{m^2}}. \tag{4-12.24}$$

The relative photon emission probability stated in (4–12.16) thus becomes

$$\frac{\alpha}{2\pi}\frac{q^2}{m^2}\int_0^1 dv(1+v^2)\;\frac{\log\!\left(\dfrac{\delta M}{2\mu}\dfrac{1-v^2}{v}\right)+\dfrac{1}{2}}{1+\dfrac{1-v^2}{4}\dfrac{q^2}{m^2}}.\tag{4–12.25}$$

It combines with the factor that gives the modified elastic cross section, Eq. (4–4.86), to produce the essentially elastic cross section, as expressed by the modification factor

$$1-\frac{\alpha}{2\pi}\frac{q^2}{m^2}\int_0^1 dv\;\frac{(1+v^2)\left(\log\!\left(\dfrac{4m}{\delta M}\dfrac{v^2}{(1-v^2)^{3/2}}\right)-\dfrac{3}{2}\right)-\dfrac{1}{6}v^4}{1+\dfrac{1-v^2}{4}\dfrac{q^2}{m^2}}.\tag{4–12.26}$$

This is what we must reproduce by direct calculation of the essentially elastic cross section. Under nonrelativistic circumstances, incidentally, where no distinction need be made between δM and $k^0_{\min}$, one expects (4–12.26) to yield (4–4.99), which it does.

Let us review the direct calculation of scattering cross sections, using the example of a local vector potential A. This is the relativistic generalization of the discussion given in Section 4–11. The vacuum amplitude representing single particle exchange in the presence of the potential is

$$i\int (dx)(dx')K_1(x)\Delta_+{}^A(x,x')K_2(x'),\tag{4–12.27}$$

where, only exhibiting terms linear and quadratic in A,

$$\Delta_+{}^A = \Delta_+ + \Delta_+(eq(pA + Ap) - e^2A^2)\Delta_+$$
$$+\ \Delta_+ eq(pA + Ap)\Delta_+ eq(pA + Ap)\Delta_+ + \cdots.\tag{4–12.28}$$

On extracting the coefficient of $(iK^*_{1pq})(iK_{2pq})$, we get the persistence probability amplitude

$$1 + i\int (dx)(dx')\phi_{pq}(x)^*[\delta(x-x')(eq(pA + Ap) - e^2A^2)(x)$$
$$+\ eq(pA + Ap)(x)\Delta_+(x-x')eq(pA + Ap)(x')]\phi_{pq}(x'),\tag{4–12.29}$$

in which

$$\phi_{pq}(x) = \varphi_q(d\omega_p)^{1/2}\exp(ipx).\tag{4–12.30}$$

The charge eigenvector φ_q can be omitted if one replaces the charge matrix with its eigenvalue. For the situation of a time-independent potential that can act upon the particle for the time interval T, this becomes

$$1 + i\, d\omega_p\, T\left[(2eqpA - e^2A^2)(0) + \int\frac{(d\mathbf{p}')}{(2\pi)^3}\frac{|e(p+p')A(\mathbf{p}-\mathbf{p}')|^2}{p'^2 + m^2 - i\varepsilon}\right], \quad (4\text{--}12.31)$$

where three-dimensional Fourier transforms are used, and $p^{0\prime} = p^0$. From the implied persistence probability,

$$1 - d\omega_p\, T\int\frac{(d\mathbf{p}')}{(2\pi)^3}\, 2\pi\delta(p'^2 + m^2)|e(p+p')A(\mathbf{p}-\mathbf{p}')|^2, \quad (4\text{--}12.32)$$

we deduce the total scattering cross section,

$$\sigma = \frac{1}{2|\mathbf{p}|}\int\frac{(d\mathbf{p}')}{(2\pi)^3}\, 2\pi\delta(p'^2 + m^2)|e(p+p')A(\mathbf{p}-\mathbf{p}')|^2, \quad (4\text{--}12.33)$$

and, by remarking that

$$\int\frac{(d\mathbf{p}')}{(2\pi)^2}\delta(p'^2 + m^2) = \int d\Omega\,\frac{2|\mathbf{p}|}{(4\pi)^2}, \quad (4\text{--}12.34)$$

infer the differential cross section

$$\frac{d\sigma}{d\Omega} = \frac{1}{(4\pi)^2}|e(p+p')A(\mathbf{p}-\mathbf{p}')|^2 = \left(\frac{p^0}{2\pi}\right)^2|eA^0(\mathbf{p}-\mathbf{p}')|^2. \quad (4\text{--}12.35)$$

The last version in (4–12.35) refers to a purely scalar potential. Using the example of the Coulomb potential, we indeed recover the familiar result stated in Eq. (3–14.8), for example.

This brings us to the consideration of the modified, nonlocal potential that is implied by the coupling with photons. We are, of course, acquainted with one aspect of it, as expressed by the form factor of Section 4–5. It is essential to note here that what is required is not the form factor applicable to free particles, where both momenta of the double spectral form are on the mass shell, as the expression goes, but the form factor with only one momentum on the mass shell. That is evident in (4–12.31) where the momentum p' is not restricted to free particle values. Thus, the form factor that multiplies both $A(\mathbf{p} - \mathbf{p}')$ and $A(\mathbf{p}' - \mathbf{p})$ is, primarily,

$$1 - \frac{\alpha}{4\pi}q^2\int\frac{x\,dx\,\tfrac{1}{2}dv}{\delta^{3/2}}(1+x)(1+(1+x)v^2)\frac{q^2+4m^2}{(M_1{}^2-m^2)(p'^2+M_2{}^2-i\varepsilon)}\cdot \quad (4\text{--}12.36)$$

It has been derived from (4–5.86) (mark the notational change $k \to q$) by using the relation (4–6.99), but it does not include the contribution involving the combination (4–6.100), which will be considered separately. On placing $p'^2 =$

— m^2, we recover the form factor expression of (4–5.98, 99). In computing the total cross section one takes the imaginary part of the square of this function multiplied by $(p'^2 + m^2 - i\varepsilon)^{-1}$, which, with a factor of π removed, is

$$\delta(p'^2 + m^2) - \frac{\alpha}{2\pi} q^2 \int \frac{x \, dx \, \tfrac{1}{2} dv}{\delta^{3/2}} (1 + x)(1 + (1 + x)v^2) \frac{q^2 + 4m^2}{(M_1{}^2 - m^2)(M_2{}^2 - m^2)}$$

$$\times \, [\delta(p'^2 + m^2) - \delta(p'^2 + M_2{}^2)]. \tag{4–12.37}$$

As in the nonrelativistic discussion, we see combined here the elastic process, indicated by $\delta(p'^2 + m^2)$, and the inelastic processes selected by $\delta(p'^2 + M_2{}^2)$. The cancellation as $M_2 \to m$ means that no infrared singularity occurs in (4–12.37). However, since it is convenient to utilize the form factor results already obtained, we shall introduce a photon mass and consider the two classes of processes separately.

The integrations in the inelastic term are constrained by the photon mass at one end, as expressed by the restriction (4–5.102) and, at the other end, by the requirement that $M_2{}^2 - m^2 < 2m \, \delta M$. Throughout this interval, $x \ll 1$, and one can simplify the inelastic coefficient in (4–12.37) to

$$\frac{\alpha}{2\pi} \frac{q^2}{m^2} \int \frac{dx}{x} \frac{1}{2} \, dv \, \frac{1 + v^2}{1 + \dfrac{1 - v^2}{4} \dfrac{q^2}{m^2}}, \tag{4–12.38}$$

where

$$\frac{\mu}{m} (1 - v^2)^{-1/2} \left(1 + \frac{q^2}{4m^2}\right)^{-1/2} < x < \frac{\delta M}{m} \frac{1}{1 + \dfrac{q^2}{4m^2} - v\left(\dfrac{q^2}{4m^2}\left(1 + \dfrac{q^2}{4m^2}\right)\right)^{1/2}}. \tag{4–12.39}$$

This gives the following value to (4–12.38),

$$\frac{\alpha}{2\pi} \frac{q^2}{m^2} \int_{-1}^{1} \frac{1}{2} \, dv \, \frac{1 + v^2}{1 + \dfrac{1 - v^2}{4} \dfrac{q^2}{m^2}} \log\left[\frac{\delta M}{\mu} \frac{(1 - v^2)^{1/2}}{\left(1 + \dfrac{q^2}{4m^2}\right)^{1/2} - v\left(\dfrac{q^2}{4m^2}\right)^{1/2}}\right], \tag{4–12.40}$$

where symmetrization between v and $- v$ induces the replacement

$$\left(1 + \frac{q^2}{4m^2}\right)^{1/2} - v\left(\frac{q^2}{4m^2}\right)^{1/2} \to \left(1 + \frac{1 - v^2}{4} \frac{q^2}{m^2}\right)^{1/2}. \tag{4–12.41}$$

Then the use of the identity [it is equivalent to (4–4.109)]

$$\int_0^1 dv(1 + v^2) \frac{\log\left(1 + \dfrac{1 - v^2}{4} \dfrac{q^2}{m^2}\right)}{1 + \dfrac{1 - v^2}{4} \dfrac{q^2}{m^2}} = \int_0^1 dv \, \frac{(1 + v^2) \log\left(\dfrac{4v^2}{1 - v^2}\right) - 2v^2}{1 + \dfrac{1 - v^2}{4} \dfrac{q^2}{m^2}} \tag{4–12.42}$$

produces

$$\frac{\alpha}{2\pi} \frac{q^2}{m^2} \int_0^1 dv \, \frac{(1+v^2) \log\left(\dfrac{\delta M}{2\mu} \dfrac{1-v^2}{v}\right) + v^2}{1 + \dfrac{1-v^2}{4} \dfrac{q^2}{m^2}}. \tag{4-12.43}$$

As far as the dominant logarithmic term is concerned, this agrees completely with (4–12.25). But, as we have stated, the above form factor consideration is incomplete. It behooves us now to examine the whole picture.

The vacuum amplitude for the causal exchange of a photon, and a particle that moves in a potential A, can be expressed as

$$e^2 \int (dx)(dx')\phi_1(2\Pi - k)(x)i \, d\omega_k \exp[ik(x - x')] \, \Delta_+{}^A(x, x')(2\Pi - k)\phi_2(x'),$$

$$\tag{4-12.44}$$

where the appearance of the gauge covariant combination

$$\Pi = p - eqA \tag{4-12.45}$$

indicates that we are now considering emission and absorption acts occurring in the region occupied by the potential. We shall again use the expansion (4–12.28) for $\Delta_+{}^A$. The initial term, Δ_+, appears under causal conditions as

$$\exp[ik(x - x')] \, \Delta_+(x - x') = i \int d\omega_p \exp[i(p + k)(x - x')]$$

$$= i \int dM^2 \, d\omega_P \, \delta((P - k)^2 + m^2) \exp[iP(x - x')].$$

$$\tag{4-12.46}$$

The corresponding vacuum amplitude contributions are

$$- e^2 \int dM^2 \, d\omega_P \, d\omega_k \, \delta(2Pk + M^2 - m^2)[\phi_1(- P)(- 2)(M^2 + m^2)\phi_2(P)$$

$$- (\phi_1 2eqA)(- P)(2P - k)\phi_2(P) - \phi_1(- P)(2P - k)(2eqA\phi_2)(P)$$

$$+ 4(\phi_1 eqA)(- P)(eqA\phi_2)(P)], \tag{4-12.47}$$

where one can introduce the effective replacement

$$k \to \frac{kP}{P^2} P = \frac{M^2 - m^2}{2M^2} P \tag{4-12.48}$$

and the integral (4–12.5). The ensuing space-time extrapolation, with appropriate contact terms, gives the action expression

$$\frac{\alpha}{4\pi} \int \frac{(dp)}{(2\pi)^4} \frac{dM^2}{M^2} [(M^2 - m^2)^2 - 4\mu^2 M^2]^{1/2} \left\{ - 2(M^2 + m^2)\tfrac{1}{2}\phi(-p) \right.$$

$$\times \left(\frac{1}{p^2 + M^2 - i\varepsilon} - \frac{1}{M^2 - m^2} + \frac{p^2 + m^2}{(M^2 - m^2)^2} \right) \phi(p)$$

$$- \left(\frac{3}{2} + \frac{m^2}{2M^2} \right) (\phi 2eqAp)(-p) \left(\frac{1}{p^2 + M^2 - i\varepsilon} - \frac{1}{M^2 - m^2} \right) \phi(p)$$

$$\left. + 2(\phi eqA)(-p) \left(\frac{1}{p^2 + M^2 - i\varepsilon} - \frac{1}{M^2 - m^2} \right) (eqA\phi)(p) \right\}. \qquad (4\text{–}12.49)$$

We recognize the additional action for a free particle with the contact terms needed to supply an effective $(p^2 + m^2)^2$ factor [essentially Eq. (4–6.40), modified for a nonzero photon mass]. There are also terms linear and quadratic in the vector potential. They have been provided with the simpler contact term that suffices to avoid modifying the primitive interaction when applied to a slowly varying potential and the fields of essentially real particles.

Since the field ϕ in (4–12.49) describes particles moving in the potential A, one can make the substitution

$$(p^2 + m^2)\phi(p) \rightarrow (2eqpA\phi)(p), \qquad (4\text{–}12.50)$$

referring to a Lorentz gauge, with the A^2 term omitted, which is sufficiently accurate for our limited purposes. The resulting action term is

$$\frac{\alpha}{4\pi} \int \frac{(dp)}{(2\pi)^4} \frac{dM^2}{M^2} [(M^2 - m^2)^2 - 4\mu^2 M^2]^{1/2} \left\{ \left[- \frac{4m^2}{(M^2 - m^2)^2} + \frac{2}{M^2 - m^2} - \frac{1}{M^2} \right] \right.$$

$$\times \tfrac{1}{2}(\phi 2eqpA)(-p) \frac{1}{p^2 + M^2 - i\varepsilon} (2eqpA\phi)(p)$$

$$\left. + 2(\phi eqA)(-p) \left(\frac{1}{p^2 + M^2 - i\varepsilon} - \frac{1}{M^2 - m^2} \right) (eqA\phi)(p) \right\}. \qquad (4\text{–}12.51)$$

It makes a contribution to the probability amplitude which, since its imaginary part contains $\delta(p^2 + M^2)$, represents inelastic processes. In view of the restriction to almost elasticity, it is only the infrared singular term of (4–12.51) that is significant. This supplement to the modification factor (4–12.43) is

$$- \frac{\alpha}{4\pi} \int \frac{dM^2}{M^2} [(M^2 - m^2)^2 - 4\mu^2 M^2]^{1/2} \frac{4m^2}{(M^2 - m^2)^2}$$

$$\cong - \frac{\alpha}{\pi} \int_\mu^{\delta M} d(M - m)((M - m)^2 - \mu^2)^{1/2} \frac{1}{(M - m)^2} = - \frac{\alpha}{\pi} \left(\log \frac{2\delta M}{\mu} - 1 \right).$$

$$(4\text{–}12.52)$$

The next term in the expansion of $\Delta_+{}^A$, containing $eq(pA + Ap) - e^2 A^2$, makes three distinct contributions in (4–12.44), all of which are expressed as double spectral forms. In the application of the A^2 term, both momenta are placed on the mass shell and no imaginary contribution ensues. Now consider the term that combines the linear A dependence of the propagation function with the A dependence of the effective emission or absorption source. Here, one momentum is on the mass shell. The corresponding imaginary part, which describes inelastic processes, has this appearance:

$$\int x \, dx (\cdots) \frac{1}{M_1{}^2 - m^2} \delta(p'^2 + M_2{}^2), \qquad (4\text{–}12.53)$$

where $(\cdots)$ indicates factors that have finite limits as $x \to 0$. Thus, in contrast with the analogous term in (4–12.37), there is only one spectral denominator, which does not yield an infrared singular structure. What remains is the double spectral form of Section 4–5, which has already been considered in its major implications. Now let us complete that discussion.

According to the complete structure of Eq. (4–5.86), and the relation (4–6.100), the following term has been omitted in the form factor (4–12.36),

$$- \frac{\alpha}{4\pi} \int \frac{x \, dx \, \tfrac{1}{2} dv}{\delta^{3/2}} (1 + x)^2 \frac{4}{x^2} \left[\frac{(M_1{}^2 - m^2)(M_2{}^2 - m^2)}{(M_1{}^2 - m^2)(p'^2 + M_2{}^2 - i\varepsilon)} - 1 \right]$$

$$= (p'^2 + m^2) \frac{\alpha}{\pi} \int \frac{dx}{x} \frac{1}{2} \, dv \frac{(1 + x)^2}{\delta^{3/2}} \frac{1}{p'^2 + M_2{}^2 - i\varepsilon}. \qquad (4\text{–}12.54)$$

Its contribution to the imaginary part of the squared form factor multiplied by $(p'^2 + m^2 - i\varepsilon)^{-1}$, and divided by π, is

$$\frac{2\alpha}{\pi} \int \frac{dx}{x} \frac{1}{2} \, dv \frac{(1 + x)^2}{\delta^{3/2}} \delta(p'^2 + M_2{}^2), \qquad (4\text{–}12.55)$$

which represents inelastic processes. Since the x integration is limited as in (4–12.39), the coefficient of interest is

$$\frac{2\alpha}{\pi} \int \frac{1}{2} \, dv \frac{dx}{x} = \frac{2\alpha}{\pi} \int_{-1}^{1} \frac{1}{2} \, dv \log \left[\frac{\delta M}{\mu} \frac{(1 - v^2)^{1/2}}{\left(1 + \frac{q^2}{4m^2}\right)^{1/2} - v \left(\frac{q^2}{4m^2}\right)^{1/2}} \right]$$

$$= \frac{2\alpha}{\pi} \left[\log \frac{\delta M}{\mu} - \frac{1}{2} \int_0^1 dv \log \frac{1}{1 - v^2} - \frac{1}{2} \int_0^1 dv \log \left(1 + \frac{1 - v^2}{4} \frac{q^2}{m^2}\right) \right], \qquad (4\text{–}12.56)$$

or, after partial integration,

$$\frac{2\alpha}{\pi}\left[\log\frac{2\delta M}{\mu}-1-\frac{q^2}{4m^2}\int_0^1 dv\,\frac{v^2}{1+\frac{1-v^2}{4}\frac{q^2}{m^2}}\right]. \qquad (4\text{-}12.57)$$

It is worth pointing out that this contribution would not occur had we adopted the provisional procedure of Eq. (4–5.90) where more elaborate contact terms are used.

We come now to the last term of the expansion (4–12.28). When inserted in the vacuum amplitude (4–12.44) it describes a process wherein two scatterings by the vector potential A occur between the photon emission and absorption acts. The causal arrangement that is most convenient for our purposes is this. A free particle emits a photon, and the resulting virtual particle is scattered by the potential to produce the real particle which, in company with the photon, is detected by an analogous arrangement. This will lead to a single spectral form associated with the mass M of the two-particle system. Its imaginary part correspondingly represents inelastic processes. We are interested only in soft photons for which the process under consideration can be factored into the elastic double scattering of the particle and the photon exchange process. Accordingly, this contribution to the modification factor for essentially elastic scattering is just the part of (4–12.4) that refers entirely to the initial momentum p_2, combining the photon emission by the initial particle with the photon detection by the final particle. It is

$$e^2\int_{(m+\mu)^2}^{(m+\delta M)^2} dM^2\,d\omega_k\,\delta(2Pk+M^2-m^2)\left(\frac{p_2}{kp_2}\right)^2$$

$$\cong -\frac{\alpha}{\pi}\int_\mu^{\delta M} d(M-m)((M-m)^2-\mu^2)^{1/2}\,\frac{1}{(M-m)^2+\mu^2\frac{q^2}{m^2}\left(1+\frac{q^2}{4m^2}\right)}$$

$$= -\frac{\alpha}{\pi}\left[\log\frac{2\delta M}{\mu}-1-\frac{q^2}{4m^2}\int_0^1 dv\,\frac{1+v^2}{1+\frac{1-v^2}{4}\frac{q^2}{m^2}}\right], \qquad (4\text{-}12.58)$$

according to the various evaluations illustrated by (4–12.8) and (4–12.23).

When one adds the three additional effects given in Eqs. (4–12.52, 57, 58), all reference to photon mass and the criterion of inelasticity disappears, leaving the following term:

$$\frac{\alpha}{\pi}\frac{q^2}{4m^2}\int_0^1 dv\,\frac{1-v^2}{1+\frac{1-v^2}{4}\frac{q^2}{m^2}}. \qquad (4\text{-}12.59)$$

Its addition to (4–12.43) yields precisely Eq. (4–12.25), which combines with the elastic factor to give (4–12.26). Note again that, while we have retained the artificial distinctions of elasticity and inelasticity for a certain convenience, the method directly supplies the modification factor for essentially elastic scattering, deriving it from the double spectral version of the form factor and requiring only relatively simple considerations to obtain a small supplementary term.

For a given δM, the modification factor (4–12.26) depends only upon the momentum transfer measure q^2. This is in contrast with (4–4.114), where an explicit energy dependence also occurs. For values of $q^2 \ll m^2$, the nonrelativistic evaluation applies,

$$\frac{q^2}{m^2} \ll 1: \quad 1 - \frac{2\alpha}{3\pi} \frac{q^2}{m^2}\left(\log \frac{m}{2\delta M} + \frac{7}{120}\right). \tag{4–12.60}$$

At the other extreme limit, we find that the modification factor for almost elasticity is

$$\frac{q^2}{m^2} \gg 1: \quad 1 - \frac{2\alpha}{\pi}\left[\left(\log \frac{q^2}{m^2} - 1\right)\left(\log \frac{4m}{\delta M} - \frac{19}{12}\right) + \frac{3}{4}\left(\log \frac{q^2}{4m^2}\right)^2\right.$$
$$\left. + 3\log 2(1 - \log 2) - \frac{31}{36}\right]. \tag{4–12.61}$$

In view of the simple relation that is known between the spin 0 and the spin $\frac{1}{2}$ problems [cf. Eq. (4–4.135) and related discussion], we can immediately supply the analogous spin $\frac{1}{2}$ results,

$$\frac{q^2}{m^2} \ll 1: \quad 1 - \frac{2\alpha}{3\pi} \frac{q^2}{m^2}\left(\log \frac{m}{2\delta M} + \frac{19}{30}\right),$$

$$\frac{q^2}{m^2} \gg 1: \quad 1 - \frac{2\alpha}{\pi}\left[\left(\log \frac{q^2}{m^2} - 1\right)\left(\log \frac{4m}{\delta M} - \frac{19}{12}\right) + \frac{3}{4}\left(\log \frac{q^2}{4m^2}\right)^2\right.$$
$$\left. + 3\log 2(1 - \log 2) - \frac{19}{36}\right]. \tag{4–12.62}$$

4–13 PHOTON-CHARGED PARTICLE SCATTERING

The preceding and following sections are concerned with different physical aspects of charged particle couplings that involve the repeated action of a Coulomb field. This section has a similar preoccupation, but with the fields of photons. The physical process is the scattering of photons by charged particles—Compton scattering—as it is modified through the mechanism of two-particle exchange. One systematic approach to that problem is provided by the additional action terms that are inferred from the vacuum amplitude for the causal exchange of a photon, Eq. (4–12.44). The terms independent of A describe the modified propagation function of the charged particle, terms linear in A represent the modification in the single photon emission and absorption mechanisms, and terms quadratic in A do a similar service for two-photon processes. The ensuing effective coupling that contains two free particle fields and two photon fields represents the process of interest. We propose to attack this problem in another, and possibly simpler, way by considering the process as a unit instead of building it up from various individual acts. Since there are no experimental results of great accuracy for electron-photon scattering, the question is primarily of methodological interest and we shall restrict the discussion to spin 0 particles.

To illustrate the possibility of a more direct calculation, let us consider the following causal arrangement. The charged particle and the photon collide to produce a new configuration of these particles. After a lapse of time the particles again collide and the end products are subsequently detected. This is a two-particle exchange process, with the scattering mechanism used both as effective production and detection sources. The action term of (3–12.92) implies the following effective emission source,

$$iK(p)J^\mu(k)\big|_{\text{eff.}} = (2\pi)^4\,\delta(k + p - k_2 - p_2)2e^2V_2^{\mu\nu}e_{2\nu}(d\omega_{p_2}\,d\omega_{k_2})^{1/2}iK_2iJ_2, \quad (4\text{–}13.1)$$

where

$$V_2^{\mu\nu} = \frac{p^\mu p_2^\nu}{p_2 k_2} - \frac{p_2^\mu p^\nu}{p_2 k} - g^{\mu\nu} \qquad (4\text{–}13.2)$$

incorporates the simplifications of a Lorentz gauge, and a reduced notation has been used for the sources of the particles that enter the collision. The analogous effective absorption source is

$$iK(-p)J^\mu(-k)\big|_{\text{eff.}}$$
$$= iK_1^*iJ_1^*(d\omega_{p_1}\,d\omega_{k_1})^{1/2}(2\pi)^4\,\delta(k_1 + p_1 - k - p)2e^2e_{1\kappa}^*V_1^{\kappa\mu}, \quad (4\text{–}13.3)$$

with

$$V_1^{\kappa\mu} = \frac{p_1^\kappa p^\mu}{p_1 k_1} - \frac{p^\kappa p_1^\mu}{p_1 k} - g^{\kappa\mu}. \qquad (4\text{–}13.4)$$

The vacuum amplitude that represents the two-particle exchange is derived from the coupling expression

$$\int iK(-p)J^\mu(-k)\big|_{\text{eff.}}\,i\,d\omega_k\,i\,d\omega_p\,iK(p)J_\mu(k)\big|_{\text{eff.}}. \qquad (4\text{–}13.5)$$

It can be expressed as the following transition matrix element (the subscript

c is appended as a reminder that this refers to a causal arrangement):

$$\langle 1|T|2\rangle_c = 2\pi i (d\omega_{p_1} \cdots d\omega_{k_2})^{1/2} 4e^4 \int d\omega_k \, d\omega_p \, (2\pi)^3 \, \delta(k + p - P) e_1^* V_1 V_2 e_2,$$

$$(4\text{--}13.6)$$

where

$$P = k_1 + p_1 = k_2 + p_2 \qquad (4\text{--}13.7)$$

is the total momentum. We shall use the rest frame of this vector to select a convenient gauge, as expressed by the conditions

$$Pe_1 = Pe_2 = 0. \qquad (4\text{--}13.8)$$

Since each polarization vector is also orthogonal to its own momentum vector, we can introduce the following substitutions:

$$p_1 e_1 = p_2 e_2 = 0, \qquad pe_2 = -ke_2, \qquad pe_1 = -ke_1. \qquad (4\text{--}13.9)$$

Accordingly,

$$e_1^* V_1 V_2 e_2 = e_1^* \left[-1 + \frac{kp_1}{(kp_1)} \right] \left[-1 + \frac{p_2 k}{(kp_2)} \right] e_2$$

$$= e_1^* \left[1 - \frac{kp_1}{(kp_1)} - \frac{p_2 k}{(kp_2)} + k \frac{(p_1 p_2)}{(kp_1)(kp_2)} k \right] e_2, \qquad (4\text{--}13.10)$$

where parentheses are used to distinguish scalar products from dyadics. One can also exercise the option of writing

$$p_1 e_2 = -k_1 e_2, \qquad p_2 e_1 = -k_2 e_1. \qquad (4\text{--}13.11)$$

The relevant terms vanish on integration, however, since the vector

$$\int d\omega_k \, d\omega_p (2\pi)^3 \, \delta(k + p - P) \frac{k}{kp_1} = \frac{1}{(4\pi)^2} \left(1 - \frac{m^2}{M^2} \right) \left\langle \frac{k}{kp_1} \right\rangle, \qquad (4\text{--}13.12)$$

for example, can only be a linear combination of the vectors P and p_1, both of which are annulled by multiplication with e_1^*. A similar remark applies to $\langle k/kp_2 \rangle$ and the polarization vector e_2. Thus the only integrals required are comprised in the tensor

$$\left\langle \frac{k^\mu k^\nu}{kp_1 kp_2} \right\rangle = \int_{-1}^{1} \frac{1}{2} \, dv \left\langle \frac{k^\mu k^\nu}{\left[k \left(\frac{1+v}{2} p_1 + \frac{1-v}{2} p_2 \right) \right]^2} \right\rangle. \qquad (4\text{--}13.13)$$

The last form makes evident that the calculation is not appreciably more complicated for $p_1 \neq p_2$, compared to $p_1 = p_2$. Nevertheless, since we shall use the result of this causal arrangement only for $p_1 = p_2$ we now adopt that specializa-

tion. In the interests of the following calculational device, we shall also introduce the symbol q to denote an arbitrary vector that is finally identified with
$p_1 = p_2$:

$$\frac{k^\mu k^\nu}{(kp_2)^2} = -\frac{\partial}{\partial q_\mu}\frac{\partial}{\partial q_\nu}\log(kq)\big|_{q=p_2}. \tag{4-13.14}$$

The necessary integration thus reduces to

$$\langle\log(kq)\rangle = \int_{-1}^{1}\frac{1}{2}\,dz\left\{\log\frac{M^2-m^2}{2M} + \log\left[\frac{qP}{M} + z\left(q^2+\left(\frac{qP}{M}\right)^2\right)^{1/2}\right]\right\} \tag{4-13.15}$$

where, in the rest frame of P, the single integration variable is the cosine of the
angle between $\mathbf{q}$ and $\mathbf{k}$. A first differentiation gives

$$\frac{\partial}{\partial q}\langle\log(kq)\rangle = \int_{-1}^{1}\frac{1}{2}\,dz\,\frac{\dfrac{P}{M}+z\left(q^2+\left(\dfrac{qP}{M}\right)^2\right)^{-1/2}\left(q+\dfrac{P}{M}\left(\dfrac{qP}{M}\right)\right)}{\dfrac{qP}{M}+z\left(q^2+\left(\dfrac{qP}{M}\right)^2\right)^{1/2}}. \tag{4-13.16}$$

But, the vector P that appears here will multiply one of the polarization vectors
and can be omitted. Accordingly, the effective value is

$$\frac{\partial}{\partial q}\langle\log(kq)\rangle \to \int_{-1}^{1}\frac{1}{2}\,dz\,z\,\frac{q}{\left(q^2+\left(\dfrac{qP}{M}\right)^2\right)^{1/2}}\,\frac{1}{\dfrac{qP}{M}+z\left(q^2+\left(\dfrac{qP}{M}\right)^2\right)^{1/2}}$$

$$= -\int_{0}^{1}dz\,z^2\,\frac{q}{\left(\dfrac{qP}{M}\right)^2-z^2\left(q^2+\left(\dfrac{qP}{M}\right)^2\right)}. \tag{4-13.17}$$

At the next and final stage of differentiation, both the vectors P and $q \to p_2$ can
be discarded. Thus the only contribution arises from the differentiation of the
vector q in the numerator. This gives

$$\left\langle\frac{k^\mu k^\nu}{(kp_2)^2}\right\rangle \to g^{\mu\nu}\int_{0}^{1}dz\,\frac{z^2}{\left(\dfrac{M^2+m^2}{2M}\right)^2-z^2\left(\dfrac{M^2-m^2}{2M}\right)^2}, \tag{4-13.18}$$

where we have made the replacements

$$q^2 = p_2{}^2 = -m^2, \qquad (qP)^2 = (p_2 P)^2 = \left(\frac{M^2+m^2}{2}\right)^2 \tag{4-13.19}$$

in which M is the mass associated with P.

Our results are expressed in terms of the following function,

$$\chi(M^2) = 1 - m^2 \int_0^1 dz \frac{z^2}{\left(\dfrac{M^2 + m^2}{2M}\right)^2 - z^2 \left(\dfrac{M^2 - m^2}{2M}\right)^2}, \qquad (4\text{-}13.20)$$

or

$$\chi(M^2) = 1 - \frac{4m^2 M^2}{M^2 - m^2} \int_{-1}^1 \frac{1}{2} dz \frac{z}{M^2 + m^2 - z(M^2 - m^2)}$$

$$= 1 - \frac{2m^2 M^2}{(M^2 - m^2)^2} \left[\frac{M^2 + m^2}{M^2 - m^2} \log \frac{M^2}{m^2} - 2 \right], \qquad (4\text{-}13.21)$$

which has simple limiting values,

$$M^2 = m^2: \quad \chi(M^2) = \tfrac{2}{3},$$

$$M^2 \gg m^2: \quad \chi(M^2) = 1. \qquad (4\text{-}13.22)$$

The causal transition matrix element is

$$\langle 1|T|2\rangle_c = 2\pi i (d\omega_{p_1} \cdots d\omega_{k_2})^{1/2} 4\alpha^2 e_1^* e_2 \left(1 - \frac{m^2}{M^2}\right) \chi(M^2). \qquad (4\text{-}13.23)$$

Its space-time extrapolation proceeds by writing

$$(2\pi)^4 \delta(k_1 + p_1 - k_2 - p_2)$$

$$= \int (2\pi)^4 \delta(k_1 + p_1 - P) \frac{(dP)}{(2\pi)^4} (2\pi)^4 \delta(P - k_2 - p_2)$$

$$= \int (dx)(dx') \exp[-i(k_1 + p_1)x] \frac{dM^2}{2\pi i} \left[i \int d\omega_P \exp[iP(x - x')] \right] \exp[i(k_2 + p_2)x'],$$

$$(4\text{-}13.24)$$

where

$$i \int d\omega_P \exp[iP(x - x')] \rightarrow \int \frac{(dp)}{(2\pi)^4} \frac{\exp[ip(x - x')]}{p^2 + M^2 - i\varepsilon}. \qquad (4\text{-}13.25)$$

One also removes the causal distinction between the fields of the initial and final particles; this introduces the crossing symmetry

$$e_1^* \leftrightarrow e_2, \qquad k_1 \leftrightarrow -k_2, \qquad (4\text{-}13.26)$$

or

$$p_1 \leftrightarrow -p_2. \qquad (4\text{-}13.27)$$

But, in order to maintain gauge invariance after these extrapolations have been performed, the polarization vectors in (4–13.23) must first be replaced by equivalent field strength combinations, as indicated by

$$e_1 \to - \frac{1}{\frac{1}{2}(M^2 - m^2)} ((p_1 k_1)e_1 - (p_1 e_1)k_1),$$

$$e_2 \to - \frac{1}{\frac{1}{2}(M^2 - m^2)} ((p_2 k_2)e_2 - (p_2 e_2)k_2). \tag{4–13.28}$$

Then, evaluated in the gauge where the polarization vectors are orthogonal to the particle (or total) momentum, the resulting transition matrix element describing forward scattering is

$$\langle 1|T|2 \rangle = - (d\omega_{p_1} \cdots d\omega_{k_2})^{1/2} 8\pi\alpha e_1^* e_2$$

$$+ (d\omega_{p_1} \cdots d\omega_{k_2})^{1/2} 4\alpha^2 e_1^* e_2 (2p_2 k_2)^2 \int_{m^2}^{\infty} \frac{dM^2}{M^2} \frac{\chi(M^2)}{M^2 - m^2}$$

$$\times \left[\frac{1}{(p_2 + k_2)^2 + M^2 - i\varepsilon} + \frac{1}{(p_2 - k_2)^2 + M^2} \right], \tag{4–13.29}$$

where the spectral integral can also be written as

$$2 \int_{m^2}^{\infty} \frac{dM^2}{M^2} \chi(M^2) \frac{1}{(M^2 - m^2)^2 - (2p_2 k_2)^2 - i\varepsilon}. \tag{4–13.30}$$

Note that we have now included the direct implication of the primitive interaction, as given in Eq. (3–12.98) and specialized to forward scattering.

There is a test of this statement, in which it is applied to predict the total cross section for photon-particle scattering. Such ideas have appeared in Sections 4–11 and 4–12. They involve computing the probability that the initial configuration of particles has persisted, despite the effect of the interaction. The discussion of Section 3–12, where the transition matrix was introduced, indicates that the vacuum persistence amplitude for the given two-particle state is (in simplified notation)

$$\langle 2|2 \rangle = 1 + iV\langle 2|T|2 \rangle, \tag{4–13.31}$$

where V measures the four-dimensional interaction volume. The implied persistence probability,

$$|\langle 2|2 \rangle|^2 = 1 - 2V \operatorname{Im}\langle 2|T|2 \rangle, \tag{4–13.32}$$

exhibits the complementary total probability of a scattering process. That probability, per unit volume, and per unit initial invariant flux F, gives the total cross section:

$$\sigma = \frac{2}{F} \, \text{Im}\langle 2|T|2\rangle. \tag{4-13.33}$$

In the situation of immediate concern, where one particle is massless, Eqs. (3–12.68, 69) supply the flux expression

$$F = d\omega_{p_2} \, d\omega_{k_2} \, 2(-2k_2 p_2), \qquad -2k_2 p_2 = M^2 - m^2, \tag{4-13.34}$$

and we deduce that

$$\sigma = 4\pi\alpha^2 \, \frac{\chi(M^2)}{M^2}. \tag{4-13.35}$$

The limiting forms supplied by (4–13.22) are indeed those stated in Eq. (3–12.118), and the total cross section produced by integrating Eq. (3–12.117) agrees with the general expression for $\chi(M^2)$, Eq. (4–13.21).

The real part of the spectral integral (4–13.30) is conveniently evaluated by inserting the integral expression for $\chi(M^2)$ given in Eq. (4–13.21) and carrying out the M^2 integration first. The result is presented (without subscripts) in

$$-P \int \frac{dM^2}{M^2} \, \frac{\chi(M^2)}{(M^2 - m^2)^2 - (2pk)^2}$$

$$= \left[\frac{1}{m^4 - (2pk)^2} + \int_{-1}^{1} dz \, \frac{z(1-z)^2}{4m^4 - (1-z)^2(2pk)^2} \right] \log\left(\frac{m^2}{-2pk}\right)$$

$$+ \int_{-1}^{1} dz \, \frac{z(1-z)^2}{4m^4 - (1-z)^2(2pk)^2} \log\left(\frac{2}{1-z}\right). \tag{4-13.36}$$

At low photon energies the right-hand side reduces to

$$-\frac{pk}{m^2} \ll 1: \quad \frac{2}{3} \frac{1}{m^4} \left[\log\left(\frac{m^2}{-2pk}\right) - \frac{1}{24} \right], \tag{4-13.37}$$

while the other limit is

$$-\frac{pk}{m^2} \gg 1: \quad \frac{1}{(2pk)^2} \left[\log\left(\frac{-2pk}{m^2}\right) - 1 \right]. \tag{4-13.38}$$

Expressed in terms of the photon energy in the particle rest frame,

$$k^0 = -pk/m, \tag{4-13.39}$$

the two limiting forms of the differential cross section for forward scattering are

$$k^0 \ll m: \quad \frac{d\sigma}{d\Omega} = \left(\frac{\alpha}{m}\right)^2 |e_1^* \cdot e_2|^2 \left[1 + \frac{16\alpha}{3\pi} \left(\frac{k^0}{m}\right)^2 \left(\log\frac{m}{2k^0} - \frac{1}{24}\right) \right] \tag{4-13.40}$$

and

$$k^0 \gg m: \quad \frac{d\sigma}{d\Omega} = \frac{\alpha^2}{2mk^0} |e_1^* \cdot e_2|^2 \left[1 + \frac{2\alpha}{\pi} \left(\log \frac{2k^0}{m} - 1 \right) \right]. \quad (4\text{--}13.41)$$

Note that there is no infrared problem here, since the charged particle is undeflected, and that the cross section is increased relative to its skeletal interaction value. The summation over final polarizations replaces $|e_1^* \cdot e_2|^2$ with unity, independently of the initial polarization.

In the discussion of photon-photon scattering, the single spectral form supplied by forward scattering considerations, taken in conjunction with a double spectral form, completely specified the transition matrix. The present situation, involving two particle and two photon fields, is more complicated since there are two possible forward scattering arrangements. The second one can be chosen as the collision of two photons to form a particle pair, which subsequently undergoes a scattering interaction. This is the two-photon analogue of the arrangement used in Section 4–4. Here, however, one considers real photons instead of the virtual photon of that discussion. The forward scattering restriction is realized, in the center of mass frame, by requiring that the oppositely moving charged particles have the same directions as the oppositely moving photons. The effective two-particle source, which is again inferred from (3–12.92), is

$$iK(p_2)K(p'_2)|_{\text{eff.}} = (2\pi)^4 \, \delta(p_2 + p'_2 - k - k') 2e^2 e_\mu V^{\mu\nu} e'_\nu (d\omega_k \, d\omega_{k'})^{1/2} i J_2 i J_{2'},$$

$$(4\text{--}13.42)$$

with (dyadic notation)

$$V = \frac{p_2 p'_2}{(p_2 k)} + \frac{p'_2 p_2}{(p_2 k')} - 1. \quad (4\text{--}13.43)$$

This is the source that is to be inserted in the Coulomb scattering part of the vacuum amplitude (4–4.4). The annihilation mechanism term is also present, of course, but it selects effective sources that are antisymmetrical in p_2 and p'_2, whereas (4–13.42) is symmetrical. The implied transition matrix element for the causal arrangement is (superfluous causal labels are omitted)

$$\langle 1|T|2 \rangle_c = - 2\pi i (d\omega_p \cdots d\omega_{k'})^{1/2} 2e^4 \int d\omega_{p_2} \, d\omega_{p'_2} (2\pi)^3 \delta(p_2 + p'_2 - k - k')$$

$$\times \frac{(p + p_2)(p' + p'_2)}{(p - p_2)^2} \, eVe'. \quad (4\text{--}13.44)$$

The integration problem conveyed by the tensor expectation value

$$\Lambda^{\mu\nu} = \left\langle \frac{(p + p_2)(p' + p'_2)}{(p - p_2)^2} V^{\mu\nu} \right\rangle \quad (4\text{--}13.45)$$

is simplified on specializing to the forward scattering circumstance that is expressed covariantly by

$$p - p' = \left(1 - \frac{4m^2}{M^2}\right)^{1/2} (k - k'),$$ (4–13.46)

where

$$M^2 = -(k + k')^2 = -(p + p')^2.$$ (4–13.47)

The tensor Λ must then be constructed from the vectors k and k', in the various combinations kk, $k'k'$, kk', $k'k$, and the unit tensor. But there are also restrictions implied by the gauge properties

$$kV = k', \qquad Vk' = k.$$ (4–13.48)

They are incorporated in the form

$$\Lambda = a\left(1 - \frac{k'k}{(kk')}\right) + bkk' + c(kk + k'k').$$ (4–13.49)

However, only the coefficient $a(M^2)$ appears in the final result:

$$\langle 1|T|2\rangle_c = 2\pi i (d\omega_p \cdots d\omega_{k'})^{1/2} 4\alpha^2 \left(1 - \frac{4m^2}{M^2}\right)^{1/2} \frac{a(M^2)}{M^2} e((kk') - k'k)e'.$$ (4–13.50)

The information needed to construct $a(M^2)$ is obtained by forming the trace of the tensor Λ:

$$-\left\langle \frac{(p + p_2)(p' + p'_2)}{(p - p_2)^2} \left[4 + \tfrac{1}{2}(M^2 - 2m^2)\left(\frac{1}{p_2 k} + \frac{1}{p_2 k'}\right)\right]\right\rangle = 3a - \tfrac{1}{2}M^2 b,$$ (4–13.51)

and by computing $k'\Lambda k$:

$$\left\langle \frac{(p + p_2)(p' + p'_2)}{(p - p_2)^2} \left[\frac{1}{2}M^2 + \frac{(p_2 k')^2}{p_2 k} + \frac{(p_2 k)^2}{p_2 k'}\right]\right\rangle = -\tfrac{1}{2}M^2 a + (\tfrac{1}{2}M^2)^2 b.$$ (4–13.52)

The relation

$$p_2 k + p_2 k' = -\tfrac{1}{2}M^2$$ (4–13.53)

enables (4–13.52) to be simplified to

$$\left\langle \frac{(p + p_2)(p' + p'_2)}{(p - p_2)^2} \left[4 + \frac{1}{2}M^2\left(\frac{1}{p_2 k} + \frac{1}{p_2 k'}\right)\right]\right\rangle = -a + \tfrac{1}{2}M^2 b,$$ (4–13.54)

and adding this to (4–13.51) gives

$$a(M^2) = \tfrac{1}{2}m^2 \left\langle \frac{(p + p_2)(p' + p'_2)}{(p - p_2)^2} \left[\frac{1}{p_2 k} + \frac{1}{p_2 k'} \right] \right\rangle. \qquad (4\text{–}13.55)$$

The equivalent integration over the scattering angle in the center of mass frame reads (inserting the photon mass μ),

$$a(M^2) = \frac{2m^2}{M^2} \int_{-1}^{1} \frac{1}{2}\, dz \left[\frac{2(M^2 - 2m^2)}{\left(\dfrac{M^2}{4} - m^2\right) 2(1 - z) + \mu^2} - 1 \right]$$

$$\times \left[\frac{1}{1 - z\left(1 - \dfrac{4m^2}{M^2}\right)^{1/2}} + \frac{1}{1 + z\left(1 - \dfrac{4m^2}{M^2}\right)^{1/2}} \right]$$

$$= 2\frac{M^2 - 2m^2}{M^2 - 4m^2} \log \frac{M^2 - 4m^2}{\mu^2} - 2 \frac{\left(1 - \dfrac{m^2}{M^2}\right)}{\left(1 - \dfrac{4m^2}{M^2}\right)^{1/2}} \log \frac{1 + \left(1 - \dfrac{4m^2}{M^2}\right)^{1/2}}{1 - \left(1 - \dfrac{4m^2}{M^2}\right)^{1/2}}.$$

$$(4\text{–}13.56)$$

The space-time extrapolation of the probability amplitude described by (4–13.50) is produced on writing

$$(2\pi)^4 \delta(p + p' - k - k') = \int (2\pi)^4 \delta(p + p' - P) \frac{(dP)}{(2\pi)^4} \delta(P - k - k')$$

$$= \int (dx)(dx') \exp[-i(p + p')x] \frac{dM^2}{2\pi i}$$

$$\times \left[i \int d\omega_P \exp[iP(x - x')] \right] \exp[i(k + k')x']$$

$$\to (2\pi)^4 \delta(p + p' - k - k') \int \frac{dM^2}{2\pi i} \frac{1}{(k + k')^2 + M^2 - i\varepsilon},$$

$$(4\text{–}13.57)$$

where additional functions of M^2 are placed under the sign of integration. The corresponding contribution to the physical transition matrix element is

$$(d\omega_p \cdots d\omega_{k'})^{1/2} 4\alpha^2 e((kk') - k'k)e' \int_{(2m)^2}^{\infty} \frac{dM^2}{M^2} \left(1 - \frac{4m^2}{M^2}\right)^{1/2}$$

$$\times a(M^2) \frac{1}{(k + k')^2 + M^2 - i\varepsilon}, \qquad (4\text{–}13.58)$$

which has been inferred directly from (4–13.50) since the latter is already in gauge invariant form.

Now we consider a causal arrangement that leads to a double spectral form. (The reader is encouraged to draw the diamond shaped causal diagram for the process to be described.)   An extended photon source, emitting the time-like momentum K_2, creates a particle-antiparticle pair. One of these particles travels to the vicinity of another extended photon source, where the space-like momentum K_a is transferred to it.   The other reaches the neighborhood of an extended particle source where, by combining this particle with a virtual (anti) particle of momentum P_a, a photon is produced. The photon and particle that have appeared at these intermediate stages subsequently join, and are detected by an extended particle source that absorbs momentum P_1, where

$$P_1 = K_2 + K_a + P_a. \tag{4–13.59}$$

This arrangement contains a real photon of momentum k, and three real particles of momenta p', p'', p'''. The identification of these momenta is indicated by the conservation statements appropriate to each interaction:

$$K_2 = p'' + p''', \qquad P_a = k - p''',$$

$$K_a = p' - p'', \qquad P_1 = k + p'. \tag{4–13.60}$$

Thus, the momentum of the particle that is detected with the photon is

$$p' = P_1 - k, \tag{4–13.61}$$

that of the particle which contributes to the production of the photon is

$$p''' = k - P_a, \tag{4–13.62}$$

and the remaining momentum is

$$p'' = K_2 + P_a - k = P_1 - k - K_a. \tag{4–13.63}$$

The process in question is one of single photon exchange, with the particle interacting successively with the fields A_2 and A_a, in the sense of the symbolically written vacuum amplitude

$$\int i \, d\omega_k \, \phi_1 eq(p \exp(ikx) + \exp(ikx) \, p) \Delta_+ eq(pA_a + A_a p) \Delta_+ eq(pA_2 + A_2 p)$$

$$\times \Delta_+ eq(p \exp(- ikx) + \exp(- ikx) \, p) \phi_a. \tag{4–13.64}$$

This is explicitly displayed as

$$e^4 \int \frac{(dP_1)}{(2\pi)^4} \frac{(dP_a)}{(2\pi)^4} \frac{(dK_a)}{(2\pi)^4} \frac{(dK_2)}{(2\pi)^4} (2\pi)^4 \delta(P_1 - P_a - K_a - K_2)$$

$$\times \phi_1(-P_1)A_a{}^\mu(K_a)A_2{}^\nu(K_2)\phi_a(P_a)I_{\mu\nu}, \qquad (4\text{-}13.65)$$

where

$$I_{\mu\nu} = \int (dk)\,\delta(k^2 + \mu^2)\,\delta((P_1 - k)^2 + m^2)\,\delta((P_1 - k - K_a)^2 + m^2)\,\delta((P_a - k)^2 + m^2)$$

$$\times (2P_1 - k)(2P_a - k)[2(P_1 - k) - K_a]_\mu [2(P_a - k) + K_2]_\nu. \qquad (4\text{-}13.66)$$

Note, in the last expression, that the momentum factor in the coupling to the field $A_a{}^\mu$ is $(p' + p'')_\mu$, since this refers to the deflection of a particle with given charge, while the field $A_2{}^\nu$, which creates a pair of oppositely charged particles, is multiplied by $(p'' - p''')_\nu$. The associated conservation, or gauge invariance statements are

$$K_a{}^\mu I_{\mu\nu} = 0, \qquad I_{\mu\nu}K_2{}^\nu = 0. \qquad (4\text{-}13.67)$$

The two independent ways of viewing excitations as proceeding through this system are indicated by the spectral masses

$$M^2 = -(P_a + K_2)^2 = -(P_1 - K_a)^2 > m^2, \qquad (4\text{-}13.68)$$

and

$$M'^2 = -(K_2 + K_a)^2 = -(P_1 - P_a)^2 > 4m^2. \qquad (4\text{-}13.69)$$

They refer to the propagation of a particle and photon, and a particle-antiparticle pair, respectively. [Of course, the first inequality reads $M^2 > (m + \mu)^2$; we have stated the physical lower limit in (4-13.68).] We also introduce the masses

$$M_1{}^2 = -P_1{}^2, \qquad M_a{}^2 = -P_a{}^2; \qquad (4\text{-}13.70)$$

they will eventually be extrapolated, from values appropriate for the causal arrangement, to the value of interest in the actual scattering process, namely, m^2. The scalars $K_a{}^2, K_2{}^2$, which later will be extrapolated to zero, complete the list of six kinematical scalar quantities (the twelve components of three independent vectors with the six parameters of the Lorentz group removed). The delta functions in (4-13.66) supply such evaluations as (the photon mass is omitted here)

$$-2kP_1 = M_1{}^2 - m^2, \qquad -2kP_a = M_a{}^2 - m^2,$$

$$-2kK_a = M_1{}^2 - M^2, \qquad -2kK_2 = M^2 - M_a{}^2. \qquad (4\text{-}13.71)$$

Other useful combinations are

$$2P_1P_a = M'^2 - M_1{}^2 - M_a{}^2, \qquad -2K_2K_a = M'^2 + K_2{}^2 + K_a{}^2 \qquad (4\text{-}13.72)$$

and

$$- 2K_2P_a = M^2 - M_a{}^2 + K_2{}^2, \qquad\qquad 2K_aP_1 = M^2 - M_1{}^2 + K_a{}^2,$$

$$- 2K_2P_1 = M^2 + M'^2 - M_a{}^2 + K_a{}^2, \qquad 2K_aP_a = M^2 + M'^2 - M_1{}^2 + K_2{}^2.$$

$$(4\text{--}13.73)$$

As an application, we note that

$$(2P_1 - k)(2P_a - k) = 2M'^2 - M_1{}^2 - M_a{}^2 - 2m^2. \qquad (4\text{--}13.74)$$

The basic integral in (4–13.66) is

$$I = \int (dk)\delta(k^2 + \mu^2)\delta(2kP_1 + M_1{}^2 - m^2)\delta(2kK_2 + M^2 - M_a{}^2)\delta(2kP_a + M_a{}^2 - m^2)$$

$$= \int (d\kappa)\,\delta\left(\kappa^2 + \mu^2 + \frac{(M_a{}^2 - M^2)^2}{4K_2{}^2}\right)\delta\left(2\kappa P_1 + M_1{}^2 - m^2 + \frac{M_a{}^2 - M^2}{K_2{}^2}K_2P_1\right)$$

$$\times\,\delta(2\kappa K_2)\,\delta\left(2\kappa P_a + M_a{}^2 - m^2 + \frac{M_a{}^2 - M^2}{K_2{}^2}K_2P_a\right), \qquad (4\text{--}13.75)$$

in which

$$k = \kappa + \frac{M_a{}^2 - M^2}{2K_2{}^2}K_2. \qquad (4\text{--}13.76)$$

This integral is essentially similar to that of (4–10.7), and, analogously,

$$I = 1/[8(-\varDelta)^{1/2}], \qquad (4\text{--}13.77)$$

where

$$(-\varDelta)^{1/2} = |\varepsilon_{\kappa\lambda\mu\nu}K_2{}^\kappa P_a{}^\lambda P_1{}^\mu \kappa^\nu|. \qquad (4\text{--}13.78)$$

The general form of the vector κ^μ is given by [cf. (4–10.13)]

$$\kappa^\mu = a\left(P_a - K_2\frac{K_2P_a}{K_2{}^2}\right)^\mu + b\left(P_1 - K_2\frac{K_2P_1}{K_2{}^2}\right)^\mu + c\varepsilon^{\mu\nu\kappa\lambda}P_{a\nu}P_{1\kappa}K_{2\lambda}. \quad (4\text{--}13.79)$$

Multiplication by the vectors P_a and P_1 supplies the information to determine a and b:

$$-\tfrac{1}{2}\left(M_a{}^2 - m^2 + \frac{M_a{}^2 - M^2}{K_2{}^2}K_2P_a\right) = a\left(-M_a{}^2 - \frac{(K_2P_a)^2}{K_2{}^2}\right)$$

$$+ b\left(P_aP_1 - \frac{P_aK_2K_2P_1}{K_2{}^2}\right),$$

$$-\tfrac{1}{2}\left(M_1{}^2 - m^2 + \frac{M_a{}^2 - M^2}{K_2{}^2}K_2P_1\right) = a\left(P_1P_a - \frac{P_1K_2K_2P_a}{K_2{}^2}\right)$$

$$+ b\left(- M_1{}^2 - \frac{(K_2 P_1)^2}{K_2{}^2}\right). \tag{4-13.80}$$

The determinant of this array of coefficients is

$$D = \left(- M_a{}^2 - \frac{(K_2 P_a)^2}{K_2{}^2}\right)\left(- M_1{}^2 - \frac{(K_2 P_1)^2}{K_2{}^2}\right) - \left(P_a P_1 - \frac{P_a K_2 P_1 K_2}{K_2{}^2}\right)^2$$

$$= M_a{}^2 M_1{}^2 - (P_a P_1)^2 + \frac{1}{K_2{}^2}[M_a{}^2(K_2 P_1)^2 + M_1{}^2(K_2 P_a)^2 + 2 P_a P_1 P_a K_2 P_1 K_2].$$

$$\tag{4-13.81}$$

The square of the combination (4–13.79) is

$$\kappa^2 = - \mu^2 - \frac{(M_a{}^2 - M^2)^2}{4 K_2{}^2}$$

$$= - \tfrac{1}{2}a\left(M_a{}^2 - m^2 + \frac{M_a{}^2 - M^2}{K_2{}^2} K_2 P_a\right)$$

$$- \tfrac{1}{2}b\left(M_1{}^2 - m^2 + \frac{M_a{}^2 - M^2}{K_2{}^2} K_2 P_1\right) + c^2(- K_2{}^2 D), \tag{4-13.82}$$

where

$$c^2(- K_2{}^2 D) > 0 \tag{4-13.83}$$

is necessary for the nonvanishing of the integral (4–13.75). The alternative form produced by multiplying (4–13.79) with κ_μ,

$$\kappa^2 = a \kappa P_a + b \kappa P_1 + c \varepsilon^{\mu\nu\kappa\lambda} \kappa_\mu P_{a\nu} P_{1\kappa} K_{2\lambda}, \tag{4-13.84}$$

shows, on comparison with (4–13.82), that

$$c^2(- K_2{}^2 D) = |c|(- \varDelta)^{1/2}, \tag{4-13.85}$$

or

$$- \varDelta = (- K_2{}^2 D)(- K_2{}^2 D c^2). \tag{4-13.86}$$

To illustrate these relations, we consider directly the situation of interest, where

$$M_1{}^2 = M_a{}^2 = m^2, \qquad K_a{}^2 = K_2{}^2 = 0, \tag{4-13.87}$$

although $K_2{}^2 = 0$ must be realized by a limiting process. For convenience, we list the values of various scalars in this limit:

$$2 P_1 P_a \rightarrow M'^2 - 2m^2; \qquad\qquad - 2 K_2 K_a \rightarrow M'^2;$$

$$2K_aP_1, -2K_2P_a \to M^2 - m^2; \qquad 2K_aP_a, -2K_2P_1 \to M^2 + M'^2 - m^2.$$
$$(4\text{-}13.88)$$

Then,

$$K_2^2D \to \tfrac{1}{4}M'^2[(M^2-m^2)^2 + M^2M'^2]$$
$$(4\text{-}13.89)$$

while the solutions of the equations in (4–13.80) with $M_1^2 = M_a^2 = m^2$,

$$K_2^2Da = -\tfrac{1}{2}(M^2-m^2)[m^2K_2P_a + K_2P_1P_aP_1],$$

$$K_2^2Db = -\tfrac{1}{2}(M^2-m^2)[m^2K_2P_1 + K_2P_aP_1P_a], \qquad (4\text{-}13.90)$$

lead to

$$a \to \frac{1}{2}\frac{(M^2-m^2)(M^2+M'^2-3m^2)}{(M^2-m^2)^2 + M^2M'^2}, \qquad b \to \frac{1}{2}\frac{(M^2-m^2)(M^2+m^2)}{(M^2-m^2)^2 + M^2M'^2}. \quad (4\text{-}13.91)$$

The condition for the existence of the integral, evaluated for $M_1^2 = M_a^2 = m^2$, is

$$-\frac{1}{2K_2^2}(M^2-m^2)[K_2P_aa + K_2P_1b + \tfrac{1}{2}(M^2-m^2)] - \mu^2 > 0. \quad (4\text{-}13.92)$$

The linear equations of (4–13.80) can be used to give two equivalent evaluations for the combination of (4–13.92),

$$\frac{1}{K_2^2}[K_2P_aa + K_2P_1b + \tfrac{1}{2}(M^2-m^2)] = \frac{1}{K_2P_a}(-m^2a + P_1P_ab)$$

$$= \frac{1}{K_2P_1}(P_1P_aa - m^2b). \quad (4\text{-}13.93)$$

The resulting limiting form of (4–13.92) is

$$\frac{1}{4}\frac{(M^2-m^2)^2(M'^2-4m^2)}{(M^2-m^2)^2 + M^2M'^2} > \mu^2. \qquad (4\text{-}13.94)$$

When it is permissible to ignore the fictitious photon mass, this inequality simply combines the independent spectral requirements

$$M^2 > m^2, \qquad M'^2 > 4m^2. \qquad (4\text{-}13.95)$$

But, if the finiteness of the photon mass is significant, neither of the lower limits in (4–13.95) can be realized. A sufficiently accurate expression of this spectral domain is

$$M^2 - m^2 > \frac{2m\mu}{[1 - (4m^2/M'^2)]^{1/2}}. \qquad (4\text{-}13.96)$$

The construction of (4–13.86) supplies the limiting form of Δ as

$$\Delta \to \frac{1}{16}(M^2 - m^2)^2 M'^2(M'^2 - 4m^2) - \tfrac{1}{4}\mu^2 M'^2[(M^2 - m^2)^2 + M^2 M'^2]. \quad (4\text{--}13.97)$$

Corresponding to the simplification used in (4–13.96), the coefficient of μ^2 is well enough approximated by replacing M^2 with m^2.

As in the photon-photon scattering discussion, the extrapolated value of Δ is positive, requiring a selection to be made between the values

$$(-\Delta)^{1/2} = \pm i\Delta^{1/2}. \quad (4\text{--}13.98)$$

Again we make this decision by comparing the causal and noncausal solutions of a related, simplified problem. We ask to evaluate the integral (here D_+ includes a photon mass)

$$J = \int (dx) \cdots (dx''') \, D_+(x''' - x) \exp(ipx) \Delta_+(x - x') \Delta_+(x^\frown - x'')$$

$$\times \Delta_+(x'' - x''') \exp(-ip'x'''), \quad (4\text{--}13.99)$$

where

$$p^2 + m^2 = 0, \quad (4\text{--}13.100)$$

first in the noncausal manner that introduces the four-dimensional constructions of these propagation functions,

$$J = (2\pi)^4 \delta(p - p') J_0,$$

$$J_0 = \int \frac{(dk)}{(2\pi)^4} \frac{1}{k^2 + \mu^2 - i\varepsilon} \left[\frac{1}{(k - p)^2 + m^2 - i\varepsilon} \right]^3, \quad (4\text{--}13.101)$$

and then with the double spectral form that gives the space-time extrapolation of a causal arrangement. The kinematical integral is (4–13.75), where we are interested only in the extrapolated situation with $K_2 = K_a = 0$, and $P_1 = P_a = p$. The introduction of double spectral forms, in the manner illustrated by the discussion leading to (4–10.36), then supplies the alternative evaluation

$$J_0 = -\int \frac{dM^2 \, dM'^2}{2\pi} \frac{1}{2\pi} \frac{1}{M^2 - m^2} \frac{1}{M'^2} \frac{1}{8(\pm i)\Delta^{1/2}}. \quad (4\text{--}13.102)$$

The first computation of J_0 proceeds by combining the parametric representations of the two propagation functions in (4–13.101),

$$J_0 = \int \frac{(dk)}{(2\pi)^4} \left[\int_0^\infty i \, ds_1 \frac{(is_1)^2}{2} \exp\{-is_1[(k - p)^2 + m^2]\} \right]$$

$$\times \left[\int_0^\infty i\, ds_2 \exp[-\,is_2(k^2 + \mu^2)] \right]. \qquad (4\text{-}13.103)$$

On changing the variables according to

$$s_1 = su, \quad s_2 = s(1 - u), \quad ds_1\, ds_2 = s\, ds\, du, \quad 0 < s < \infty, \quad 0 < u < 1, \quad (4\text{-}13.104)$$

this reads

$$J_0 = \int \tfrac{1}{2}\, du\, u^2\, ds\, s^3 \int \frac{(dk)}{(2\pi)^4} \exp\{-\,is[k^2 - 2kpu + \mu^2(1 - u)]\}. \qquad (4\text{-}13.105)$$

The momentum integration is performed by completing the square,

$$k^2 - 2kpu = (k - pu)^2 + m^2 u^2, \qquad (4\text{-}13.106)$$

and then applying the standard integral (4–8.57). This gives

$$J_0 = -\frac{i}{32\pi^2} \int_0^1 du\, u^2 \int_0^\infty ds\, s \exp\{-\,is[m^2 u^2 + \mu^2(1 - u)]\}$$

$$= \frac{i}{32\pi^2} \int_0^1 du\, \frac{u^2}{[m^2 u^2 + \mu^2(1 - u)]^2}. \qquad (4\text{-}13.107)$$

Since the integral that appears here is positive, a comparison with (4–13.102) settles the question of $\pm\, i$ in favor of $+\, i$. But let us also verify that the two numerical factors are the same, at least for $\mu/m \ll 1$.

Inasmuch as the μ^2 term in the denominator is only significant when u is quite small, the $1 - u$ factor can be replaced by unity. Then, introducing the variable

$$x = (m/\mu)u, \qquad 0 < x < (m/\mu) \sim \infty, \qquad (4\text{-}13.108)$$

we get, effectively,

$$J_0 = \frac{i}{32\pi^2} \frac{1}{m^3\mu} \int_0^\infty dx\, \frac{x^2}{(x^2 + 1)^2} = \frac{i}{128\pi} \frac{1}{m^3\mu}. \qquad (4\text{-}13.109)$$

The alternative computation is

$$J_0 = i \int \frac{dM^2}{2\pi} \frac{dM'^2}{2\pi} \frac{1}{M^2 - m^2} \frac{1}{M'^2} \frac{1}{2} \frac{1}{M'(M'^2 - 4m^2)^{1/2}}$$

$$\times \frac{1}{\left[(M^2 - m^2)^2 - 4\mu^2 m^2 \dfrac{M'^2}{M'^2 - 4m^2} \right]^{1/2}}, \qquad (4\text{-}13.110)$$

or, in terms of the variable y defined by

$$M^2 - m^2 = 2\mu m \frac{M'}{(M'^2 - 4m^2)^{1/2}} y, \qquad 1 < y < \infty, \tag{4-13.111}$$

$$J_0 = \frac{i}{16\pi^2} \frac{1}{m\mu} \int_1^\infty \frac{dy}{y} \frac{1}{(y^2 - 1)^{1/2}} \int_{(2m)^2}^\infty \frac{dM'^2}{(M'^2)^2} = \frac{i}{128\pi} \frac{1}{m^3\mu}. \tag{4-13.112}$$

The explicit construction of the photon momentum vector is obtained from (4–13.76) and (4–13.79) as

$$k^\mu = aP_a{}^\mu + bP_1{}^\mu + c\varepsilon^{\mu\nu\kappa\lambda}P_{a\nu}P_{1\kappa}K_{2\lambda} + dK_2{}^\mu, \tag{4-13.113}$$

where

$$d = -\frac{1}{K_2{}^2}[K_2P_a a + K_2P_1 b + \tfrac{1}{2}(M^2 - m^2)] \to \frac{I}{2} \frac{(M^2 - m^2)(M'^2 - 4m^2)}{(M^2 - m^2)^2 + M^2M'^2}, \tag{4-13.114}$$

according to (4–13.93). Note that the various limiting values are related by

$$d = a - b. \tag{4-13.115}$$

Let us also recall that only the magnitude of c is fixed and therefore an average of both signs is to be used (the basic integral I already contains the factor of two). The extrapolated value of the square of c is derived from (4–13.86) as

$$-c^2 \to \left(1 - \frac{4m^2}{M'^2}\right) \frac{(M^2 - m^2)^2}{[(M^2 - m^2)^2 + M^2M'^2]^2}. \tag{4-13.116}$$

The vector combinations that occur in (4–13.66) can be presented in this way:

$$[2(P_1 - k) - K_a]^\mu = (1 - a - b)(P_1 + P_a)^\mu + (a - b)K_a{}^\mu + (1 - a + b)K_2{}^\mu$$
$$+ 2c\varepsilon^{\mu\nu\kappa\lambda}P_{a\nu}P_{1\kappa}K_{a\lambda},$$

$$[2(P_a - k) + K_2]^\mu = (1 - a - b)(P_1 + P_a)^\mu - (a - b)K_2{}^\mu - (1 - a + b)K_a{}^\mu$$
$$- 2c\varepsilon^{\mu\nu\kappa\lambda}P_{a\nu}P_{1\kappa}K_{2\lambda}, \tag{4-13.117}$$

which make use of such rearrangements as

$$P_1 = \tfrac{1}{2}(P_1 + P_a) + \tfrac{1}{2}(K_2 + K_a). \tag{4-13.118}$$

There is a relation between a and b that is demanded by the gauge invariance conditions,

$$0 = K_a[2(P_1 - k) - K_a] = K_2[2(P_a - k) + K_2] = (M^2 - m^2)(1 - a - b) - M'^2 b. \tag{4-13.119}$$

It is indeed satisfied since

$$1 - a - b = \frac{1}{2} \frac{M'^2(M^2 + m^2)}{(M^2 - m^2)^2 + M^2 M'^2}.$$ (4-13.120)

The space-time extrapolations that we are about to perform will not maintain this gauge invariance, apart from the terms having the coefficient c, unless we make it explicit at the causal stage. That can be done in various ways, as in the photon-photon scattering discussion. One possibility is described by the following gauge transformation:

$$A_a(K_a) \rightarrow \frac{1}{\frac{1}{2}M'^2} [-(K_a K_2) + K_a(K_2 + K_a)]A_a(K_a),$$

$$A_2(K_2) \rightarrow \frac{1}{\frac{1}{2}M'^2} [-(K_a K_2) + K_2(K_2 + K_a)]A_2(K_2).$$ (4-13.121)

The new structures vanish on multiplication with $K_2 + K_a$. If we then adopt a Lorentz gauge, scalar multiplication by either photon momentum annuls these combinations. This also enables us to replace $P_1 + P_a$, when multiplying the vector potentials, by either $2P_1$ or $2P_a$. As a result, we have

$$A_a(K_2)[2(P_1 - k) - K_a][2(P_a - k) + K_2]A_2(K_2)$$

$$= \frac{16}{M'^4}(1 - a - b)^2[P_1 K_a K_2 A_a(K_a) - K_2 K_a P_1 A_a(K_a)][P_a K_2 K_a A_2(K_2)$$

$$- K_a K_2 P_a A_2(K_2)] - 4c^2 \varepsilon^{\mu\nu\kappa\lambda} K_{a\mu} A_{a\nu}(K_a) K_{2\kappa} P_{1\lambda} \varepsilon^{\mu\nu\kappa\lambda} K_{2\mu} A_{2\nu}(K_2) K_{a\kappa} P_{a\lambda},$$ (4-13.122)

where terms linear in c have been omitted. Note that the two terms involve field strengths and dual field strengths, respectively.

As in the discussion that leads from (4-10.34) to (4-10.36), the space-time extrapolation is performed by introducing the double spectral form structure

$$-\frac{dM^2}{2\pi} \frac{dM'^2}{2\pi} \frac{1}{(P_a + K_2)^2 + M^2 - i\varepsilon} \frac{1}{(K_a + K_2)^2 + M'^2 - i\varepsilon}.$$ (4-13.123)

In order to present our results with a minimum repetition of complicated structures, we shall state directly the double spectral part of the physical transition matrix, which is produced by the extrapolation

$$P_a \rightarrow p_2, \qquad P_1 \rightarrow p_1, \qquad K_a \rightarrow -k_1, \qquad K_2 \rightarrow k_2,$$ (4-13.124)

supplemented by the crossing transformation

$$k_2 \leftrightarrow -k_1, \qquad e_2 \leftrightarrow e_1^*,$$ (4-13.125)

or the equivalent one,

$$p_2 \leftrightarrow - p_1. \tag{4-13.126}$$

It is

$$(d\omega_{p_1} \cdots d\omega_{k_2})^{1/2} 4\alpha^2 \int \frac{dM^2\, dM'^2}{\Delta^{1/2}} (M'^2 - 2m^2) \left[\frac{1}{(p_2 + k_2)^2 + M^2 - i\varepsilon} \right.$$

$$+ \frac{1}{(p_2 - k_1)^2 + M^2} \left] \frac{1}{(k_1 - k_2)^2 + M'^2} \left\{ f_+ (k_1 k_2 p_1 e_1{}^* - k_1 p_1 k_2 e_1{}^*)(k_1 k_2 p_2 e_2 \right.$$

$$- k_2 p_2 k_1 e_2) + \left(1 - \frac{4m^2}{M'^2}\right) f_- \varepsilon^{\mu\nu\kappa\lambda} k_{1\mu} e_{1\nu}^* k_{2\kappa} p_{1\lambda} \varepsilon^{\mu\nu\kappa\lambda} k_{2\mu} e_{2\nu} k_{1\kappa} p_{2\lambda} \right\}, \tag{4-13.127}$$

where

$$f_\pm = \frac{(M^2 \pm m^2)^2}{[(M^2 - m^2)^2 + M^2 M'^2]^2}. \tag{4-13.128}$$

The procedure we have just followed to make gauge invariance explicit is straightforward, but has the disadvantage that the resulting numerator momentum dependence is not irreducible. In this situation there are two basic combinations, involving the polarization vectors, that are gauge invariant and crossing symmetric, namely

$$G_1 = k_1 p_2 k_2 p_2 e_1{}^* e_2 + k_2 p_2 p_2 e_1{}^* p_1 e_2 - k_1 p_2 p_1 e_1{}^* p_2 e_2, \tag{4-13.129}$$

which gives the structure of the skeletal scattering process, and

$$G_2 = k_1 k_2 e_1{}^* e_2 - k_2 e_1{}^* k_1 e_2. \tag{4-13.130}$$

Written in terms of these, we have

$$(k_1 k_2 p_1 e_1{}^* - k_1 p_1 k_2 e_1{}^*)(k_1 k_2 p_2 e_2 - k_2 p_2 k_1 e_2) = k_1 k_2 G_1 - k_1 p_2 k_2 p_2 G_2 \tag{4-13.131}$$

and

$$\varepsilon^{\mu\nu\kappa\lambda} k_{1\mu} e_{1\nu}^* k_{2\kappa} p_{1\lambda} \varepsilon^{\mu\nu\kappa\lambda} k_{2\mu} e_{2\nu} k_{1\kappa} p_{2\lambda} = k_1 k_2 G_1 + (k_1 p_2 k_2 p_2 + m^2 k_1 k_2) G_2. \tag{4-13.132}$$

These statements can be verified by algebraic rearrangement or by comparing the two sides in some coordinate system such as the p_2 rest frame. Since it is only the demand of gauge invariance that motivates the appearance of momentum factors in the numerator, superfluous ones should be removed by transforming them into added single spectral forms. In doing this $k_1 k_2$ is replaced by $\frac{1}{2} M'^2$, while $- k_2 p_2$ and $k_1 p_2$, multiplying the corresponding denominator, are replaced by $\frac{1}{2}(M^2 - m^2)$. The outcome is this double spectral form:

$$(d\omega_{p_1} \cdots)^{1/2} \alpha^2 (2G_1 + m^2 G_2) \int \frac{dM^2\, dM'^2}{\Delta^{1/2}} (M'^2 - 2m^2) M'^2 \left[f_+ + \left(1 - \frac{4m^2}{M'^2}\right) f_- \right]$$

$$\times \frac{1}{(k_1 - k_2)^2 + M'^2} \left[\frac{1}{(p_2 + k_2)^2 + M^2 - i\varepsilon} + \frac{1}{(p_2 - k_1)^2 + M^2} \right]$$

$$- (d\omega_{p_1} \cdots)^{1/2} 4\alpha^2 m^2 G_2 \int \frac{dM^2 \, dM'^2}{\Delta^{1/2}} \left(1 - \frac{2m^2}{M'^2} \right) \frac{1}{(k_1 - k_2)^2 + M'^2}$$

$$\times \left[\frac{1}{(p_2 + k_2)^2 + M^2 - i\varepsilon} + \frac{1}{(p_2 - k_1)^2 + M^2} \right], \qquad (4\text{-}13.133)$$

which uses the relation

$$[(M^2 - m^2)^2 + M^2 M'^2] \left[f_+ - \left(1 - \frac{4m^2}{M'^2} \right) f_- \right] = \frac{4m^2}{M'^2}. \qquad (4\text{-}13.134)$$

The added single spectral forms are determined by forward scattering information. We first construct the single spectral form that gives the generalization of (4–13.29) to arbitrary scattering angles. As one can recognize from the gauge invariant structures (4–13.28), the space-time version of this coupling involves fields in the combination

$$\tfrac{1}{2} \int (dx)(dx') \, \partial_\mu \, \phi(x) F^{\mu\nu}(x) \, \Delta_+(x - x', M^2) F_{\nu\lambda}(x') \, \partial'^\lambda \, \phi(x'), \qquad (4\text{-}13.135)$$

which is to be integrated over M^2 with the appropriate weight factor. The implied single spectral form contains the reducible gauge invariant combinations

$$[(p_1 k_1)e_1^* - (p_1 e_1^*)k_1][(p_2 k_2)e_2 - (p_2 e_2)k_2] = G_1 + k_2 p_2 G_2,$$

$$[(p_2 k_1)e_1^* - (p_2 e_1^*)k_1][(p_1 k_2)e_2 - (p_1 e_2)k_2] = G_1 - k_2 p_1 G_2. \qquad (4\text{-}13.136)$$

The reduced version is

$$(d\omega_{p_1} \cdots)^{1/2} 16\alpha^2 G_1 \int_{m^2}^{\infty} \frac{dM^2}{M^2} \frac{\chi(M^2)}{M^2 - m^2} \left[\frac{1}{(p_2 + k_2)^2 + M^2 - i\varepsilon} + \frac{1}{(p_2 - k_1)^2 + M^2} \right]$$

$$- (d\omega_{p_1} \cdots)^{1/2} 8\alpha^2 G_2 \int_{m^2}^{\infty} \frac{dM^2}{M^2} \chi(M^2) \left[\frac{1}{(p_2 + k_2)^2 + M^2 - i\varepsilon} + \frac{1}{(p_2 - k_1)^2 + M^2} \right],$$

$$(4\text{-}13.137)$$

which has no infrared singularity in the forward scattering situation of actual interest, where $p_2 k_2 = p_2 k_1$. This is to be compared with the forward scattering limit of the double spectral form, which is selected by imposing the kinematical restriction $(k_1 - k_2)^2 = 0$.

One of the resulting spectral integrals is $(\mu = 0)$

$$\int_{(2m)^2}^{\infty} \frac{dM'^2}{\Delta^{1/2}} (M'^2 - 2m^2) \left[f_+ + \left(1 - \frac{4m^2}{M'^2} \right) f_- \right]$$

$$= \frac{16m^2}{M^2 - m^2} \int_0^1 dv(1 + v^2) \frac{(M^2 + m^2)^2 + v^2(M^2 - m^2)^2}{[(M^2 + m^2)^2 - v^2(M^2 - m^2)^2]^2} = \frac{8}{M^2} \frac{\chi(M^2)}{M^2 - m^2}.$$

$$(4\text{–}13.138)$$

The function $\chi(M^2)$ that appears here is

$$\chi(M^2) = (M^2 + m^2)^2 \int_0^1 dv \frac{1 - v^2}{(M^2 + m^2)^2 - v^2(M^2 - m^2)^2}, \qquad (4\text{–}13.139)$$

which definition is equivalent to that of (4–13.20). We also need the integral

$$\int_{(2m)^2}^{\infty} \frac{dM'^2}{\varDelta^{1/2}} \frac{M'^2 - 2m^2}{(M'^2)^2} = \frac{1}{m^2} \frac{1}{M^2 - m^2} \int_0^1 dv(1 + v^2) = \frac{4}{3m^2} \frac{1}{M^2 - m^2}. \quad (4\text{–}13.140)$$

The ensuing single spectral form is

$$(d\omega_{p_1} \cdots)^{1/2} 16\alpha^2 G_1 \int \frac{dM^2}{M^2} \frac{\chi(M^2)}{M^2 - m^2} [\ ] \div (d\omega_{p_1} \cdots)^{1/2} 8\alpha^2 G_2 \int \frac{dM^2}{M^2} \chi(M^2)[\ ]$$

$$+ (d\omega_{p_1} \cdots)^{1/2} 8\alpha^2 G_2 \int dM^2 \frac{\chi(M^2) - \frac{2}{3}}{M^2 - m^2} [\ ], \qquad (4\text{–}13.141)$$

where the brackets indicate the propagation function combination that is displayed in (4–13.137). The first two terms that appear above reproduce (4–13.137). Should we conclude that an additional single spectral form is needed to cancel the last term of (4–13.141)?

The answer to this question is negative; no single spectral form involving the momentum combinations $(p_2 + k_2)^2$ and $(p_2 - k_1)^2$ is required. We have allowed this apparently paradoxical situation to develop in order to emphasize a subtlety in the study of special kinematical circumstances. Does "forward scattering" mean $k_1 = k_2$, or $(k_1 - k_2)^2 = 0$? The calculation from which (4–13.137) was inferred used the $k_1 = k_2$ characterization; the reduction of the double spectral form to (4–13.141) exploited $(k_1 - k_2)^2 = 0$. The distinction between the two approaches appears on noting that G_2 vanishes for $k_1 = k_2$; only the single spectral form with G_1 as a factor can really be inferred in this way. What we must do is return to the calculation of (4–13.13) that culminated in (4–13.18) and retain the additional numerator structure that introduces G_2, while continuing to use the denominator simplifications that express $(k_1 - k_2)^2 = 0$. Then, with the notation

$$q = \frac{1 + v}{2} p_1 + \frac{1 - v}{2} p_2, \qquad (4\text{–}13.142)$$

we have

$$-\frac{\partial}{\partial q_\mu}\frac{\partial}{\partial q_\nu}\langle\log(kq)\rangle = g^{\mu\nu}\int_0^1 dz\,\frac{z^2}{[(M^2+m^2)/2M]^2-z^2[(M^2-m^2)/2M]^2}$$

$$+2q^\mu q^\nu\int_0^1 dz\,\frac{z^4}{[((M^2+m^2)/2M)^2-z^2((M^2-m^2)/2M)^2]^2}\,,$$

$$(4\text{-}13.143)$$

where, in the context of multiplication by $e_{1\mu}^*$ and $e_{2\nu}$,

$$q^\mu q^\nu \to \frac{1-v^2}{4}\,k_2^\mu k_1^\nu \to \tfrac{1}{8}k_2^\mu k_1^\nu. \qquad (4\text{-}13.144)$$

The last step gives the result of the v integration that is required in (4–13.13). After performing a partial integration in the additional term of (4–13.143), we find that (4–13.10) is now evaluated as

$$e_1{}^*e_2\chi(M^2) - e_1{}^*k_2k_1e_2\,\frac{2M^2}{(M^2-m^2)^2}\,(\chi(M^2)-\tfrac{2}{3}), \qquad (4\text{-}13.145)$$

or, with the gauge invariance substitution (4–13.28),

$$\frac{4}{(M^2-m^2)^2}\,[G_1\chi(M^2) - G_2\tfrac{1}{2}(M^2-m^2)\chi(M^2) + G_2\tfrac{1}{2}M^2(\chi(M^2)-\tfrac{2}{3})]. \quad (4\text{-}13.146)$$

We recognize, in the relative coefficients of these three terms, precisely the combination of (4–13.141).

In order to apply the double spectral form in the pair creation regime we make the crossing substitutions

$$k_1, e_1^* \to -k, e, \qquad p_2 \to -p', \qquad (4\text{-}13.147)$$

and also write k', e', p instead of k_2, e_2, p_1. As a result,

$$G_1 \to -kpkp'ee' + kpep'e'p + kp'epe'p',$$

$$G_2 \to -kk'ee' + ek'e'k. \qquad (4\text{-}13.148)$$

In the forward scattering circumstance expressed by

$$p+p' = k+k', \qquad p-p' = \left[1+\frac{4m^2}{(k+k')^2}\right]^{1/2}(k-k'), \quad (4\text{-}13.149)$$

we find that

$$2G_1 + m^2G_2 \to 0, \qquad (4\text{-}13.150)$$

while

$$\frac{1}{M^2 - m^2 + 2p_2k_2} + \frac{1}{M^2 - m^2 - 2p_2k_1} \rightarrow \frac{1}{M^2 - m^2 - 2pk} + \frac{1}{M^2 - m^2 - 2pk'}$$

$$= \frac{2}{M^2} \frac{M^2 - m^2 - kk'}{(k - k')^2 + [(M^2 - m^2)^2/M^2]}.$$

$$(4\text{-}13.151)$$

The combination that occurs in the double spectral form is

$$\frac{1}{(k + k')^2 + M'^2 - i\varepsilon} \frac{2}{M^2} \frac{M^2 - m^2 - kk'}{(k - k')^2 + [(M^2 - m^2)^2/M^2]}$$

$$= \frac{2(M^2 - m^2) + M'^2}{(M^2 - m^2)^2 + M^2M'^2} \frac{1}{(k + k')^2 + M'^2 - i\varepsilon}$$

$$+ \frac{1}{M^2} \frac{(M^2 - m^2)(M^2 + m^2)}{(M^2 - m^2)^2 + M^2M'^2} \frac{1}{(k - k')^2 + [(M^2 - m^2)^2/M^2]}. \quad (4\text{-}13.152)$$

The function $[(k - k')^2 + (M^2 - m^2)^2/M^2]^{-1}$ describes the transfer of a spacelike excitation, which is distinct from the pair-exchange mechanism represented by $[(k + k')^2 + M'^2 - i\varepsilon]^{-1}$. The contribution of the latter is

$$- (d\omega_{p_1} \cdots)^{1/2} 4\alpha^2 m^2 G_2 \int dM'^2 \frac{1 - (2m^2/M'^2)}{(k + k')^2 + M'^2 - i\varepsilon} \int \frac{dM^2}{\Delta^{1/2}} \frac{2(M^2 - m^2) + M'^2}{(M^2 - m^2)^2 + M^2M'^2}.$$

$$(4\text{-}13.153)$$

The M^2 integral that occurs here can be evaluated by decomposing it at a value of $M^2 - m^2 = X$ such that

$$2\mu m \ll X \ll m^2. \quad (4\text{-}13.154)$$

This enables one to write it as

$$\frac{4}{m^2M'^2}\left(1 - \frac{4m^2}{M'^2}\right)^{-1/2}\left[\int_1^Y \frac{dy}{(y^2 - 1)^{1/2}} + m^2 \int_X^\infty \frac{d(M^2 - m^2)}{M^2 - m^2} \frac{2(M^2 - m^2) + M'^2}{(M^2 - m^2)^2 + M^2M'^2}\right],$$

$$(4\text{-}13.155)$$

where

$$Y = \frac{X}{2\mu m}\left(1 - \frac{4m^2}{M'^2}\right)^{1/2}, \quad (4\text{-}13.156)$$

with the result

$$\frac{2}{m^2M'^2}\left(1 - \frac{4m^2}{M'^2}\right)^{-1/2}\left[\log\frac{M'^2 - 4m^2}{\mu^2} - \left(1 - \frac{4m^2}{M'^2}\right)^{1/2}\log\frac{1 + [1 - (4m^2/M'^2)]^{1/2}}{1 - [1 - (4m^2/M'^2)]^{1/2}}\right].$$

$$(4\text{-}13.157)$$

The comparison with (4–13.56, 58) indicates that the following single spectral form, which is stated for the photon scattering situation, is needed to supplement the double spectral form,

$$(d\omega_{p_1}\cdots)^{1/2}4\alpha^2 G_2 \int_{(2m)^2}^{\infty} \frac{dM'^2}{M'^2}\frac{2m^2}{M'^2}\log\frac{1+[1-(4m^2/M'^2)]^{1/2}}{1-[1-(4m^2/M'^2)]^{1/2}}\frac{1}{(k_1-k_2)^2+M'^2}.$$

$$(4\text{–}13.158)$$

One question remains, however. The forward scattering condition (4–13.149) leaves undetermined any structure involving $G_1 + \frac{1}{2}m^2 G_2$. Does the double spectral form specify the latter correctly, or is an additional single spectral form required? As one might anticipate from the earlier discussion concerning G_2, the double spectral form does contain the necessary information. In order to verify this, we return to the calculation in (4–13.44) and consider, in the center of mass frame, an arrangement in which the particle momenta $\mathbf{p} = -\mathbf{p}'$ make an angle α with the photon momenta $\mathbf{k} = -\mathbf{k}'$. We also subtract the results for the two arrangements where the polarization vectors $\mathbf{e} = \mathbf{e}'$ lie in the plane of $\mathbf{p}$ and $\mathbf{k}$, and are perpendicular to it, respectively. This isolates the G_1 dependence, since G_2 assumes the same value for both polarization choices, whereas

$$G_1|_{\parallel} - G_1|_{\perp} = -kk'(\mathbf{e}\cdot\mathbf{p})^2|_{\parallel} = \frac{M^2}{2}\left(\frac{M^2}{4} - m^2\right)\sin^2\alpha. \qquad (4\text{–}13.159)$$

The integral that now appears is

$$-\int\frac{\sin\theta\,d\theta\,d\phi}{4\pi}\left[\frac{2(M^2-2m^2)}{((M^2/4)-m^2)2(1-\cos\theta)+\mu^2}-1\right]\left(\frac{M^2}{4}-m^2\right)\left(\frac{1}{-p_2k}+\frac{1}{-p_2k'}\right)$$

$$\times\,[(-\sin\alpha\cos\theta+\cos\alpha\sin\theta\cos\phi)^2-(\sin\theta\sin\phi)^2], \qquad (4\text{–}13.160)$$

where

$$\frac{1}{-p_2k}+\frac{1}{-p_2k'}=\frac{8}{M^2}\frac{1}{1-(1-(4m^2/M^2))(\cos\alpha\cos\theta+\sin\alpha\sin\theta\cos\phi)^2}, \qquad (4\text{–}13.161)$$

which employs a spherical coordinate system that selects $\mathbf{p}$ as the z-axis while $\mathbf{k}$ lies in the xz-plane. It should be noted that the -1 term of the first bracket makes no contribution since, lacking reference to the $\mathbf{p}$ vector, the two polarization choices are equivalent, and cancel.

It suffices to consider the limit $\alpha \to 0$, after differentiating twice. The result of this operation on (4–13.160), in which the azimuthal integration has been performed, is

$$- 16\left(1 - \frac{2m^2}{M^2}\right)\int_{-1}^{1} \frac{1}{2}\, dz\, \frac{2((M^2/4) - m^2)}{2((M^2/4) - m^2)(1 - z) + \mu^2}\, \frac{1}{D}$$

$$+ 16\left(1 - \frac{2m^2}{M^2}\right)\int_{0}^{1} dz\, \left\{\frac{1}{2}\frac{1}{D} + \frac{1}{D^2} + \left(1 - \frac{4m^2}{M^2}\right)\left[\frac{z^2}{D^2} + \frac{3}{4}\frac{1 - z^2}{D^2} - \frac{1 - z^2}{D^3}\right]\right\},$$

$$\tag{4-13.162}$$

which employs the abbreviation

$$D = 1 - \left(1 - \frac{4m^2}{M^2}\right)z^2. \tag{4-13.163}$$

Completing the integration gives

$$-\frac{2}{m^2}(M^2 - 4m^2)\rho(M^2), \tag{4-13.164}$$

where

$$\rho(M^2) = \frac{M^2 - 2m^2}{M^2 - 4m^2}\left[\log\frac{M^2 - 4m^2}{\mu^2} - \frac{3}{2} - \left(1 - \frac{m^2}{M^2}\right)\right.$$

$$\left. \times \frac{1}{(1 - (4m^2/M^2))^{1/2}}\log\frac{1 + (1 - (4m^2/M^2))^{1/2}}{1 - (1 - (4m^2/M^2))^{1/2}}\right]. \tag{4-13.165}$$

The implied spectral form is

$$(d\omega_{p_1}\cdots)^{1/2}16\frac{\alpha^2}{m^2}(G_1 + \tfrac{1}{2}m^2 G_2)\int_{(2m)^2}^{\infty}\frac{dM'^2}{M'^2}\left(1 - \frac{4m^2}{M'^2}\right)^{1/2}\rho(M'^2)\frac{1}{(k + k')^2 + M'^2 - i\varepsilon}.$$

$$\tag{4-13.166}$$

Comparison must now be made with the single spectral form that is extracted from the first term of (4-13.133) by imposing a scalar version of the forward scattering condition. As one can recognize from the coefficients of ee' in (4-13.148), this statement is

$$2kpkp' + m^2 kk' = 0; \tag{4-13.167}$$

it suffices to bring about the reduction given in Eq. (4-13.151). The spectral form can be written in the notation of (4-13.166), with

$$\rho(M'^2) = \tfrac{1}{8}m^2(M'^2)^2\frac{M'^2 - 2m^2}{(1 - (4m^2/M'^2))^{1/2}}\int\frac{dM^2}{\Delta^{1/2}}\frac{2(M^2 - m^2) + M'^2}{(M^2 - m^2)^2 + M^2 M'^2}$$

$$\times \left[f_+ + \left(1 - \frac{4m^2}{M'^2}\right)f_-\right]. \tag{4-13.168}$$

The outcome of the integration is just (4-13.165). We conclude that the modification in particle-photon scattering is described completely by the double spectral form (4–13.133) and the single spectral form (4–13.158).

Harold interjects a question.

H.: Before you go on, please tell me this. Why has there been no mention of single particle exchange? After all, the skeletal Compton scattering process can be inferred from a causal arrangement in which a particle is scattered by an electromagnetic field and the resulting real particle detected by another such scattering process. We would be concerned with the dynamical modification of each scattering act.

S.: In general, you are right; single particle exchange must also be considered. Spin 0 is an exception. If one considers the scattering of a real particle by an appropriate field and then extrapolates to the field of a photon, the charge form factor $F(k)$ is finally evaluated for $k^2 = 0$, and there it remains unity. Single particle exchange continues to be described by the skeletal coupling, for spin 0.

Only the double spectral form involves the photon mass. There is a simple test of this dependence, since it must balance the additional scattering that is accompanied by a soft photon, the so-called double Compton scattering. Our concern now is only with the coefficients of log $1/\mu$ and not with additive constants. The relevant contribution of the double spectral form, which comes from the neighborhood $M^2 \sim m^2$, is

$$(d\omega_{p_1} \cdots)^{1/2} 8\alpha^2 G_1 \frac{(k_1 - k_2)^2}{k_1 p_2 k_2 p_2} \log \frac{1}{\mu} \int_{(2m)^2}^{\infty} \frac{dM'^2}{M'^2} \frac{1 - (2m^2/M'^2)}{(1 - (4m^2/M'^2))^{1/2}} \frac{1}{(k_1 - k_2)^2 + M'^2},$$

(4–13.169)

according to the relation

$$\frac{1}{2k_2 p_2} - \frac{1}{2k_1 p_2} = \frac{1}{4} \frac{(k_1 - k_2)^2}{k_1 p_2 k_2 p_2}.$$

(4–13.170)

This is superimposed on the amplitude of the skeletal scattering process [Eq. (3–12.98)] to give

$$\langle 1|T|2 \rangle = - (d\omega_{p_1} \cdots)^{1/2} 8\pi\alpha \frac{G_1}{k_1 p_2 k_2 p_2}$$

$$\times \left[1 - \frac{\alpha}{4\pi} \log \frac{1}{\mu} \frac{(k_1 - k_2)^2}{m^2} \int_0^1 dv \frac{1 + v^2}{1 + [(k_1 - k_2)^2/4m^2](1 - v^2)} \right].$$

(4–13.171)

The implied fractional modification of the elastic scattering cross section, in its dependence on the photon mass,

$$1 - \frac{\alpha}{2\pi} \log \frac{1}{\mu} \frac{(k_1 - k_2)^2}{m^2} \int_0^1 dv \, \frac{1 + v^2}{1 + [(k_1 - k_2)^2/4m^2](1 - v^2)}, \quad (4\text{-}13.172)$$

does indeed combine with the soft photon emission probability stated in (4–12.16), where

$$q^2 = (p_1 - p_2)^2 = (k_1 - k_2)^2, \quad (4\text{-}13.173)$$

to produce the total elimination of the fictitious photon mass.

The low energy limit of the scattering process is characterized, in part, by the condition

$$(k_1 - k_2)^2 \ll 4m^2, \quad (4\text{-}13.174)$$

and therefore resembles the forward scattering discussion that produced (4–13.141), except that attention must be paid to the infrared phenomena near $M^2 = m^2$. The latter consideration would repeat the photon mass discussion just given, subject to (4–13.174), except that we are now interested in the constant added to $\log(1/\mu)$. Let the first integral of (4–13.141) be terminated at the lower limit $M^2 = m^2 + X$, where X obeys the restrictions of (4–13.154). Then, added to it is

$$(d\omega_{p_1} \cdots)^{1/2} 8\alpha^2 G_1 \frac{(k_1 - k_2)^2}{k_1 p_2 k_2 p_2} \int_{(2m)^2}^{\infty} \frac{dM'^2}{(M'^2)^2} \frac{1 - (2m^2/M'^2)}{(1 - (4m^2/M'^2))^{1/2}} \int_1^Y \frac{dy}{(y^2 - 1)^{1/2}}, \quad (4\text{-}13.175)$$

where Y is defined by (4–13.156). The double integral here is evaluated as

$$\frac{1}{4m^2} \int_0^1 dv(1 + v^2) \log\left(\frac{X}{\mu m} v\right) = \frac{1}{3m^2}\left(\log \frac{X}{\mu m} - \frac{5}{6}\right). \quad (4\text{-}13.176)$$

The soft photon emission that removes the μ dependence is represented by (4–12.25) where, for the nonrelativistic situation that occurs in the rest frame of the initial particle, δM can be identified as $k^0_{\min}$, the minimum detectable frequency of an additional soft photon. The dependence on these photon parameters is contained in the integral

$$\int_0^1 dv(1 + v^2)\left[\log\left(\frac{k^0_{\min}}{2\mu} \frac{1 - v^2}{v}\right) + \frac{1}{2}\right] = \frac{4}{3}\left[\log \frac{2k^0_{\min}}{\mu} - \frac{5}{6}\right]. \quad (4\text{-}13.177)$$

Thus, the effective value of the infrared sensitive term is

$$(d\omega_{p_1} \cdots)^{1/2} \frac{8}{3} \frac{\alpha^2}{m^2} G_1 \frac{(k_1 - k_2)^2}{k_1 p_2 k_2 p_2} \log \frac{X}{2mk^0_{\min}}. \quad (4\text{-}13.178)$$

To find the low energy limit of the contribution stated in (4–13.141), we apply an integral, which is related to (4–13.36),

$$\int_{m^2}^{\infty} \frac{dM^2}{M^2} \chi(M^2) \frac{1}{M^2 - m^2 + x} = \left[\frac{1}{m^2 - x} + \int_{-1}^{1} dz \frac{z(1-z)}{2m^2 - (1-z)x} \right] \log \left(\frac{m^2}{x} \right)$$

$$+ \int_{-1}^{1} dz \frac{z(1-z)}{2m^2 - (1-z)x} \log \left(\frac{2}{1-z} \right). \quad (4\text{-}13.179)$$

As stated, it refers to $x > 0$. With $x < 0$, the corresponding proper value integral is obtained on writing $|x|$ in the argument of the first logarithm. When one extracts the part of (4-13.179) that is odd in x, the formula (4-13.36) is recovered. The leading terms of an expansion for small x are displayed in

$$\int_{m^2}^{\infty} \frac{dM^2}{M^2} \chi(M^2) \frac{1}{M^2 - m^2 + x} = \frac{2}{3m^2} \left[\left(1 + \frac{x}{m^2} \right) \log \left(\frac{m^2}{x} \right) + \frac{1}{12} \left(1 - \frac{1}{2} \frac{x}{m^2} \right) + \cdots \right].$$

$$(4\text{-}13.180)$$

The type of integral that appears in the G_1 term of (4-13.141) is

$$\int_{m^2+x}^{\infty} \frac{dM^2}{M^2} \frac{\chi(M^2)}{M^2 - m^2} \frac{1}{M^2 - m^2 + x} = \frac{1}{x} \int_{m^2+x}^{\infty} \frac{dM^2}{M^2} \frac{\chi(M^2)}{M^2 - m^2}$$

$$- \frac{1}{x} \int_{m^2}^{\infty} \frac{dM^2}{M^2} \frac{\chi(M^2)}{M^2 - m^2 + x}. \quad (4\text{-}13.181)$$

The first of these integrals on the right-hand side is obtained from (4-13.180) by considering $x \ll X \ll m^2$:

$$\int_{m^2+x}^{\infty} \frac{dM^2}{M^2} \frac{\chi(M^2)}{M^2 - m^2} + \int_{m^2}^{m^2+X} \frac{dM^2}{m^2} \frac{2}{3} \frac{1}{M^2 - m^2 + x} = \frac{2}{3m^2} \left(\log \frac{m^2}{x} + \frac{1}{12} \right),$$

$$(4\text{-}13.182)$$

namely

$$\int_{m^2+x}^{\infty} \frac{dM^2}{M^2} \frac{\chi(M^2)}{M^2 - m^2} = \frac{2}{3m^2} \left(\log \frac{m^2}{X} + \frac{1}{12} \right). \quad (4\text{-}13.183)$$

This is the term that combines with (4-13.178) to eliminate the arbitrary parameter X.

In the low energy limit, the last of the three terms of (4-13.141) supplies the integral

$$K = \int_{m^2}^{\infty} dM^2 \frac{\chi(M^2) - \frac{2}{3}}{M^2 - m^2} \frac{2}{M^2 - m^2}, \quad (4\text{-}13.184)$$

the existence of which is confirmed by rewriting (4-13.139) as

$$\chi(M^2) - \frac{2}{3} = \left(\frac{M^2 - m^2}{M^2 + m^2} \right)^2 \int_0^1 dv \frac{v^2(1-v^2)}{1 - v^2[(M^2 - m^2)/(M^2 + m^2)]^2}. \quad (4\text{-}13.185)$$

On introducing the variable

$$u = \frac{M^2 - m^2}{M^2 + m^2}, \qquad 0 < u < 1, \qquad (4\text{-}13.186)$$

we get

$$K = \frac{1}{m^2} \int_0^1 du \int_0^1 dv \, \frac{v^2(1 - v^2)}{1 - u^2 v^2} = \frac{1}{2m^2} \int_0^1 dv \, v(1 - v^2) \log \frac{1 + v}{1 - v} = \frac{1}{6m^2}. \quad (4\text{-}13.187)$$

The last of the needed integrals is supplied by the single spectral form (4–13.158). It is

$$\int_{(2m)^2}^{\infty} \frac{dM'^2}{(M'^2)^2} \frac{2m^2}{M'^2} \log \frac{1 + (1 - (4m^2/M'^2))^{1/2}}{1 - (1 - (4m^2/M'^2))^{1/2}} = \frac{1}{2} K. \quad (4\text{-}13.188)$$

The various contributions are put together to give the low energy form of the transition matrix:

$$\langle 1|T|2 \rangle = - (d\omega_{p_1} \cdots)^{1/2} 8\pi\alpha \, \frac{G_1}{k_1 p_2 k_2 p_2}$$

$$+ (d\omega_{p_1} \cdots)^{1/2} \frac{32}{3} \frac{\alpha^2}{m^2} \left[\frac{1}{4} \frac{(k_1 - k_2)^2}{k_1 p_2 k_2 p_2} G_1 \left(\log \frac{|k_2 p_2|}{m k^0_{\min}} - 1 \right) \right.$$

$$\left. - 2 \frac{G_1}{m^2} \left(\log \frac{m^2}{2|k_2 p_2|} - \frac{1}{24} \right) - G_2 \left(\log \frac{m^2}{2|k_2 p_2|} - \frac{7}{96} \right) \right], \quad (4\text{-}13.189)$$

omitting the imaginary term since it has negligible effect in the computation of the differential cross section, which is our present concern. Considered in the rest frame of the initial particle, and in the gauge where polarization vectors are orthogonal to this momentum ($p_2 e = 0$), our result reads

$$\langle 1|T|2 \rangle = - (d\omega_{p_1} \cdots)^{1/2} 8\pi\alpha e_1{}^* \cdot e_2 + (d\omega_{p_1} \cdots)^{1/2} \frac{32}{3} \alpha^2 \left(\frac{k^0}{m} \right)^2$$

$$\times \left[\tfrac{1}{2}(1 - \cos\theta) e_1{}^* \cdot e_2 \left(\log \frac{k^0}{k^0_{\min}} - 1 \right) - 2 e_1{}^* \cdot e_2 \left(\log \frac{m}{2k^0} - \frac{1}{24} \right) \right.$$

$$\left. + ((1 - \cos\theta) e_1{}^* \cdot e_2 + n_2 \cdot e_1{}^* n_1 \cdot e_2) \left(\log \frac{m}{2k^0} - \frac{7}{96} \right) \right]. \quad (4\text{-}13.190)$$

Here, $n_{1,2}$ are the unit photon propagation vectors,

$$n_1 \cdot n_2 = \cos\theta, \qquad (4\text{-}13.191)$$

and k^0 represents the photon energy, which is essentially unaltered in this low energy collision.

The modified differential cross section for polarized photons is obtained immediately from (4–13.190), and generalizes the result of Eq. (4–13.40). We shall only state the modification in the differential cross section for unpolarized photons. The required summations over final polarizations and averages over initial polarizations are given by [cf. Eq. (3–14.101)]

$$\tfrac{1}{2} \sum |e_1{}^* \cdot e_2|^2 = \tfrac{1}{2}(1 + \cos^2\theta), \tag{4–13.192}$$

and

$$\tfrac{1}{2} \sum e_2{}^* \cdot e_1 n_2 \cdot e_1{}^* n_1 \cdot e_2 = -\tfrac{1}{4} \cos\theta(1 - \cos^2\theta). \tag{4–13.193}$$

The conclusion is that

$$\delta\left(\frac{d\sigma}{d\Omega}\right) = \left(\frac{\alpha}{m}\right)^2 \frac{4\alpha}{3\pi} \left(\frac{k^0}{m}\right)^2 \left[(1 + \cos\theta)^2 \left(\log\frac{m}{2k^0} - \frac{1}{24} \right) + \frac{1}{32}(1 - \cos\theta)^2 \right.$$
$$\left. - \tfrac{1}{2}(1 - \cos\theta)(1 + \cos^2\theta)\left(\log\frac{k^0}{k^0_{\min}} - 1 \right) \right]. \tag{4–13.194}$$

In the particular situation of back scattering this becomes

$$\theta = \pi: \quad \delta\left(\frac{d\sigma}{d\Omega}\right) = -\left(\frac{\alpha}{m}\right)^2 \frac{8\alpha}{3\pi} \left(\frac{k^0}{m}\right)^2 \left(\log\frac{k^0}{k^0_{\min}} - \frac{17}{16} \right), \tag{4–13.195}$$

while the modification in the total cross section is

$$\delta\sigma = \frac{64}{9} \frac{\alpha^3}{m^2} \left(\frac{k^0}{m}\right)^2 \left[\log\frac{m}{2k^0} - \frac{1}{96} - \frac{1}{2}\left(\log\frac{k^0}{k^0_{\min}} - 1 \right) \right], \tag{4–13.196}$$

which can be an increase or a decrease, depending upon the quantitative relation between $k^0/k^0_{\min}$ and $(m/2k^0)^2$. We shall not trouble to give details about the high energy behavior, except to note that the fractional modification in the differential cross section is of order α, multiplied by logarithms that vary in form with the particular angular region [an example is Eq. (4–13.41)].

Historical note: The energy- and angle-dependent factor of Eq. (4–13.194), in an equivalent form, was stated in 1948 by E. Corinaldesi and R. Jost, *Helv. Phys. Acta* **21**, 183. These authors used the unitary transformation method that also gave the first results concerning the electron magnetic moment, energy displacements, and Coulomb scattering modifications.

4–14 NONCAUSAL METHODS

Occasional use has been made of noncausal calculational techniques, most notably in the treatment of low frequency scattering of light by light and related questions.

The calculation of the probability that an electron-positron pair be created by a strong homogeneous electric field was particularly striking since no finite number of single scattering encounters can produce this act. We now want to recognize the special ability of noncausal methods to handle the problem of bound state energy displacements, which similarly involve an unlimited number of interactions. This is welcome since the causal methods have not suggested any very elegant solution to the problem of finding a unified treatment of high and low energy phenomena. Indeed, much time and energy were expended on one such calculation of the $Z\alpha$ modification to the energy displacement in a Coulomb field before the unnecessary complexity of the procedure was admitted and the attempt discarded. It is worth remarking here that the freedom to choose between, or to combine, causal and noncausal calculational methods emphasizes the unification that source theory has brought about between the causally oriented analytic S-matrix theory (by removing the hypothesis of analyticity) and the noncausal operator field theory (by removing the operator fields).

To indicate the kind of approach now to be studied, let us return to the spin 0 vacuum amplitude of Eq. (4–12.44). There the particle propagation function retains its general space-time form, but that of the photon has been specialized to its causal version. We shall remove this restriction through the replacement

$$i \, d\omega_k \to \frac{(dk)}{(2\pi)^4} \frac{1}{k^2},$$ (4–14.1)

$(-i\varepsilon$ is omitted) and no longer insist that ϕ_1 and ϕ_2 be in causal relationship. They are, however, still required not to overlap. Thus, the vacuum amplitude now reads, in a symbolic notation,

$$e^2 \int \frac{(dk)}{(2\pi)^4} \phi_1 (2\Pi - k) \frac{1}{k^2} \frac{1}{(\Pi - k)^2 + m^2} (2\Pi - k) \phi_2.$$ (4–14.2)

A useful rearrangement is stated by

$$(2\Pi - k) \frac{1}{k^2} \frac{1}{(\Pi - k)^2 + m^2} (2\Pi - k) = (2\Pi - k)^2 \cdot \frac{1}{k^2} \frac{1}{(\Pi - k)^2 + m^2}$$

$$- 2 \left[\Pi, \left[\Pi, \frac{1}{k^2} \frac{1}{(\Pi - k)^2 + m^2} \right] \right],$$ (4–14.3)

where the dot indicates symmetrized multiplication,

$$A.B = \tfrac{1}{2}\{A, B\}.$$ (4–14.4)

We have, furthermore,

$$(2\Pi - k)^2 = 2[(\Pi - k)^2 + m^2] - k^2 + 2\Pi^2 - 2m^2,$$ (4–14.5)

in which the first term on the right can be omitted since the corresponding local structure in (4–14.3) vanishes under the nonoverlapping circumstances being considered. Accordingly, our actual starting point is the vacuum amplitude expression

$$
e^2 \int \frac{(dk)}{(2\pi)^4} \, \phi_1 \left\{ - \frac{1}{(\Pi - k)^2 + m^2} + (2\Pi^2 - 2m^2) \cdot \frac{1}{k^2} \frac{1}{(\Pi - k)^2 + m^2} \right.
$$
$$
\left. - 2 \left[\Pi, \left[\Pi, \frac{1}{k^2} \frac{1}{(\Pi - k)^2 + m^2} \right] \right] \right\} \phi_2. \tag{4–14.6}
$$

We shall use the propagation function representations [cf. Eq. (4–8.34)]

$$
\frac{1}{(\Pi - k)^2 + m^2} = i \int_0^\infty ds \exp\{- is[(\Pi - k)^2 + m^2]\},
$$
$$
\frac{1}{k^2} = i \int_0^\infty ds \exp(- isk^2), \tag{4–14.7}
$$

and the implied product representation

$$
\frac{1}{k^2} \frac{1}{(\Pi - k)^2 + m^2} = - \int_0^\infty ds_1 \, ds_2 \exp\{- is_1[(\Pi - k)^2 + m^2] - is_2 k^2\}. \tag{4–14.8}
$$

In the latter, the parameter transformation

$$
s_1 = su, \qquad s_2 = s(1 - u), \qquad ds_1 \, ds_2 = s \, ds \, du \tag{4–14.9}
$$

produces the form

$$
\frac{1}{k^2} \frac{1}{(\Pi - k)^2 + m^2} = - \int_0^\infty ds \, s \int_0^1 du \exp[- is\chi(u)], \tag{4–14.10}
$$

where

$$
\chi(u) = u[(\Pi - k)^2 + m^2] + (1 - u)k^2
$$
$$
= (k - u\Pi)^2 + u(1 - u)\Pi^2 + m^2 u. \tag{4–14.11}
$$

Note that if we want to consider a "photon" of mass μ, the term $\mu^2(1 - u)$ should be appended to $\chi(u)$.

Let us illustrate the use of the product representation in the simple situation without an electromagnetic field where, according to the commutation relation [Eq. (4–8.44)]

$$
[\Pi, \Pi] = ieqF, \tag{4–14.12}
$$

the components of Π are commutative. Then, the last term of (4–14.6) vanishes. Also, the permissible redefinition of the integration variable in the first term

$k - \Pi \to k$, exhibits that contribution as a local one, which can be omitted. The analogous transformation in $\chi(u)$, $k - u\Pi \to k$, yields [a photon mass term is included],

$$\int \frac{(dk)}{(2\pi)^4} \frac{1}{k^2} \frac{1}{(\Pi - k)^2 + m^2}$$

$$= -\int_0^\infty ds\, s \int_0^1 du \exp\{- is[u(1-u)\Pi^2 + m^2 u + \mu^2(1-u)]\} \int \frac{(dk)}{(2\pi)^4} \exp(-isk^2)$$

$$= \frac{i}{(4\pi)^2} \int \frac{ds}{s} du \exp\{- is[u(1-u)\Pi^2 + m^2 u + \mu^2(1-u)]\}, \qquad (4\text{-}14.13)$$

according to the momentum integral (4–8.57). It is convenient to carry out a partial integration on u:

$$\int_0^1 du \exp(- is[\ ])$$

$$= u \exp(- is[\ ])|_0^1 + is \int_0^1 du\, u((1 - 2u)\Pi^2 + m^2 - \mu^2) \exp(- is[\ ]). \qquad (4\text{-}14.14)$$

The first term on the right contributes, at $u = 1$, the value $\exp(-ism^2)$ which, being local, vanishes in Eq. (4–14.6). The s integration is now performed, and (4–14.6) becomes

$$i\, \frac{\alpha}{4\pi} \int_0^1 du\, u\phi_1 \frac{(2\Pi^2 - 2m^2)((1 - 2u)\Pi^2 + m^2 - \mu^2)}{u(1-u)\Pi^2 + m^2 u + \mu^2(1-u)} \phi_2. \qquad (4\text{-}14.15)$$

The continued reference to nonoverlapping conditions enables one to reduce Π^2 in the numerator according to the substitution

$$\Pi^2 \to -\frac{m^2}{1-u} - \frac{\mu^2}{u}. \qquad (4\text{-}14.16)$$

This yields [for simplicity, μ^2 is set equal to zero in the numerator]

$$-i\, \frac{\alpha}{2\pi} \int du\, \frac{u}{(1-u)^2} \left(1 + \frac{1}{1-u}\right) m^4 \phi_1 \frac{1}{\Pi^2 + (m^2/1 - u) + (\mu^2/u)} \phi_2. \qquad (4\text{-}14.17)$$

It is at this stage that the full space-time extrapolation is performed by adding contact terms, in the known manner,

$$\frac{1}{\Pi^2 + M^2} \to \frac{1}{\Pi^2 + M^2} - \frac{1}{M^2 - m^2} + \frac{\Pi^2 + m^2}{(M^2 - m^2)^2}. \qquad (4\text{-}14.18)$$

When μ is set equal to zero, and thus

$$M^2 = \frac{m^2}{1 - u}, \tag{4-14.19}$$

the spectral weight factor that appears in (4–14.17) is

$$\frac{du}{1-u}\frac{u}{1-u}\left(1 + \frac{1}{1-u}\right)m^4 = \frac{dM^2}{M^2}(M^2 - m^2)(M^2 + m^2), \tag{4-14.20}$$

in complete agreement with (4–6.40). For finite photon mass, the spectral representation that is provided by (4–14.17), in which

$$M^2 = \frac{m^2}{1-u} + \frac{\mu^2}{u}, \tag{4-14.21}$$

uses the parametrization already encountered in Eq. (4–1.32). In this connection, it should be noted that

$$\int_0^1 du\, \eta\left(M^2 - \frac{m_a^2}{u} - \frac{m_b^2}{1-u}\right) = \int_0^1 du\, u\left(\frac{m_b^2}{(1-u)^2} - \frac{m_a^2}{u^2}\right)\delta\left(M^2 - \frac{m_a^2}{u} - \frac{m_b^2}{1-u}\right). \tag{4-14.22}$$

We are going to make extensive use of expansions of $\exp[-is\chi(u)]$ that are patterned after the quantum mechanical perturbation expansion

$$\exp[-it(H_0 + H_1)]$$

$$= \exp(-itH_0) - i\int_0^t dt_1 \exp[-i(t - t_1)H_0]\, H_1 \exp(-it_1H_0)$$

$$+ (-i)^2 \int_0^t dt_1 \int_0^t dt_2\, \eta(t_1 - t_2) \exp[-i(t - t_1)H_0]\, H_1$$

$$\times \exp[-i(t_1 - t_2)H_0]\, H_1 \exp(-it_2H_0) + \cdots \tag{4-14.23}$$

[for a quantum action principle derivation, see *Quantum Kinematics and Dynamics*, W. A. Benjamin, Inc., Menlo Park, 1970, Section 7–6, although the time dependence is there left implicit]. It is convenient to use fractional elapsed times as the integration variables, and to give them appropriately symmetrical forms. Thus, for the situation of interest, with

$$\chi = \chi_0 + \chi_1, \tag{4-14.24}$$

we write

$$\exp(-is\chi)$$

$$= \exp(-is\chi_0) - is\int_{-1}^1 \tfrac{1}{2}\, dv \exp\{-is[(1+v)/2]\chi_0\}\, \chi_1 \exp\{-is[(1-v)/2]\chi_0\}$$

$$-s^2 \int_{-1}^{1} \tfrac{1}{2}\, dv \int_0^1 dw\, w \exp\{-is[(1+v)/2]w\chi_0\}\, \chi_1 \exp[-is(1-w)\chi_0]\, \chi_1$$

$$\times \exp\{-is[(1-v)/2]w\chi_0\} + \cdots. \tag{4-14.25}$$

If χ_1 is an infinitesimal quantity, the expansion terminates with the linear χ_1 term. That is the situation when $\chi(u)$ is subjected to an infinitesimal variation, as illustrated by

$$[\Pi, \exp(-is\chi)] = -is \int_{-1}^{1} \tfrac{1}{2}\, dv \exp\{-is[(1+v)/2]\chi\}\, [\Pi, \chi]\, \exp\{-is[(1-v)/2]\chi\}. \tag{4-14.26}$$

In this example we have

$$[\Pi, \chi(u)] = u[\Pi, (\Pi - k)^2] = 2uieqF.(\Pi - k). \tag{4-14.27}$$

Let us also discuss here the evaluation of the double commutator $[\Pi, [\Pi, \exp(-is\chi)]]$, in which the scalar product of the Π vectors is understood. Consider, for that purpose, the transformation

$$\exp(\lambda\Pi) \exp(-is\chi) \exp(-\lambda\Pi) = \exp[-is \exp(\lambda\Pi) \chi \exp(-\lambda\Pi)], \tag{4-14.28}$$

where λ is an arbitrary constant vector. We shall compare the expansions of both sides, based upon the general commutator expansion

$$e^A B e^{-A} = B + [A, B] + \frac{1}{2!}[A, [A, B]] + \cdots, \tag{4-14.29}$$

which can be verified through successive differentiations of a scale parameter in A. Thus, up to terms quadratic in λ, we have

$$\exp(-is\chi) + [\lambda\Pi, \exp(-is\chi)] + \tfrac{1}{2}[\lambda\Pi, [\lambda\Pi, \exp(-is\chi)]]$$

$$= \exp\{-is\chi - is[\lambda\Pi, \chi] - is\tfrac{1}{2}[\lambda\Pi, [\lambda\Pi, \chi]]\}. \tag{4-14.30}$$

The expansion of Eq. (4–14.25) presents the right-hand side as

$$\exp(-is\chi) - is \int_{-1}^{1} \tfrac{1}{2}\, dv \exp\{-is[(1+v)/2]\chi\}\, [\lambda\Pi, \chi]\, \exp\{-is[(1-v)/2]\chi\}$$

$$- is\tfrac{1}{2} \int_{-1}^{1} \tfrac{1}{2}\, dv \exp\{-is[(1+v)/2]\chi\}\, [\lambda\Pi, [\lambda\Pi, \chi]]\, \exp\{-is[(1-v)/2]\chi\}$$

$$- s^2 \int_{-1}^{1} \tfrac{1}{2}\, dv \int_0^1 dw\, w \exp\{-is[(1+v)/2]w\chi\}\, [\lambda\Pi, \chi]\, \exp[-is(1-w)\chi]\, [\lambda\Pi, \chi]$$

$$\times \exp\{-is[(1-v)/2]w\chi\}. \tag{4-14.31}$$

Comparison with the left side then restates (4–14.26) and gives the desired expression:

$$[\Pi, [\Pi, \exp(-is\chi)]] = -is \int_{-1}^{1} \tfrac{1}{2} dv \exp\{-is[(1+v)/2]\chi\} [\Pi, [\Pi, \chi]]$$

$$\times \exp\{-is[(1-v)/2]\chi\} - 2s^2 \int_{-1}^{1} \tfrac{1}{2} dv \int_{0}^{1} dw\, w$$

$$\times \exp\{-is[(1+v)/2]w\chi\} [\Pi, \chi] \exp[-is(1-w)\chi] [\Pi, \chi]$$

$$\times \exp\{-is[(1-v)/2]w\chi\}, \tag{4–14.32}$$

where

$$[\Pi, [\Pi, \chi(u)]] = -2ueq(\Pi - k).J - 2ue^2 F^{\mu\nu} F_{\mu\nu} \tag{4–14.33}$$

and we have introduced

$$J^\mu = \partial_\nu F^{\mu\nu}. \tag{4–14.34}$$

Note that the symmetrization in the associated term is unnecessary, since

$$i[\Pi_\mu, J^\mu] = \partial_\mu J^\mu = 0. \tag{4–14.35}$$

The technical problem before us is to carry out the k integration when the substitution $k - u\Pi \to k$ cannot be made owing to the noncommutativity of the components of Π in the presence of an electromagnetic field. To that end we propose a device suggested by the following quantum mechanical considerations. In a system of n q- and p-variables, the expectation value $\langle q'|f(p)|q'\rangle$ has the evaluation

$$\langle q'|f(p)|q'\rangle = \int \langle q'|p'\rangle (dp') f(p') \langle p'|q'\rangle = \int \frac{(dp')}{(2\pi)^n} f(p'), \tag{4–14.36}$$

which is independent of q'. Accordingly, let ξ be the (four-vector) coordinate complementary to k, and write

$$\int \frac{(dk)}{(2\pi)^4} f(k) = \langle \xi' = 0|f(k)|\xi' = 0\rangle. \tag{4–14.37}$$

The advantage offered by this reformulation is the possibility of introducing canonical transformations that do not affect the expectation value, but alter its form in a useful manner. Thus, we shall bring about the nearest correct version of the invalid substitution $k - u\Pi \to k$ through an operator transformation. Indeed, when the field vanishes and the components of Π commute, we have

$$\exp(-iu\xi\Pi) f(k - u\Pi) \exp(iu\xi\Pi) = f(k) \tag{4–14.38}$$

and

$$\langle \xi' = 0 | f(k - u\Pi) | \xi' = 0 \rangle = \langle \xi' = 0 | f(k) | \xi' = 0 \rangle. \qquad (4\text{--}14.39)$$

The simplest procedure is to use the same transformation in the presence of the field. We therefore seek to evaluate the transformed quantities

$$\mathring{k} = \exp(- iu\xi\Pi) \, k \exp(iu\xi\Pi), \qquad \mathring{\Pi} = \exp(- iu\xi\Pi) \, \Pi \exp(iu\xi\Pi). \qquad (4\text{--}14.40)$$

One can exploit the presence of the variable u to produce differential equations:

$$\frac{d}{du} \mathring{k} = \exp(- iu\xi\Pi) \, i[k, \xi\Pi] \exp(iu\xi\Pi) = \mathring{\Pi},$$

$$\frac{d}{du} \mathring{\Pi} = \exp(- iu\xi\Pi) \, i[\Pi, \xi\Pi] \exp(iu\xi\Pi) = - eq\mathring{F}\xi, \qquad (4\text{--}14.41)$$

where

$$\mathring{F} = F(\mathring{x}), \qquad \mathring{x} = \exp(- iu\xi\Pi) \, x \exp(iu\xi\Pi). \qquad (4\text{--}14.42)$$

The chain of transformations stops with $\mathring{x}$, since

$$\frac{d}{du} \mathring{x} = \exp(- iu\xi\Pi) \, i[x, \xi\Pi] \exp(iu\xi\Pi) = - \xi, \qquad (4\text{--}14.43)$$

in virtue of the commutativity of the ξ components among themselves. Hence,

$$\mathring{x} = x - u\xi \qquad (4\text{--}14.44)$$

and

$$\frac{d}{du} \mathring{\Pi} = - eqF(x - u\xi)\xi, \qquad (4\text{--}14.45)$$

which is integrated to give

$$\mathring{\Pi} = \Pi - eq \int_0^u du' \, F(x - u'\xi)\xi. \qquad (4\text{--}14.46)$$

Integration of the differential equation for $\mathring{k}$ then produces

$$\mathring{k} = k + u\Pi - eq \int_0^u du' \int_0^{u'} du'' \, F(x - u''\xi)\xi$$

$$= k + u\Pi - eq \int_0^u du'(u - u')F(x - u'\xi)\xi. \qquad (4\text{--}14.47)$$

We also note the combinations

$$\mathring{k} - u\mathring{\Pi} = k + eq \int_0^u du' \, u'F(x - u'\xi)\xi, \qquad (4\text{--}14.48)$$

and

$$\hat{\Pi} - \hat{k} = (1 - u)\Pi - k + eq \int_0^u du'(u - 1 - u')F(x - u'\xi)\xi. \quad (4\text{-}14.49)$$

The transformed version of $\chi(u)$ is

$$\hat{\chi}(u) = \left[k + eq \int_0^u du' \, u' F(x - u'\xi)\xi \right]^2 + u(1 - u)\left[\Pi - eq \int_0^u du' \, F(x - u'\xi)\xi \right]^2 + m^2 u,$$

$$(4\text{-}14.50)$$

or

$$\hat{\chi}(u) = k^2 + u(1 - u)\Pi^2 + m^2 u + 2eq \int_0^u du' \, u'k.F(x - u'\xi)\xi$$

$$- 2equ(1 - u) \int_0^u du' \, \Pi.F(x - u'\xi)\xi + e^2 \left[\int_0^u du' \, u' F(x - u'\xi)\xi \right]^2$$

$$+ e^2 u(1 - u) \left[\int_0^u du' \, F(x - u'\xi)\xi \right]^2. \quad (4\text{-}14.51)$$

As a first application, we shall extract just the terms that are explicitly linear in the electromagnetic field. When specialized to particle fields that obey $(\Pi^2 + m^2)\phi = 0$, the result should imply the very well-known form factor for real particles. In this situation we must include a photon mass. Let us begin with the last term of (4-14.6) and use the analysis of $[\Pi, [\Pi, \exp(-is\chi)]]$ given in Eq. (4-14.32). To avoid higher powers of F than the first, we retain only the first term on the right side of (4-14.33) and simplify (4-14.32) to

$$[\Pi, [\Pi, \exp(-is\chi)]] = is \int_{-1}^1 \tfrac{1}{2} \, dv \, \exp\{-is[(1 + v)/2]\chi\} \, 2ueq(\Pi - k)J$$

$$\times \exp\{-is[(1 - v)/2]\chi\}. \quad (4\text{-}14.52)$$

When the transformation signaled by $\hat{\ }$ is performed, we have, to the desired accuracy

$$\chi(u) \to k^2 + u(1 - u)\Pi^2 + m^2 u, \quad \Pi - k \to (1 - u)\Pi - k, \quad J(x) \to J(x - u\xi).$$

$$(4\text{-}14.53)$$

Now consider a typical harmonic component of $J(x)$, $\exp(ipx)$, and examine the ξ, k operator structure

$$\exp\{-is[(1 + v)/2]k^2\} \, \exp(-iup\xi) \, \exp\{-is[(1 - v)/2]k^2\}, \quad (4\text{-}14.54$$

or

$$\exp\{-is[(1+v)/2]k^2\}\exp\{-iu[(1-v)/2]p\xi\}\exp\{-iu[(1+v)/2]p\xi\}$$

$$\times\exp\{-is[(1-v)/2]k^2\}$$

$$=\exp\{-iu[(1-v)/2]p\xi\}\exp\left\{-is[(1+v)/2]\left[k-u\frac{1-v}{2}p\right]^2\right\}$$

$$\times\exp\left\{-is[(1-v)/2]\left[k+u\frac{1+v}{2}p\right]^2\right\}\exp\{-iu[(1+v)/2]p\xi\}$$

$$=\exp\{-iu[(1-v)/2]p\xi\}\,[\exp(-isk^2)\exp\{-isu^2[(1-v^2)/4]p^2\}]$$

$$\times\exp\{-iu[(1+v)/2]p\xi\}. \tag{4-14.55}$$

This rearrangement uses only the momentum translation property of the ξ exponential factors. When the $\xi' = 0$ diagonal matrix element is extracted, these factors, in their final positions, are replaced by unity, leaving just the factor in brackets. There also occurs in (4-14.52), with the substitutions of (4-14.53), a linear k term which, as written, would become $k - u[(1-v)/2]p$ after the translations indicated in (4-14.55) are performed. But $pJ = 0$, and the residual odd function of k vanishes on integration. Accordingly, to terms linear in F,

$$\int\frac{(dk)}{(2\pi)^4}\,[\Pi,[\Pi,\exp\{-is\chi(u)\}]]$$

$$=\frac{1}{(4\pi)^2}\frac{2u(1-u)}{s}\int_{-1}^{1}\frac{1}{2}\,dv\,\exp\{-is[(1+v)/2]\psi\}$$

$$\times\exp\{-isu^2[(1-v^2)/4]p^2\}\,eq\Pi J\,\exp\{-is[(1-v)/2]\psi\}, \tag{4-14.56}$$

where

$$\psi(u) = u(1-u)\Pi^2 + m^2 u + \mu^2(1-u), \tag{4-14.57}$$

and the Gaussian function of p is defined through its multiplicative action on the Fourier components of $J(x)$. When the specialization of particle fields indicated by $\Pi^2 \to -m^2$ is introduced, we get for the last term of (4-14.6),

$$-2e^2\int\frac{(dk)}{(2\pi)^4}\left[\Pi,\left[\Pi,\frac{1}{k^2}\frac{1}{(\Pi-k)^2+m^2}\right]\right]$$

$$=\frac{\alpha}{\pi}\int_{-1}^{1}\frac{1}{2}\,dv\int_{0}^{1}du\,u(1-u)\int_{0}^{\infty}ds\,\exp\left\{-is\left[m^2u^2+\mu^2(1-u)\right.\right.$$

$$\left.\left.+u^2\frac{1-v^2}{4}p^2\right]\right\}eq\Pi J$$

$$=-i\frac{\alpha}{\pi}\int_{0}^{1}dv\int_{0}^{1}du\,u(1-u)\frac{1}{m^2u^2+\mu^2(1-u)+u^2[(1-v^2)/4]p^2}\,eq\Pi J. \tag{4-14.58}$$

Note that, with $\mu = 0$, an infrared singularity appears in the limit $u \to 0$.

Next, consider the middle term of (4–14.6) for which we need the expansion

$$\exp(- is\hat{\chi}) = \exp[- is(k^2 + \psi)] - is \int_{-1}^{1} \tfrac{1}{2}\, dv \exp\{- is[(1 + v)/2](k^2 + \psi)\}$$

$$\times 2eq \int_{0}^{u} du'\, u'k.F(x - u'\xi)\xi \exp\{- is[(1 - v)/2](k^2 + \psi)\}$$

$$+ isu(1 - u) \int_{-1}^{1} \tfrac{1}{2}\, dv \exp\{- is[(1 + v)/2](k^2 + \psi)\}$$

$$\times 2eq \int_{0}^{u} du'\, \Pi.F(x - u'\xi)\xi \exp\{- is[(1 - v)/2](k^2 + \psi)\},$$

$$(4\text{–}14.59)$$

which is limited to the first power of F. The combination

$$k.F\xi = \tfrac{1}{2}k_v\xi_\mu F^{v\mu} + \tfrac{1}{2}F^{v\mu}\xi_\mu k_v \qquad (4\text{–}14.60)$$

exhibits the angular momentum structure $\xi_\mu k_v - \xi_v k_\mu$. This commutes with any function of k^2, and annuls the rotationally invariant states $\langle \xi' = 0 |, | \xi' = 0 \rangle$. That restricts attention to the last term of (4–14.59), where we encounter

$$\exp\{- is[(1 + v)/2]k^2\} \exp(- iu'p\xi)\, \xi \exp\{- is[(1 - v)/2]k^2\}$$

$$= \frac{i}{u'}\frac{\partial}{\partial p} \left(\exp\{- is[(1 + v)/2]k^2\} \exp(- iu'p\xi) \exp\{- is[(1 - v)/2]k^2\}\right)$$

$$\to 2su' \frac{1 - v^2}{4}\, p \exp(- isk^2) \exp\{- isu'^2[(1 - v^2)/4]p^2\}. \qquad (4\text{–}14.61)$$

After the real particle specialization ($\Pi^2 \to - m^2$), and with only the term linear in F exhibited, we have

$$- 4m^2e^2 \int \frac{(dk)}{(2\pi)^4} \frac{1}{k^2} \frac{1}{(\Pi - k)^2 + m^2}$$

$$= - i\frac{\alpha}{2\pi} \int_0^1 dv \int_0^1 du\, u(1 - u)(1 - v^2)$$

$$\times \int_0^{u^2} du'^2\, m^2 \int_0^\infty ds\, s \exp\left\{- is\left[m^2u^2 + \mu^2(1 - u) + u'^2 \frac{1 - v^2}{4}p^2\right]\right\} eq\Pi J.$$

$$(4\text{–}14.62)$$

The latter double integral can be performed in either order: for example,

$$- \int_0^{u^2} du'^2 \frac{m^2}{[m^2u^2 + \mu^2(1 - u) + u'^2((1 - v^2)/4)p^2]^2}$$

$$= - \frac{m^2 u^2}{m^2 u^2 + \mu^2(1 - u)} \frac{1}{m^2 u^2 + \mu^2(1 - u) + u^2[(1 - v^2)/4]p^2} . \quad (4\text{-}14.63)$$

Concerning the first term of (4–14.6), we observe that its exponential form contains

$$\chi(1) = (\Pi - k)^2 + m^2, \quad (4\text{-}14.64)$$

and correspondingly the form of the ξ transformation with $u = 1$ is to be used. As one recognizes from the factor $1 - u$ in the last term of (4–14.59), no linear term in F results from this contribution.

In this way we get, as the operator standing between $i\phi_1$ and ϕ_2, the following:

$$- \frac{\alpha}{2\pi} \int_0^1 dv \int_0^1 du\, u(1 - u) \left[1 + v^2 + (1 - v^2) \frac{\mu^2}{m^2 u^2 + \mu^2} \right]$$

$$\times \frac{1}{(m^2 + [(1 - v^2)/4]p^2)u^2 + \mu^2}\, eq\Pi J, \quad (4\text{-}14.65)$$

in which we have introduced the permissible simplification $(\mu \ll m)$

$$\mu^2(1 - u) \simeq \mu^2. \quad (4\text{-}14.66)$$

The u^2 term in $u(1 - u)$ gives a well-defined integral as $\mu \to 0$; here, the photon mass can be set equal to zero. The u term supplies these integrals:

$$\int_0^1 du\, u \frac{1}{(m^2 + [(1 - v^2)/4]p^2)u^2 + \mu^2}$$

$$= \frac{1}{2} \frac{\log(m^2/\mu^2) + \log(1 + (1 - v^2)(p^2/4m^2))}{m^2 + (1 - v^2)(p^2/4)},$$

$$(1 - v^2) \int_0^1 du\, u \frac{\mu^2}{m^2 u^2 + \mu^2} \frac{1}{(m^2 + [(1 - v^2)/4]p^2)u^2 + \mu^2}$$

$$= \frac{2}{p^2} \log\left(1 + (1 - v^2) \frac{p^2}{4m^2} \right). \quad (4\text{-}14.67)$$

With the aid of the identity (4–12.42), and of the relation

$$\int_0^1 dv\, \frac{2}{p^2} \log\left(1 + (1 - v^2) \frac{p^2}{4m^2} \right) = \int_0^1 dv\, \frac{v^2}{m^2 + [(1 - v^2)/4]p^2} , \quad (4\text{-}14.68)$$

we arrive directly at this result for (4–14.65),

$$- \frac{\alpha}{2\pi} \int_0^1 dv(1 + v^2) \frac{\log((4m^2/\mu^2)[v^2/(1 - v^2)]) - 2}{4m^2 + (1 - v^2)p^2}\, 2eq\Pi J. \quad (4\text{-}14.69)$$

When set between particle fields ϕ, and multiplied by $\frac{1}{2}$, it constitutes an addition to the action. Since $J = p^2 A$, in a Lorentz gauge, we recognize here, as the coefficient of $2eq\Pi A$, the anticipated expression for $F(p) - 1$.

For a relatively simple example of effects that are quadratic in F, consider a field far from its source so that $J = 0$. This describes the situation of photon scattering where we shall restrict attention to forward scattering. That enables us to discard the field combination $F^{\mu\nu}F_{\mu\nu}$ [it produces the polarization vector combination G_2 of Eq. (4–13.130)]. In this situation both terms of $[\Pi, [\Pi, \chi]]$ disappear and

$$[\Pi, [\Pi, \exp(-is\chi)]] \rightarrow 8s^2 u^2 e^2 \int \tfrac{1}{2}\, dv\, dw\, w \exp\{-is[(1+v)/2]w\chi\}\, F^{\mu\nu}(\Pi - k)_\nu$$

$$\times \exp[-is(1-w)\chi]\, F_{\mu\lambda}(\Pi - k)^\lambda \exp\{-is[(1-v)/2]w\chi\}.$$

$$(4\text{–}14.70)$$

Since the required powers of F are already in evidence, the calculation is particularly easy; the ξ transformation is simply

$$k \rightarrow k + u\Pi, \qquad x \rightarrow x - u\xi. \qquad (4\text{–}14.71)$$

The restriction to forward scattering means that no net momentum is supplied by the two fields. Consider, then, the combination

$$\exp\{-is[(1+v)/2]wk^2\} \exp(iup\xi) \exp[-is(1-w)k^2] \exp(-iup\xi)$$

$$\times \exp\{-is[(1-v)/2]wk^2\}$$

$$\rightarrow \exp\{-is[(1+v)/2]w(k+up)^2\} \exp[-is(1-w)k^2]$$

$$\times \exp\{-is[(1-v)/2]w(k+up)^2\} \qquad (4\text{–}14.72)$$

which, with the integration variable transformation

$$k \rightarrow k - uwp, \qquad (4\text{–}14.73)$$

becomes

$$\exp(-isk^2) \exp[-isu^2 w(1-w)p^2]. \qquad (4\text{–}14.74)$$

In the application to photon scattering, where $p^2 = 0$, the last factor reduces to unity. The transformation (4–14.73) has no effect on the linear k factors of (4–14.70), since $J = 0$. We also encounter the momentum integral

$$\int \frac{(dk)}{(2\pi)^4} k_\nu k_\lambda \exp(-isk^2) = \int \frac{(dk)}{(2\pi)^4} k_\nu \left(\frac{i}{2s}\right) \frac{\partial}{\partial k^\lambda} \exp(-isk^2)$$

$$= -\frac{1}{2} g_{\nu\lambda} \frac{1}{(4\pi)^2} \frac{1}{s^3}, \qquad (4\text{–}14.75)$$

but, since this produces the field combination $F^{\mu\nu}F_{\mu\nu}$, it does not contribute to our limited objective. In this way, we arrive at

$$e^2 \int \frac{(dk)}{(2\pi)^4} [\Pi, [\Pi, \exp(-is\chi)]] = -i8\alpha^2 u^2(1-u)^2 \exp(-ism^2u^2) \int_0^1 dw\, w F^{\mu\nu}\Pi_\nu$$

$$\times \exp[-is(1-w)u(1-u)(\Pi^2+m^2)]\, F_{\mu\lambda}\Pi^\lambda.$$

$$(4\text{–}14.76)$$

It is convenient to perform a partial integration on w according to

$$\int_0^1 dw\, w \exp[-is(1-w)\mathcal{H}] = -\frac{i}{s}\frac{1}{\mathcal{H}}\left[1 - \int_0^1 dw \exp[-is(1-w)\mathcal{H}]\right], \quad (4\text{–}14.77)$$

where

$$\mathcal{H} = u(1-u)(\Pi^2+m^2).$$

$$(4\text{–}14.78)$$

Then, when the s integration is performed, we get

$$-2e^2 \int \frac{(dk)}{(2\pi)^4} \left[\Pi, \left[\Pi, \frac{1}{k^2}\frac{1}{(\Pi-k)^2+m^2}\right]\right]$$

$$= i\frac{16\alpha^2}{m^2} \int_0^1 du(1-u)^2 \int_0^1 dw(1-w)F^{\mu\nu}\Pi_\nu$$

$$\times \frac{1}{m^2u^2+(1-w)u(1-u)(\Pi^2+m^2)}\, F_{\mu\lambda}\Pi^\lambda. \qquad (4\text{–}14.79)$$

Now consider the expansion of $\exp(-is\hat{\chi})$, where only the term containing $F\Pi$ quadratically is of interest to us. When simplified by the restriction $(\Pi^2 + m^2)\phi = 0$, it is

$$-s^2e^24u^4(1-u)^2 \exp(-ism^2u^2)\int \tfrac{1}{2}\, dv\, dw\, w \exp\{-is[(1+v)/2]wk^2\}\xi\bar{F}\Pi$$

$$\times \exp\{-is(1-w)[k^2+u(1-u)(\Pi^2+m^2)]\}\, \xi\bar{F}\Pi \exp\{-is[(1-v)/2]wk^2\},$$

$$(4\text{–}14.80)$$

where

$$\bar{F} = \frac{1}{u}\int_0^u du'\, F(x-u'\xi).$$

$$(4\text{–}14.81)$$

The k, ξ operator structure appearing here is reduced by liberal use of the properties $J = 0$, $p^2 = 0$:

$$\exp\{-is[(1+v)/2]wk^2\}\, \xi_\nu\, \exp(iu'p\xi) \exp[-is(1-w)k^2]\, \xi_\lambda$$

$$\times \exp(-iu''p\xi) \exp\{-is[(1-v)/2]wk^2\}$$

$$\rightarrow \exp\{-is[(1+v)/2]w(k+u'p)^2\}\, \xi_\nu\, \exp[-is(1-w)k^2]\, \xi_\lambda$$

$$\times \exp\{-is[(1-v)/2]w(k+u''p)^2\}$$

$$\rightarrow -s^2(1-v^2)w^2k_\nu k_\lambda \exp[-isk^2 - isw((1+v)u' + (1-v)u'')kp]$$

$$\rightarrow -s^2(1-v^2)w^2k_\nu k_\lambda \exp(-isk^2), \tag{4-14.82}$$

where the last step employs the integration variable transformation

$$k \rightarrow k - w\left(\frac{1+v}{2}u' + \frac{1-v}{2}u''\right)p. \tag{4-14.83}$$

The disappearance of the parameters u', u'' makes $\bar{F}$ effectively equal to F. Accordingly, with only the quadratic F term stated,

$$e^2 \int \frac{(dk)}{(2\pi)^4} \exp[-is\chi(u)] = -\frac{4}{3}\alpha^2 su^4(1-u)^2 \exp(-ism^2u^2) \int_0^1 dw\, w^3 F^{\mu\nu}\Pi_\nu$$

$$\times \exp[-is(1-w)u(1-u)(\Pi^2+m^2)]\, F_{\mu\lambda}\Pi^\lambda. \tag{4-14.84}$$

A partial integration on w is again advisable:

$$\int_0^1 dw\, w^3 \exp[-is(1-w)\mathscr{H}]$$

$$= -\frac{i}{s\mathscr{H}} + \frac{3}{s^2\mathscr{H}^2}\left[1 - 2\int_0^1 dw\, w \exp[-is(1-w)\mathscr{H}]\right]. \tag{4-14.85}$$

The outcome of the s integration is given by

$$-4m^2e^2 \int \frac{(dk)}{(2\pi)^4} \frac{1}{k^2} \frac{1}{(\Pi-k)^2+m^2}$$

$$= -i\frac{32\alpha^2}{m^2} \int_0^1 du(1-u)^2 \int_0^1 dw\, w(1-w)^2$$

$$\times F^{\mu\nu}\Pi_\nu \frac{1}{m^2u^2+(1-w)u(1-u)(\Pi^2+m^2)} F_{\mu\lambda}\Pi^\lambda. \tag{4-14.86}$$

Concerning the first term of (4–14.6), we note that placing $u = 1$ in (4–14.84) removes the contribution of interest.

The result of combining (4–14.79) and (4–14.86) is the following expression for the operator that stands between $i\phi_1$ and ϕ_2:

$$16\alpha^2 \int_{m^2}^\infty \frac{dM^2}{M^2} \frac{\chi(M^2)}{M^2-m^2} F^{\mu\nu}\Pi_\nu \frac{1}{\Pi^2+M^2} F_{\mu\lambda}\Pi^\lambda, \tag{4-14.87}$$

where

$$\frac{1}{M^2}\frac{\chi(M^2)}{M^2 - m^2} = \frac{1}{m^2}\int_0^1 du \int_0^1 dw \frac{1-u}{u}(1 - 2w(1-w))$$

$$\times \delta\left(M^2 - m^2 - m^2\frac{u}{(1-u)(1-w)}\right). \qquad (4\text{-}14.88)$$

On eliminating the variable u, and writing

$$w = \tfrac{1}{2}(1 + z), \qquad (4\text{-}14.89)$$

the function called $\chi(M^2)$ becomes

$$\chi(M^2) = 2M^2m^2 \int_{-1}^1 \frac{1}{2} dz \frac{1 + z^2}{(M^2 + m^2 - z(M^2 - m^2))^2}$$

$$= 1 - \frac{4M^2m^2}{M^2 - m^2}\int_{-1}^1 \frac{1}{2} dz \frac{z}{M^2 + m^2 - z(M^2 - m^2)}. \qquad (4\text{-}14.90)$$

This will be recognized as the causally derived function of Eq. (4–13.21), which is as it should be.

We now face the major problem before us: how to handle neatly the repeated interactions that characterize the low momentum excitations of a bound system. An indication of the proper course comes by comparing the evaluation of the double commutator given in (4–14.56) for weak fields with the form obtained by refraining from exercising the second commutator. For the latter we return to (4–14.26) and (4–14.27),

$$[\Pi, \exp(-is\chi)] = 2su \int_{-1}^1 \tfrac{1}{2} dv \exp\{-is[(1 + v)/2]\chi\} eqF.(\Pi - k)$$

$$\times \exp\{-is[(1 - v)/2]\chi\}, \qquad (4\text{-}14.91)$$

and, retaining only terms linear in the field, perform the transformations indicated in (4–14.53). This gives

$$[\Pi, \exp(-is\chi)] \rightarrow 2su \int_{-1}^1 \tfrac{1}{2} dv \exp\{-is[(1 + v)/2](k^2 + \psi)\}$$

$$\times eqF(x - u\xi).((1 - u)\Pi - k) \exp\{-is[(1 - v)/2](k^2 + \psi)\}, \qquad (4\text{-}14.92)$$

where

$$\psi(u) = u(1 - u)(\Pi^2 + m^2) + m^2u^2. \qquad (4\text{-}14.93)$$

The reduction of Eq. (4–14.55) is now employed, together with

$$\exp\{-is[(1 + v)/2]k^2\} \exp(-iup\xi).k \exp\{-is[(1 - v)/2]k^2\}$$

$$\rightarrow \exp(-isk^2)(k + \tfrac{1}{2}uvp)\,\exp\{-isu^2[(1-v^2)/4]p^2\}, \tag{4-14.94}$$

which produces

$$e^2 \int \frac{(dk)}{(2\pi)^4}\,[\Pi, \exp(-is\chi)]$$

$$= -i\frac{\alpha}{2\pi}\frac{u}{s}\int \frac{1}{2}\,dv\,\exp\{-is[(1+v)/2]\psi\}\,\exp\{-isu^2[(1-v^2)/4]p^2\}$$

$$\times\,[(1-u)eqF.\Pi + \tfrac{1}{2}iuveqJ]\,\exp\{-is[(1-v)/2]\psi\}. \tag{4-14.95}$$

At this point, we write out the two terms of the second commutator with Π and, where Π^2 acts directly on the particle fields, replace it with $-m^2$. That has the following consequence,

$$e^2 \int \frac{(dk)}{(2\pi)^4}\,[\Pi, [\Pi, \exp(-is\chi)]]$$

$$= -i\frac{\alpha}{2\pi}\frac{u}{s}\exp(-ism^2u^2)\int \frac{1}{2}\,dv\,\exp\{-isu^2[(1-v^2)/4]p^2\}$$

$$\times\{\Pi^\mu \exp\{-is[(1+v)/2]u(1-u)(\Pi^2+m^2)\}\,[(1-u)eqF_{\mu\nu}.\Pi^\nu + \tfrac{1}{2}iuveqJ_\mu]$$

$$-\,[(1-u)eqF_{\mu\nu}.\Pi^\nu + \tfrac{1}{2}iuveqJ_\mu]\,\exp\{-is[(1-v)/2]u(1-u)(\Pi^2+m^2)\}\,\Pi^\mu\}, \tag{4-14.96}$$

and then

$$-2e^2 \int \frac{(dk)}{(2\pi)^4}\left[\Pi, \left[\Pi, \frac{1}{k^2}\frac{1}{(\Pi-k)^2+m^2}\right]\right]$$

$$= -\frac{\alpha}{\pi}\int_{-1}^{1}\frac{1}{2}\,dv\int_{0}^{1}du\,u$$

$$\times\left\{\Pi^\mu\frac{1}{m^2u^2 + u^2[(1-v^2)/4]p^2 + [(1+v)/2]u(1-u)(\Pi^2+m^2)}\right.$$

$$\times\,[(1-u)eqF_{\mu\nu}.\Pi^\nu + \tfrac{1}{2}iuveqJ_\mu] - [(1-u)eqF_{\mu\nu}.\Pi^\nu + \tfrac{1}{2}iuveqJ_\mu]$$

$$\left.\times\,\frac{1}{m^2u^2 + u^2[(1-v^2)/4]p^2 + [(1-v)/2]u(1-u)(\Pi^2+m^2)}\,\Pi^\mu\right\}. \tag{4-14.97}$$

If $\Pi^2 + m^2$ is set equal to zero in these denominators, we regain (4–14.58). But retaining it provides a natural "cut-off" for the infrared singularity at $u = 0$, which has otherwise been produced artificially by invoking a photon mass.

Let us also record a version of (4–14.97) that is applicable to slowly varying fields ($Z\alpha \ll 1$). To that end we omit p^2 in the denominators, and drop the J term. Furthermore, we decompose the u integral at a value of $u = u_0$ such that

$$u_0 \sim Z\alpha \ll 1. \tag{4-14.98}$$

Then

$$\int_0^1 du \, \frac{u(1-u)}{m^2u^2 + [(1+v)/2]u(1-u)(\Pi^2 + m^2)}$$

$$\cong \int_0^{u_0} du \, \frac{1}{m^2u + [(1+v)/2](\Pi^2 + m^2)} + \int_{u_0}^1 du \, \frac{1-u}{m^2u}$$

$$\cong \frac{1}{m^2}\left[\log\frac{m^2}{\Pi^2 + m^2} - \log\frac{1+v}{2} - 1\right], \tag{4-14.99}$$

where the $- i\varepsilon$ that has been left implicit in the denominator tells us that

$$\Pi^2 + m^2 < 0: \quad \log\frac{m^2}{\Pi^2 + m^2} = \log\frac{m^2}{|\Pi^2 + m^2|} + \pi i. \tag{4-14.100}$$

This simplification of (4-14.97) is expressed by

$$- 2e^2\int\frac{(dk)}{(2\pi)^4}\left[\Pi,\left[\Pi,\frac{1}{k^2}\frac{1}{(\Pi-k)^2 + m^2}\right]\right]$$

$$\cong - \frac{\alpha}{\pi}\frac{1}{m^2}\left[\Pi^\mu \log\frac{m^2}{\Pi^2 + m^2} eqF_{\mu v}.\Pi^v - eqF_{\mu v}.\Pi^v \log\frac{m^2}{\Pi^2 + m^2}\Pi^\mu\right], \tag{4-14.101}$$

which uses the integral

$$\int_{-1}^1 \frac{1}{2} dv\left(\log\frac{1+v}{2} + 1\right) = 0. \tag{4-14.102}$$

According to the commutator

$$[\Pi_\mu, \Pi^2 + m^2] = 2ieqF_{\mu v}.\Pi^v, \tag{4-14.103}$$

the result can also be presented as

$$- i\frac{\alpha}{\pi}\frac{1}{m^2}\Pi^\mu(\Pi^2 + m^2)\log\frac{m^2}{\Pi^2 + m^2}\Pi_\mu. \tag{4-14.104}$$

Now we must produce a similar improvement to the calculations of Eqs. (4-14.62, 63). This will first be directed toward the situation of slowly varying fields. Then the important field-dependent term in $\hat{\chi}(u)$ is the one containing Π, together with its quadratic partner, the last term of Eq. (4-14.51). The others either vanish or lead to quadratic field terms without infrared singularities. These significant contributions to $\hat{\chi}(u)$ are:

$$\hat{\chi}(u) = k^2 + u(1-u)\Pi^2 + m^2u - 2equ^2(1-u)\Pi.F\xi + equ^3(1-u)\Pi^\mu.\partial_\lambda F_{\mu v}\xi^v\xi^\lambda$$

$$- e^2u^3(1-u)\xi^\mu F_{\mu\lambda}F^{\lambda v}\xi_v, \tag{4-14.105}$$

where only the first two terms in a ξ power series expansion of $F(x - u'\xi)$ have been retained. With the aid of the reduction

$$\exp\{- is[(1 + v)/2]k^2\} \, \xi_\mu \xi_\nu \exp\{- is[(1 - v)/2]k^2\}$$

$$\rightarrow \exp\{- is[(1 + v)/2]k^2\} \, (- s(1 + v)k_\mu)(s(1 - v)k_\nu) \exp\{- is[(1 - v)/2]k^2\},$$

$$(4\text{--}14.106)$$

and the integral (4–14.75), one gets the following linear and quadratic field terms,

$$e^2 \int \frac{(dk)}{(2\pi)^4} \exp[- is\chi(u)] = - i \, \frac{\alpha}{12\pi} u^3(1 - u) \exp(- ism^2u^2) \, (eq\Pi J + e^2 F^{\mu\nu} F_{\mu\nu})$$

$$+ \frac{\alpha}{12\pi} su^4(1 - u)^2 \exp(- ism^2u^2) \int_0^1 dw \, w^3 2ieq F^{\mu\nu}.\Pi_\nu$$

$$\times \exp[- is(1 - w)u(1 - u)(\Pi^2 + m^2)] \, 2ieq F_{\mu\lambda}.\Pi^\lambda.$$

$$(4\text{--}14.107)$$

The commutator of Eq. (4–14.103) can be used to give alternative forms to the last factor:

$$\int dw \, w^3 2ieq F^{\mu\nu}.\Pi_\nu \exp[- is(1 - w)u(1 - u)(\Pi^2 + m^2)] \, 2ieq F_{\mu\lambda}.\Pi^\lambda$$

$$= - \frac{i}{su(1 - u)} \int dw \, w^3 \Pi^\mu \frac{d}{dw} \exp[- is(1 - w)u(1 - u)(\Pi^2 + m^2)] \, 2ieq F_{\mu\lambda}.\Pi^\lambda$$

$$= \frac{i}{su(1 - u)} \int dw \, w^3 2ieq F^{\mu\nu}.\Pi_\nu \frac{d}{dw} \exp[- is(1 - w)u(1 - u)(\Pi^2 + m^2)] \, \Pi_\mu.$$

$$(4\text{--}14.108)$$

We shall average the two forms, after partial integration. Then the commutator that appears at $w = 1$,

$$[\Pi^\mu, ieq F_{\mu\nu}.\Pi^\nu] = - eq\Pi J - e^2 F^{\mu\nu} F_{\mu\nu}, \qquad (4\text{--}14.109)$$

gives a contribution that cancels the first term on the right side of (4–14.107). What remains is [recall the definition of $\mathscr{H}$ in Eq. (4–14.78)]

$$- \frac{\alpha}{4\pi} u^3(1 - u) \exp(- ism^2u^2) \int_0^1 dw \, w^2 [\Pi^\mu \exp\{- is(1 - w)\mathscr{H}\} \, eq F_{\mu\nu}.\Pi^\nu$$

$$- eq F_{\mu\nu}.\Pi^\nu \exp\{- is(1 - w)\mathscr{H}\} \, \Pi^\mu], \qquad (4\text{--}14.110)$$

where we shall introduce another partial integration:

$$\int_0^1 dw \, w^2 \exp[-is(1-w)\mathcal{H}] = \frac{1}{is\mathcal{H}} \left[1 - 2 \int_0^1 dw \, w \exp[-is(1-w)\mathcal{H}] \right].$$

$$(4\text{-}14.111)$$

The s integration then gives, for the field-dependent terms,

$$-4m^2e^2 \int \frac{(dk)}{(2\pi)^4} \frac{1}{k^2} \frac{1}{(\Pi-k)^2+m^2}$$

$$= \frac{2\alpha}{\pi} \int_0^1 du \, u(1-u) \int_0^1 dw \, w(1-w) \left[\Pi^\mu \frac{1}{m^2u^2 + (1-w)\mathcal{H}} eqF_{\mu\nu}.\Pi^\nu \right.$$

$$\left. - eqF_{\mu\nu}.\Pi^\nu \frac{1}{m^2u^2 + (1-w)\mathcal{H}} \Pi^\mu \right]. \qquad (4\text{-}14.112)$$

The u integral that appears here is the same as (4–14.99), with $1-w$ replacing $\frac{1}{2}(1+v)$. But the remaining parametric integrals are different,

$$\int_0^1 dw \, w(1-w) = \int_{-1}^1 \frac{1}{2} dv \, \frac{1-v^2}{4} = \frac{1}{6},$$

$$\int_0^1 dw \, w(1-w)(\log(1-w) + 1) = \int_{-1}^1 \frac{1}{2} dv \, \frac{1-v^2}{4} \left(\log \frac{1+v}{2} + 1 \right) = \frac{1}{36}.$$

$$(4\text{-}14.113)$$

This simplified version of (4–14.112) can be written as

$$i \frac{\alpha}{3\pi} \frac{1}{m^2} \Pi^\mu (\Pi^2 + m^2) \left(\log \frac{m^2}{\Pi^2 + m^2} - \frac{1}{6} \right) \Pi_\mu. \qquad (4\text{-}14.114)$$

Again we note that, since (4–14.110), for example, vanishes at $u = 1$, there is no such contribution from the first term of (4–14.6). When (4–14.104) and (4–14.114) are added, and the factor of i removed, we get the following operator to stand between ϕ_1 and ϕ_2,

$$-\frac{2\alpha}{3\pi} \frac{1}{m^2} \Pi^\mu (\Pi^2 + m^2) \left(\log \frac{m^2}{\Pi^2 + m^2} + \frac{1}{12} \right) \Pi_\mu. \qquad (4\text{-}14.115)$$

The implied addition to the action is produced by forming a scalar product with the particle field ϕ and multiplying by $\frac{1}{2}$.

The energy displacement predicted by this addition to the action can be inferred by standard perturbation theory from the modified field equation, or, by considering a causal arrangement. Emission and detection sources separated by the time interval T, exchange a particle with the associated field

$$\phi(x) = \phi(\mathbf{x}) \exp(-ip^0x^0). \qquad (4\text{-}14.116)$$

The vacuum amplitude conveyed by (4–14.115) then becomes

$$iK_1^*[-iT\delta E]iK_2 \qquad (4\text{–}14.117)$$

where, using three-dimensional scalar product notation,

$$\delta E = \phi^* \frac{2\alpha}{3\pi} \frac{1}{m^2} \Pi(\Pi^2 + m^2)\left(\log\frac{m^2}{\Pi^2 + m^2} + \frac{1}{12}\right)\Pi\phi \qquad (4\text{–}14.118)$$

is the energy displacement of the state. We have only to show that, with the neglect of terms of relative order $Z\alpha$ and higher, this is the result found in Section 4.11 by combining nonrelativistic and relativistic calculations. To that accuracy, we have

$$\Pi(\Pi^2 + m^2)\left(\log\frac{m^2}{\Pi^2 + m^2} + \frac{1}{12}\right)\Pi$$

$$\to -(m + E - V)2m(H - E)\left(\log\frac{\frac{1}{2}m}{H - E} + \frac{1}{12}\right)(m + E - V)$$

$$+ \mathbf{p} \cdot 2m(H - E)\left(\log\frac{\frac{1}{2}m}{H - E} + \frac{1}{12}\right)\mathbf{p}$$

$$\to 2m\mathbf{p} \cdot (H - E)\int_0^K dk^0 \frac{1}{k^0 + H - E}\mathbf{p}, \qquad (4\text{–}14.119)$$

where

$$\log K = \log \tfrac{1}{2}m + \tfrac{1}{12}. \qquad (4\text{–}14.120)$$

The omission of the first term on the right side of (4–14.119) combines the neglect of the piece that is quadratic in V with the recognition that ϕ is an eigenvector of H with the eigenvalue E. Now, if we write

$$(H - E)\int_0^K dk^0 \frac{1}{k^0 + H - E} = \int_0^K dk^0\, k^0 \left[\frac{1}{E - H - k^0} + \frac{1}{k^0}\right]$$

$$= \int_0^K dk^0\, k^0 \left[\frac{1}{E - T - k^0} + \frac{1}{k^0}\right.$$

$$+ \frac{1}{E - T - k^0}V\frac{1}{E - T - k^0}$$

$$+ \left. \frac{1}{E - T - k^0}V\frac{1}{E - H - k^0}V\frac{1}{E - T - k^0}\right], \qquad (4\text{–}14.121)$$

and note that the factor $2m$ converts ϕ into ψ, the nonrelativistic wave function we have just the structure of Eq. (4–11.59) [taking into account the cancellation of the contact terms exhibited there], with the upper limit of integration fixed according to (4–14.120) at just the value given in Eq. (4–11.54). Here, indeed, is the unified derivation we have been seeking. The related one for spin $\frac{1}{2}$ will be given later. Incidentally, the effect of vacuum polarization, which has not been considered explicitly, is to alter the additive constant $\frac{1}{12}$. As indicated in Eq. (4–11.135), this is given by

$$\frac{1}{12} \to \frac{1}{12} - \frac{1}{40}. \tag{4–14.122}$$

4-15 H-PARTICLE ENERGY DISPLACEMENTS. SPIN 0 RELATIVISTIC THEORY

Our objective here is to find the modification of these energy displacement calculations that is of relative order $Z\alpha$. Perhaps one should first show that there are such relativistic corrections of order $Z\alpha$, rather than the generally smaller effects suggested by the characteristic fine-structure factor $(Z\alpha)^2$. The simplest place to look is the vacuum polarization calculation of Section 4–3, where the approximation of replacing $|\psi(\mathbf{x})|^2$ by $|\psi(0)|^2$ was introduced. As an improvement, one includes the variation of $\psi(\mathbf{x})$ for $|\mathbf{x}| \ll a_0 = (mZ\alpha)^{-1}$. For our restricted accuracy, this is adequately described by the Schrödinger equation,

$$\left[- |E| + \frac{1}{2m} \nabla^2 + \frac{Z\alpha}{|\mathbf{x}|} \right] \psi(\mathbf{x}) = 0. \tag{4–15.1}$$

The behavior near the origin is insensitive to the energy of the s-state, leading to the solution

$$|\mathbf{x}| \ll a_0: \quad \psi(\mathbf{x}) \doteq \psi(0) \left[1 - \frac{|\mathbf{x}|}{a_0} + \cdots \right]. \tag{4–15.2}$$

Accordingly, the energy displacement formula (4–3.53) is altered into

$$\delta E \cong - 4\pi Z\alpha |\psi(0)|^2 \int (d\mathbf{x}) \, \delta \mathscr{D}(\mathbf{x}) + 8\pi Z^2\alpha^2 |\psi(0)|^2 \int (d\mathbf{x}) \, m|\mathbf{x}| \, \delta \mathscr{D}(\mathbf{x}), \tag{4–15.3}$$

where the additional term is indeed of relative order $Z\alpha$. The new integral appearing here is, for spin 0,

$$\int (d\mathbf{x}) m|\mathbf{x}| \, \delta \mathscr{D}(\mathbf{x}) = \frac{\alpha}{6\pi} m \int_{(2m)^2}^{\infty} \frac{dM^2}{(M^2)^{5/2}} \left(1 - \frac{4m^2}{M^2} \right)^{3/2}$$

$$= \frac{\alpha}{6\pi} \frac{1}{(2m)^2} \int_0^1 dv(1 - v^2)^{1/2} v^4 = \frac{\alpha}{192} \frac{1}{(2m)^2}. \tag{4–15.4}$$

The effect is conveyed, in the additive constants of (4-14.122), by the substitution

$$-\frac{1}{40} \rightarrow -\frac{1}{40} + \frac{1}{128}\,\pi Z\alpha. \qquad (4\text{-}15.5)$$

As a first step we shall present another treatment of the integral

$$I = \int \frac{(dk)}{(2\pi)^4}\, \exp[-\,is\chi(u)], \qquad (4\text{-}15.6)$$

which will be advantageous for our present purpose. It is based on the related integral

$$I(\lambda^2) = \int \frac{(dk)}{(2\pi)^4}\, \exp[-\,is\chi_\lambda(u)], \qquad (4\text{-}15.7)$$

where

$$\chi_\lambda(u) = (k - \lambda u \Pi)^2 + u(1 - u)\Pi^2 + m^2 u. \qquad (4\text{-}15.8)$$

We have written $I(\lambda^2)$ since the integral is clearly an even function of λ. The first point is this. The quantity

$$I(0) = \int \frac{(dk)}{(2\pi)^4}\, \exp(-\,isk^2)\, \exp[-\,is\psi(u)] \qquad (4\text{-}15.9)$$

is the result that is obtained for $I = I(1)$ in the absence of an electromagnetic field, where the transformation $k - u\Pi \rightarrow k$ can be used. Accordingly, the rearrangement

$$I(1) = I(0) + \int_0^1 d\lambda^2\, \frac{d}{d\lambda^2}\, I(\lambda^2) \qquad (4\text{-}15.10)$$

separates, in the last term, the explicitly field-dependent part. Second, the additional rearrangement

$$\int_0^1 d\lambda^2\, \frac{d}{d\lambda^2}\, I(\lambda^2) = \frac{d}{d\lambda^2}\, I(\lambda^2)\big|_{\lambda=0} + \int_0^1 d\lambda^2 (1 - \lambda^2) \left(\frac{d}{d\lambda^2}\right)^2 I(\lambda^2) \qquad (4\text{-}15.11)$$

isolates, in the $\lambda = 0$ component, just the infrared sensitive part that is exhibited in (4-14.110), thus permitting a simpler handling of the residual term.

Let us begin with the derivative expression

$$\frac{dI(\lambda^2)}{d\lambda} = \int \frac{(dk)}{(2\pi)^4} \frac{dv}{2}\, \exp\{-\,is\chi_\lambda[(1 + v)/2]\}\, 2isu(k\Pi - \lambda u \Pi^2)\, \exp\{-\,is\chi_\lambda[(1 - v)/2]\}, \qquad (4\text{-}15.12)$$

which we proceed to rewrite by using the fact that

$$\int \frac{(dk)}{(2\pi)^4} \frac{dv}{2} \frac{\partial}{\partial k_\mu} [\exp\{- is\chi_\lambda[(1+v)/2]\} \Pi_\mu \exp\{- is\chi_\lambda[(1-v)/2]\}] = 0. \quad (4\text{-}15.13)$$

Differentiation of the individual exponentials gives

$$0 = \int \frac{(dk)}{(2\pi)^4} \int \frac{dv\,dv'}{2\;\;2} \left[\exp\{- is\chi_\lambda[(1+v)/2][(1+v')/2]\} \frac{1+v}{2}(k-\lambda u\Pi) \right.$$

$$\times \exp\{- is\chi_\lambda[(1+v)/2][(1-v')/2]\} \Pi \exp\{- is\chi_\lambda[(1-v)/2]\}$$

$$+ \exp\{- is\chi_\lambda[(1+v)/2]\} \Pi \exp\{- is\chi_\lambda[(1-v)/2][(1+v')/2]\} \frac{1-v}{2}(k-\lambda u\Pi)$$

$$\left. \times \exp\{- is\chi_\lambda[(1-v)/2][(1-v')/2]\} \right], \quad (4\text{-}15.14)$$

or

$$\int \frac{(dk)}{(2\pi)^4} \frac{1}{2} dv \exp\{- is\chi_\lambda[(1+v)/2]\} k\Pi \exp\{- is\chi_\lambda[(1-v)/2]\}$$

$$= 2\lambda u \int \frac{(dk)}{(2\pi)^4} \int_{-1}^{1} \frac{1}{2} dv \int_{0}^{1} dw\, w \exp\{- is\chi_\lambda[(1+v)/2]w\} \Pi$$

$$\times \exp[- is\chi_\lambda(1-w)] \Pi \exp\{- is\chi_\lambda[(1-v)/2]w\}. \quad (4\text{-}15.15)$$

In the latter version the three parameters of unit sum have been given the more symmetrical form used in (4–14.25). The transformation of the first term in (4–15.14) induces

$$\frac{1+v}{2} \frac{dv}{2} \frac{dv'}{2} \rightarrow \frac{1}{2} dv\, dw\, w, \quad (4\text{-}15.16)$$

and the second one is analogous, giving an identical contribution. This identity converts (4–15.12) into

$$\frac{dI(\lambda^2)}{d\lambda^2} = isu^2 \int \frac{(dk)}{(2\pi)^4} \left[2 \int \tfrac{1}{2} dv\, dw\, w \exp\{- is\chi_\lambda[(1+v)/2]w\} \Pi \exp[- is\chi_\lambda(1-w)] \right.$$

$$\times \Pi \exp\{- is\chi_\lambda[(1-v)/2]w\}$$

$$\left. - \int \tfrac{1}{2} dv \exp\{- is\chi_\lambda[(1+v)/2]\} \Pi^2 \exp\{- is\chi_\lambda[(1-v)/2]\} \right]. \quad (4\text{-}15.17)$$

To verify the remark about the $\lambda = 0$ value of this derivative, we have only to simplify it for use with fields obeying $(\Pi^2 + m^2)\phi = 0$:

$$e^2 \frac{dI(\lambda^2)}{d\lambda^2} \bigg|_{\lambda=0} \rightarrow \frac{\alpha}{4\pi} \frac{u^2}{s} \exp(- ism^2 u^2) 2 \int_{0}^{1} dw\, w\Pi[\exp\{- is(1-w)\mathcal{H}\} - 1]\Pi, \quad (4\text{-}15.18)$$

which will be recognized as (4–14.110), after inserting the identity (4–14.111) and using the commutator (4–14.103).

The situation is even simpler for the double commutator term, where

$$2 \int ds\, s\, du\, e^2 \int \frac{(dk)}{(2\pi)^4} [\Pi, [\Pi, \exp\{- is\chi_\lambda(u)\}]]_{\lambda=0} \qquad (4\text{–}15.19)$$

produces just the leading approximation to (4–14.97), in which J and p^2 are discarded. Beginning with

$$e^2 \int \frac{(dk)}{(2\pi)^4} \exp[- is\chi_0(u)] = - i\frac{\alpha}{4\pi}\frac{1}{s^2} \exp[- is\psi(u)], \qquad (4\text{–}15.20)$$

we get for (4–15.19):

$$- i\frac{\alpha}{2\pi} \int \frac{ds}{s}\, du\, \frac{dv}{2} \exp(- ism^2 u^2)[\Pi, \exp\{- is[(1+v)/2]\mathcal{H}\} 2s\dot{u}(1-u)$$

$$\times eqF.\Pi \exp\{- is[(1-v)/2]\mathcal{H}\}], \qquad (4\text{–}15.21)$$

which reduces $[(\Pi^2 + m^2)\phi = 0]$ to

$$- \frac{\alpha}{\pi} \int_{-1}^{1} \frac{1}{2}\, dv \int_0^1 du\, u(1-u) \left[\Pi \frac{1}{m^2 u^2 + \dfrac{1+v}{2}\mathcal{H}} eqF.\Pi \right.$$

$$\left. - eqF.\Pi \frac{1}{m^2 u^2 + \dfrac{1-v}{2}\mathcal{H}} \Pi \right]. \qquad (4\text{–}15.22)$$

This is indeed the result extracted from (4–14.97) in the manner mentioned.

Our study of the $Z\alpha$ modifications starts with the errors introduced through the approximations used in Eqs. (4–14.99) and (4–14.119). Let us compare the (incomplete) energy shift expression [combining the approximate version of (4–14.97), and (4–14.112)]

$$\frac{\alpha}{\pi} \int_0^1 du\, u(1-u) \int_{-1}^{1} \frac{1}{2} dv \frac{1+v^2}{2} \phi^* \Pi \frac{\Pi^2 + m^2}{m^2 u^2 + \dfrac{1+v}{2}\mathcal{H}} \Pi\phi \qquad (4\text{–}15.23)$$

with the further simplified one of Eq. (4–14.118):

$$\frac{2\alpha}{3\pi}\frac{1}{m^2} \phi^* \Pi(\Pi^2 + m^2)\left(\log\frac{m^2}{\Pi^2 + m^2} + \frac{1}{12}\right)\Pi\phi. \qquad (4\text{–}15.24)$$

To that end we evaluate the u integral

$$\int_0^1 du \,\frac{1-u}{m^2 u + \dfrac{1+v}{2}(1-u)(\varPi^2 + m^2)}$$

$$= \frac{m^2}{\left(m^2 - \dfrac{1+v}{2}(\varPi^2 + m^2)\right)^2} \log\left(\frac{m^2}{\dfrac{1+v}{2}(\varPi^2 + m^2)}\right) - \frac{1}{m^2 - \dfrac{1+v}{2}(\varPi^2 + m^2)}$$

$$= -m^2 \frac{d}{dm^2}\left[\frac{1}{m^2 - \dfrac{1+v}{2}(\varPi^2 + m^2)} \log \frac{m^2}{\dfrac{1+v}{2}(\varPi^2 + m^2)}\right], \qquad (4\text{-}15.25)$$

where the differentiation does not extend to the m^2 in $\varPi^2 + m^2$. The simplified version is obtained by neglecting $\varPi^2 + m^2$ in the denominator, relative to m^2. Note, in this connection, that

$$\frac{3}{2}\int_{-1}^1 \frac{1}{2}\,dv\,\frac{1+v^2}{2}\left(\log\frac{1+v}{2}+1\right) = -\frac{1}{12}. \qquad (4\text{-}15.26)$$

The difference of the two,

$$\int_0^1 du \,\frac{1-u}{m^2 u + \dfrac{1+v}{2}(1-u)(\varPi^2 + m^2)} - \left(\frac{1}{m^2}\log\frac{m^2}{\dfrac{1+v}{2}(\varPi^2 + m^2)} - \frac{1}{m^2}\right)$$

$$= -m^2 \frac{d}{dm^2}\left[\frac{1}{m^2} \frac{\dfrac{1+v}{2}(\varPi^2 + m^2)}{m^2 - \dfrac{1+v}{2}(\varPi^2 + m^2)} \log \frac{m^2}{\dfrac{1+v}{2}(\varPi^2 + m^2)}\right], \qquad (4\text{-}15.27)$$

can again be represented as

$$-m^2 \frac{d}{dm^2}\left[\frac{1}{m^2}\int_0^1 du\,\frac{1}{m^2 u + \dfrac{1+v}{2}(1-u)(\varPi^2 + m^2)}\right]\frac{1+v}{2}(\varPi^2 + m^2). \qquad (4\text{-}15.28)$$

The appearance of two factors of $\varPi^2 + m^2$ in the numerator of the resulting expression means that it is explicitly quadratic in the fields,

$$\phi^*\varPi^\mu(\varPi^2 + m^2)(\ )(\varPi^2 + m^2)\varPi_\mu\phi = 4e^2\phi^*F^{\mu\nu}.\varPi_\nu(\ )F_{\mu\lambda}.\varPi^\lambda\phi. \qquad (4\text{-}15.29)$$

As a high energy effect, only the behavior of the field ϕ near the spatial origin is significant. Accordingly, we replace $\phi(x)$ with

$$\varphi = \phi(\mathbf{x} = 0), \qquad (4\text{-}15.30)$$

which is equivalent to neglecting the spatial momentum dependence of this function. Now, one can write (4-15.29) out as

$$4e^2\varphi^*\tfrac{1}{2}[F^{0k}, p_k](\ )\tfrac{1}{2}[p_l, F_{0l}]\varphi + 4e^2\varphi^*F^{k0}(-m)(\ )F_{k0}m\varphi$$

$$= -e^2\varphi^*J^0(\ )J^0\varphi + 4m^2e^2\varphi^*F^{0k}(\ )F^{0k}\varphi, \tag{4-15.31}$$

in which Π^0 has been approximated by m. The quantities occurring here are

$$eJ^0(\mathbf{x}) = 4\pi Z\alpha \int \frac{(d\mathbf{p})}{(2\pi)^3} \exp(i\mathbf{p}\cdot\mathbf{x}),$$

$$eF^{0k}(\mathbf{x}) = 4\pi Z\alpha \int \frac{(d\mathbf{p})}{(2\pi)^3} \frac{(-i\mathbf{p})^k}{\mathbf{p}^2} \exp(i\mathbf{p}\cdot\mathbf{x}), \tag{4-15.32}$$

while $\Pi^2 + m^2$, in the denominator of (4–15.28), is approximated by $\mathbf{p}^2$, since $\Pi^0 \cong m$.

The energy shift thus obtained is

$$16\pi Z^2\alpha^3|\varphi|^2 \int_{-1}^{1} \frac{1}{2}\,dv\,\frac{1+v^2}{2}\frac{1+v}{2}\int \frac{(d\mathbf{p})}{(2\pi)^3}\left(\frac{4m^2}{\mathbf{p}^2}-1\right)\left(-m^2\frac{d}{dm^2}\right)\frac{1}{m^2}\int_0^1 du$$

$$\times \frac{1}{m^2 u + \dfrac{1+v}{2}(1-u)\mathbf{p}^2}. \tag{4-15.33}$$

But, it is essential to realize that the comparison of interest is not with the simplified expression (4–15.24), but with its nonrelativistic reduction (4–14.119). The difference resides in the time component of the scalar product, where $\Pi^0 \cong m + V$ gives the additional contribution

$$-\frac{2\alpha}{3\pi}\frac{1}{m^2}\varphi^*V\mathbf{p}^2\left(\log\frac{m^2}{\mathbf{p}^2}+\frac{1}{12}\right)V\varphi$$

$$= -16\pi Z^2\alpha^3|\varphi|^2\frac{1}{m^2}\frac{2}{3}\int\frac{(d\mathbf{p})}{(2\pi)^3}\frac{1}{\mathbf{p}^2}\left(\log\frac{m^2}{\mathbf{p}^2}+\frac{1}{12}\right). \tag{4-15.34}$$

Now observe that

$$\left(-m^2\frac{d}{dm^2}\right)\frac{1}{m^2}\int_0^1 du\,\frac{\left(-\dfrac{1+v}{2}\right)}{m^2 u + \dfrac{1+v}{2}(1-u)\mathbf{p}^2}$$

$$= \left(-m^2\frac{d}{dm^2}\right)\frac{1}{m^2}\frac{\left(-\dfrac{1+v}{2}\right)\mathbf{p}^2}{m^2 - \dfrac{1+v}{2}\mathbf{p}^2}\frac{1}{\mathbf{p}^2}\log\frac{m^2}{\dfrac{1+v}{2}\mathbf{p}^2}$$

$$= -\left(-m^2 \frac{d}{dm^2}\right) \frac{1}{\mathbf{p}^2} \int_0^1 du \, \frac{1}{m^2 u + \dfrac{1+v}{2}(1-u)\mathbf{p}^2} + \frac{1}{m^2 \mathbf{p}^2}\left(\log \frac{m^2}{\dfrac{1+v}{2}\mathbf{p}^2} - 1\right),$$

$$(4\text{-}15.35)$$

where it has been convenient to return, temporarily, to the explicit function of $\mathbf{p}^2$. The contribution to (4-15.33) of the last term in (4-15.35) precisely cancels (4-15.34), and the remainder gives the partial energy shift

$$\delta_1 E = 16\pi Z^2 \alpha^3 |\varphi|^2 \int_{-1}^1 \frac{1}{2} \, dv \, \frac{1+v^2}{2} \int \frac{(d\mathbf{p})}{(2\pi)^3} \frac{1}{\mathbf{p}^2} \left[\frac{1+v}{2} \, 4m^2 \left(-m^2 \frac{d}{dm^2}\right) \frac{1}{m^2} \int_0^1 du \right.$$

$$\times \frac{1}{m^2 u + \dfrac{1+v}{2}(1-u)\mathbf{p}^2} - \left(-m^2 \frac{d}{dm^2}\right) \int_0^1 du \, \frac{1}{m^2 u + \dfrac{1+v}{2}(1-u)\mathbf{p}^2} \right].$$

$$(4\text{-}15.36)$$

The momentum integral is

$$\int \frac{(d\mathbf{p})}{(2\pi)^3} \frac{1}{\mathbf{p}^2} \, \frac{1}{m^2 u + \dfrac{1+v}{2}(1-u)\mathbf{p}^2} = \frac{1}{4\pi} \frac{1}{\left[m^2 \dfrac{1+v}{2} u(1-u)\right]^{1/2}}, \quad (4\text{-}15.37)$$

and the subsequent u integration gives

$$\tfrac{1}{4}(m^2)^{-1/2}\left(\frac{1+v}{2}\right)^{-1/2}, \qquad (4\text{-}15.38)$$

as a simple example of the Γ-function relations

$$\int_0^1 du \, u^{a-1}(1-u)^{b-1} = \frac{\Gamma(a)\Gamma(b)}{\Gamma(a+b)},$$

$$\Gamma(a)\Gamma(1-a) = \frac{\pi}{\sin \pi a}. \qquad (4\text{-}15.39)$$

The result is

$$\delta_1 E = 8\pi Z^2 \alpha^3 |\varphi|^2 \frac{1}{m} \int_{-1}^1 \frac{1}{2} \, dv \, \frac{1}{2} v^2 \left[3\left(\frac{1+v}{2}\right)^{1/2} - \frac{1}{4}\left(\frac{1+v}{2}\right)^{-1/2}\right]$$

$$= \left(8 - \frac{44}{105}\right)\pi Z^2 \alpha^3 |\varphi|^2 \frac{1}{m}. \qquad (4\text{-}15.40)$$

In what follows, the complete neglect of spatial momentum in the particle field, as given in (4-15.30), is not always permissible. One needs, for some purposes, the short distance behavior stated in (4–15.2). We now want to show that the latter can be regarded as the initial result of an iterative solution of the homogeneous field equation

$$(p^2 + m^2)\phi = 2eqp.A\phi - e^2A^2\phi, \tag{4-15.41}$$

one that begins with the field

$$\varphi(x^0) = \varphi \exp(-imx^0). \tag{4-15.42}$$

With only the term linear in A retained, the first iteration gives

$$\phi = \varphi + \frac{1}{p^2 + m^2} 2eqp.A\varphi, \tag{4-15.43}$$

or, inserting the Coulomb field and the charge relationship necessary for a bound state,

$$\phi = \varphi + \frac{1}{\mathbf{p}^2} \frac{2mZ\alpha}{|\mathbf{x}|} \varphi. \tag{4-15.44}$$

To be quite precise, what is claimed involves a specific interpretation of the three-dimensional Green's function symbolically represented by $(\mathbf{p}^2)^{-1}$. It is stated in

$$\frac{1}{\mathbf{p}^2} f(\mathbf{x}) = \mathrm{Lim}\, P \int \frac{(d\mathbf{p})}{(2\pi)^3} \frac{\exp[i\mathbf{p} \cdot (\mathbf{x} - \mathbf{x}')]}{\mathbf{p}^2 - \varepsilon^2} f(\mathbf{x}')(d\mathbf{x}'), \tag{4-15.45}$$

where the Cauchy principal value is employed. We verify this directly [$\mathrm{Lim}\, P$ is understood],

$$\frac{1}{\mathbf{p}^2} \frac{1}{|\mathbf{x}|} = \int \frac{(d\mathbf{p})}{(2\pi)^3} \frac{\exp(i\mathbf{p} \cdot \mathbf{x})}{\mathbf{p}^2 - \varepsilon^2} \frac{4\pi}{\mathbf{p}^2} = \frac{2}{\pi} \frac{1}{|\mathbf{x}|} \int_0^\infty \frac{dp}{p} \frac{\sin p|\mathbf{x}|}{p^2 - \varepsilon^2}, \tag{4-15.46}$$

in which the last integral can be presented as

$$\mathrm{Im}\, \frac{1}{2} \int_{-\infty}^\infty \frac{dp}{p} \frac{\exp(ip|\mathbf{x}|) - 1}{p^2 - \varepsilon^2} = \frac{\pi}{2} \frac{\cos \varepsilon|\mathbf{x}| - 1}{\varepsilon^2} \to -\frac{\pi}{4} |\mathbf{x}|^2. \tag{4-15.47}$$

The evaluation uses the complex computation of a principal value integral, averaging the results supplied by the two contours drawn above and below the singularity. Of the four contours that interlace the points ε and $-\varepsilon$, the one traced entirely above the real axis vanishes, while the analogous one drawn below the real axis duplicates the sum of the other two contributions, each of which effectively encircles one singular point. We have shown that

$$\frac{1}{\mathbf{p}^2} \frac{1}{|\mathbf{x}|} = -\frac{1}{2} |\mathbf{x}|, \tag{4-15.48}$$

which differs from the elementary inverse statement

$$\nabla^2(\tfrac{1}{2}|\mathbf{x}| + \text{const.}) = \frac{1}{|\mathbf{x}|} \qquad (4\text{-}15.49)$$

through the special interpretation of the Green's function, preventing the appearance of an added constant. This confirms that the construction of Eqs. (4-15.43, 44, 45) produces

$$\phi(x) = \left(1 - \frac{|\mathbf{x}|}{a_0} + \cdots\right)\varphi(x^0). \qquad (4\text{-}15.50)$$

Next, we consider the errors introduced in replacing $I(1)$ by $I(0)$, or by $I(0) + dI/d\lambda^2|_{\lambda=0}$. Since these residuals refer exclusively to high energy phenomena we shall resort to an elementary method: expansion in powers of the vector potential. Accordingly, we now write $\chi_\lambda(u)$ [Eq. (4-15.8)] as

$$\chi_\lambda(u) = \zeta_\lambda(u) + 2\lambda u(k - \lambda up).eqA - 2u(1 - u)p.eqA + (\lambda^2 u^2 + u(1 - u))e^2 A^2,$$
$$(4\text{-}15.51)$$

where

$$\zeta_\lambda(u) = (k - \lambda up)^2 + u(1 - u)p^2 + m^2 u. \qquad (4\text{-}15.52)$$

After performing the k integration, the leading term of $I(\lambda^2)$,

$$\int \frac{(dk)}{(2\pi)^4} \exp[-is\zeta_\lambda(u)] = \frac{1}{(4\pi)^2}\frac{1}{is^2}\exp\{-is[u(1 - u)p^2 + m^2 u]\}, \qquad (4\text{-}15.53)$$

has become independent of λ and therefore does not contribute to either of the differences of interest. The term linear in A is

$$I(\lambda^2)_A = -is\int\frac{(dk)}{(2\pi)^4}\frac{dv}{2}\exp\{-is[(1 + v)/2]\zeta_\lambda\}$$
$$\times [2\lambda u(k - \lambda up).eqA - 2u(1 - u)p.eqA]\exp\{-is[(1 - v)/2]\zeta_\lambda\}.$$
$$(4\text{-}15.54)$$

For a typical matrix element, referring to momenta p' and p'', one encounters the momentum integrals

$$\int\frac{(dk)}{(2\pi)^4}\exp\left\{-is\left[\frac{1 + v}{2}(k - \lambda up')^2 + \frac{1 - v}{2}(k - \lambda up'')^2\right]\right\}$$
$$= \frac{1}{(4\pi)^2}\frac{1}{is^2}\exp\left\{-is\lambda^2 u^2\left[\frac{1 - v^2}{4}\right](p' - p'')^2\right\} \qquad (4\text{-}15.55)$$

and

$$\int \frac{(dk)}{(2\pi)^4}\left(k - \lambda u \frac{p' + p''}{2}\right)\exp\left\{-is\left[\frac{1+v}{2}(k-\lambda up')^2 + \frac{1-v}{2}(k-\lambda up'')^2\right]\right\}$$

$$= \frac{1}{(4\pi)^2}\frac{1}{is^2}\lambda u \frac{v}{2}(p'-p'')\exp\left\{-is\lambda^2 u^2\left[\frac{1-v^2}{4}\right](p'-p'')^2\right\}, \qquad (4\text{-}15.56)$$

which evaluations use the transformation

$$k - \lambda u\left(\frac{1+v}{2}p' + \frac{1-v}{2}p''\right) \to k. \qquad (4\text{-}15.57)$$

If a Lorentz gauge is adopted, so that

$$(p'-p'')A = 0, \qquad (4\text{-}15.58)$$

the latter integral will not contribute in (4–15.54), and one gets, for the matrix element:

$$-\int ds\, s\, du\, e^2 I(\lambda^2)_A \to i\frac{\alpha}{4\pi}\int \frac{1}{2}\, dv\, du\, u(1-u)\frac{(p'+p'')eqA}{D_\lambda}, \qquad (4\text{-}15.59)$$

where

$$D_\lambda = m^2 u^2 + u(1-u)\frac{1+v}{2}(p'^2+m^2) + u(1-u)\frac{1-v}{2}(p''^2+m^2)$$

$$+ \lambda^2 u^2 \frac{1-v^2}{4}(p'-p'')^2. \qquad (4\text{-}15.60)$$

Should we evaluate (4–15.59) in the null momentum state φ, the quantity D_λ would become independent of λ and no contribution to the λ difference appears. We must use the once iterated field (4–15.43). Retaining only the cross terms between the two parts of ϕ produces

$$-\int ds\, s\, du\, e^2 \phi^* I(\lambda^2)_A \phi = i\, 32\pi Z^2 \alpha^3 |\varphi|^2 m^2 \int \frac{1}{2}\, dv\, du\, u(1-u)$$

$$\times \int \frac{(d\mathbf{p})}{(2\pi)^3}\frac{1}{\mathbf{p}^2 - \varepsilon^2}\left(\frac{1}{\mathbf{p}^2}\right)^2 \frac{1}{D_\lambda}, \qquad (4\text{-}15.61)$$

where D_λ has become

$$D_\lambda = m^2 u^2 + \left[u(1-u)\frac{1+v}{2} + \lambda^2 u^2 \frac{1-v^2}{4}\right]\mathbf{p}^2. \qquad (4\text{-}15.62)$$

There are two applications for this structure, or something very similar. The first one refers to the middle term of (4–14.6) where, according to (4–15.10) and (4–15.11), we must form λ differences that amount to rejecting the linear λ^2 dependence of $I(1)$. Thus we encounter

$$\frac{1}{D_1} - \frac{1}{D_0} - \frac{d}{d\lambda^2} \frac{1}{D_\lambda}\bigg|_{\lambda=0} = \frac{\left(u^2 \dfrac{1-v^2}{4} \mathbf{p}^2\right)^2}{D_0^2 D_1}, \qquad (4\text{-}15.63)$$

with

$$D_0 = m^2 u^2 + u(1-u)\frac{1+v}{2}\mathbf{p}^2, \qquad D_1 = m^2 u^2 + \left(u\frac{1+v}{2} - \left(u\frac{1+v}{2}\right)^2\right)\mathbf{p}^2. \qquad (4\text{-}15.64)$$

This contribution to the energy shift is

$$\delta_2 E = 128\pi Z^2 \alpha^3 |\varphi|^2 m^4 \int \tfrac{1}{2}\, dv\, du\, u(1-u)\left(u^2\frac{1-v^2}{4}\right)^2 \int \frac{(d\mathbf{p})}{(2\pi)^3} \frac{1}{\mathbf{p}^2 - \varepsilon^2} \frac{1}{D_0^2 D_1}. \qquad (4\text{-}15.65)$$

The principal value interpretation of $(\mathbf{p}^2 - \varepsilon^2)^{-1}$ is not involved in the elementary $\mathbf{p}$ integration, which yields

$$\delta_2 E = 32 Z^2 \alpha^3 |\varphi|^2 \frac{1}{m} \int_0^1 dw\, w^{3/2} \int_0^1 du\, \frac{1-u}{u^{5/2}}\left[(1-uw)^{3/2} - (1-u)^{3/2}\right.$$

$$\left. - \tfrac{3}{2}u(1-u)^{1/2}(1-w)\right], \qquad (4\text{-}15.66)$$

where

$$w = \tfrac{1}{2}(1+v). \qquad (4\text{-}15.67)$$

After performing two partial integrations with respect to w, this simplifies to

$$\delta_2 E = 32 Z^2 \alpha^3 |\varphi|^2 \frac{1}{m} \frac{3}{35} \int_0^1 dw\, w^{7/2} \int_0^1 du\, u^{-1/2}(1-u)(1-uw)^{-1/2}. \qquad (4\text{-}15.68)$$

The general integral of this type is $(a,\, b > 0,\, c > -1)$

$$\int_0^1 du \int_0^1 dw\, u^{a-1} w^{b-1}(1-uw)^{c-1} = \frac{1}{a-b}\left[\frac{\Gamma(b)\Gamma(c)}{\Gamma(b+c)} - \frac{\Gamma(a)\Gamma(c)}{\Gamma(a+c)}\right]. \qquad (4\text{-}15.69)$$

Using it, we get

$$\delta_2 E = \left(\frac{1}{16} + \frac{8}{35}\right)\pi Z^2 \alpha^3 |\varphi|^2 \frac{1}{m}. \qquad (4\text{-}15.70)$$

The second application of (4-15.61), in a related version, occurs in the double commutator term of (4-14.6). When one ignores the vector potential in Π, the double commutator introduces the additional factor of $(p' - p'')^2$ into a matrix element. In the context of the expectation value computed from ϕ, where one spatial momentum is zero and the other $\pm \mathbf{p}$, this supplies an extra factor of $\mathbf{p}^2$ in the integrand of (4–15.61). The residual effect in the double commutator term is just the difference between $\lambda = 1$ and $\lambda = 0$. That produces the combination

$$\frac{1}{D_1} - \frac{1}{D_0} = - \frac{u^2 \dfrac{1 - v^2}{4} \mathbf{p}^2}{D_0 D_1}, \qquad (4\text{–}15.71)$$

and the energy shift from this source is

$$\delta_3 E = - 64\pi Z^2 \alpha^3 |\varphi|^2 m^2 \int \tfrac{1}{2}\, dv\, du\, u(1 - u)u^2 \frac{1 - v^2}{4} \int \frac{(d\mathbf{p})}{(2\pi)^3} \frac{1}{\mathbf{p}^2} \frac{1}{D_0 D_1}$$

$$= - 16 Z^2 \alpha^3 |\varphi|^2 \frac{1}{m} \int_0^1 dw\, w^{1/2} \int_0^1 du\, \frac{1 - u}{u^{3/2}} [(1 - uw)^{1/2} - (1 - u)^{1/2}]. \quad (4\text{–}15.72)$$

A single partial integration produces the form

$$\delta_3 E = - 16 Z^2 \alpha^3 |\varphi|^2 \frac{1}{m} \frac{1}{3} \int_0^1 dw\, w^{3/2} \int_0^1 du\, u^{-1/2}(1 - u)(1 - uw)^{-1/2}, \quad (4\text{–}15.73)$$

and

$$\delta_3 E = - \pi Z^2 \alpha^3 |\varphi|^2 \frac{1}{m}. \qquad (4\text{–}15.74)$$

Concerning the first term of (4–14.6), we note that $I(\lambda^2)_A$ vanishes for $u = 1$; no analogous contribution is forthcoming here.

We must finally examine the terms that are explicitly quadratic in A. They have the simplifying feature that $\phi = \varphi$ suffices. As a consequence, no contributions of this type are obtained from the double commutator expression $[\Pi, [\Pi, I]]$. Such terms are

$$[p, [p, I_{A^2}]] - [p, [eqA, I_A]] - [eqA, [p, I_A]] \qquad (4\text{–}15.75)$$

where I_{A^2} states the part, in an expansion of $I(\lambda^2)$, that is quadratic in A. Since we are now taking a diagonal matrix element for a state of definite momentum, the first two commutators vanish. Concerning the third commutator, the identity of Jacobi [Eq. (1–1.22)] tells us that

$$[eqA, [p, I_A]] = [p, [eqA, I_A]] - [[p, eqA], I_A]. \qquad (4\text{–}15.76)$$

In the Lorentz gauge that is being systematically employed, the last term above also vanishes.

The explicit expression for $I(\lambda^2)_{A^2}$ is

$$I(\lambda^2)_{A^2} = -s^2 \int \frac{(dk)}{(2\pi)^4} \int \frac{1}{2} dv\, dw\, w \exp\{-is[(1+v)/2]w\zeta_\lambda\}$$

$$\times [2\lambda u(k - \lambda up).eqA - 2u(1-u)p.eqA]$$

$$\times \exp[-is(1-w)\zeta_\lambda] [2\lambda u(k - \lambda up).eqA - 2u(1-u)p.eqA]$$

$$\times \exp\{-is[(1-v)/2]w\zeta_\lambda\}$$

$$- is \int \frac{(dk)}{(2\pi)^4} \int \frac{1}{2} dv \exp\{-is[(1+v)/2]\zeta_\lambda\}$$

$$\times (\lambda^2 u^2 + u(1-u))e^2 A^2 \exp\{-is[(1-v)/2]\zeta_\lambda\}. \tag{4-15.77}$$

We need only the diagonal matrix element in the state φ of momentum p' ($p'^0 = m$, $\mathbf{p}' = 0$). Thus the momentum integral in the last term of (4-15.77) becomes

$$\int \frac{(dk)}{(2\pi)^4} \exp[-is(k - \lambda up')^2] = \int \frac{(dk)}{(2\pi)^4} \exp(-isk^2) = \frac{1}{(4\pi)^2} \frac{1}{is^2}. \tag{4-15.78}$$

As a result, the last term is a linear function of λ^2 and does not contribute to the λ difference of interest. The basic momentum integral in the first term of (4-15.77) has the form [it is (4-15.55), with $\frac{1}{2}(1+v) \rightarrow w$]

$$\int \frac{(dk)}{(2\pi)^4} \exp[-isw(k - \lambda up')^2] \exp[-is(1-w)(k - \lambda up'')^2]$$

$$= \frac{1}{(4\pi)^2} \frac{1}{is^2} \exp[-is\lambda^2 u^2 w(1-w)(p' - p'')^2]. \tag{4-15.79}$$

There also occurs an integral with the additional factor of $k - \lambda up'$, which, as in (4-15.56), results in a multiple of $p' - p''$. This gives a vanishing contribution in the Lorentz gauge. And, there is an integral with two $k - \lambda up'$ factors. The use of the Lorentz gauge again produces a simplification, effectively reducing the integral to (4-14.75). In consequence,

$$e^2 I(\lambda^2)_{A^2} \rightarrow i\frac{\alpha}{4\pi} \int dw\, w \exp(-ism^2 u^2) \Big[4\lambda^2 u^2 (-i/2s)eqA$$

$$\times \exp[-is(1-w)u(1-u)(p''^2 + m^2)] eqA$$

$$+ 4u^3(1-u)^2 p'eqA \exp[-is(1-w)u(1-u)(p''^2 + m^2)] p'eqA \Big]$$

$$\times \exp[-is\lambda^2 u^2 w(1-w)(p' - p'')^2], \tag{4-15.80}$$

and

$$e^2 \phi^* I(\lambda^2)_{A^2}\phi = i\, 8\pi Z^2 \alpha^3 |\varphi|^2 \int_{-1}^{1} \frac{1}{2} dv \frac{1-v}{2} \int \frac{(dp)}{(2\pi)^3} \left(\frac{1}{\mathbf{p}^2}\right)^2$$

$$\times \left[i\lambda^2 \frac{u^2}{s} + 2m^2u^2(1 - u)^2 \right] \exp(- isD_\lambda), \tag{4-15.81}$$

where we have elected to write

$$w = \tfrac{1}{2}(1 - v), \tag{4-15.82}$$

so that D_λ has the structure given in (4-15.62). When the s integration is performed, we get

$$- \int ds\, s\, du\, e^2\phi^* I(\lambda^2)_{A^2}\phi$$

$$= i\, 8\pi Z^2\alpha^3 |\varphi|^2 \int \frac{1}{2}\, dv\, \frac{1 - v}{2}\, du$$

$$\times \int \frac{(d\mathbf{p})}{(2\pi)^3} \left(\frac{1}{\mathbf{p}^2}\right)^2 \left[-\lambda^2 u^2 \frac{1}{D_\lambda} + 2(1 - u)^2\left(-m^2 \frac{d}{dm^2}\right)\frac{1}{D_\lambda}\right]. \tag{4-15.83}$$

The λ differences that are needed here are those of Eqs. (4-15.63) and (4-15.71). The expression thus obtained for the energy shift contribution is

$$\delta_4 E = 32\pi Z^2\alpha^3 |\varphi|^2 m^2 \int \frac{1}{2}\, dv\, \frac{1 - v}{2}\, du \int \frac{(d\mathbf{p})}{(2\pi)^3}\left[\frac{1}{\mathbf{p}^2}\frac{u^4\dfrac{1 - v^2}{4}}{D_0 D_1}\right.$$

$$\left. + 2(1 - u)^2\left(u^2 \frac{1 - v^2}{4}\right)^2\left(-m^2\frac{d}{dm^2}\right)\frac{1}{D_0^2 D_1}\right], \tag{4-15.84}$$

or

$$\delta_4 E = \delta'_4 E + \delta''_4 E, \tag{4-15.85}$$

with

$$\delta'_4 E = 8Z^2\alpha^3 |\varphi|^2 \frac{1}{m}\int_0^1 dw(1 - w)w^{1/2}\int_0^1 du\, u^{-1/2}[(1 - uw)^{1/2} - (1 - u)^{1/2}] \tag{4-15.86}$$

and

$$\delta''_4 E = 24Z^2\alpha^3 |\varphi|^2 \frac{1}{m}\int_0^1 dw(1 - w)w^{1/2}\int_0^1 du\, \frac{(1 - u)^2}{u^{5/2}}[- (1 - uw)^{1/2}$$

$$+ (1 - u)^{1/2} + \tfrac{1}{2}u(1 - u)^{-1/2}(1 - w)], \tag{4-15.87}$$

where we have seen fit to return, by means of Eq. (4-15.67), to a variable called w. A single partial integration converts $\delta'_4 E$ into

$$\delta'_4 E = 8Z^2\alpha^3|\varphi|^2 \frac{1}{m} \int_0^1 dw \left(\tfrac{1}{3}w^{3/2} - \tfrac{1}{5}w^{5/2}\right) \int_0^1 du \, u^{1/2}(1 - uw)^{-1/2}$$

$$= \left(\frac{1}{3} - \frac{3}{20}\right)\pi Z^2\alpha^3|\varphi|^2 \frac{1}{m}, \tag{4-15.88}$$

according to (4–15.69). As for $\delta''_4 E$, two partial integrations give

$$\delta''_4 E = 24Z^2\alpha^3|\varphi|^2 \frac{1}{m} \int_0^1 dw \left(\frac{1}{15}w^{5/2} - \frac{1}{35}w^{7/2}\right) \int_0^1 du \, u^{-1/2}(1 - u)^2(1 - uw)^{-3/2}$$

$$= \left(\frac{2}{35} + \frac{3}{40}\right)\pi Z^2\alpha^3|\varphi|^2 \frac{1}{m}. \tag{4-15.89}$$

The sum of these two parts is

$$\delta_4 E = \left(\frac{41}{105} - \frac{3}{40}\right)\pi Z^2\alpha^3|\varphi|^2 \frac{1}{m}. \tag{4-15.90}$$

Remaining to be considered is the first term of (4–14.6), which has not done anything for us, so far. We set $u = 1$ in Eq. (4–15.81):

$$e^2\phi^* I(\lambda^2, u = 1)_{A^2}\phi = i \, 4\pi Z^2\alpha^3|\varphi|^2 \int_0^1 dv \int \frac{(d\mathbf{p})}{(2\pi)^3} \left(\frac{1}{\mathbf{p}^2}\right)^2 \frac{i\lambda^2}{s} \exp(-isD_\lambda), \tag{4-15.91}$$

where

$$D_\lambda(u = 1) = m^2 + \lambda^2 \frac{1 - v^2}{4} \mathbf{p}^2 \tag{4-15.92}$$

is an even function of v; accordingly, the term linear in v has been discarded and the v integral suitably rewritten. Let us carry out a partial integration with respect to v,

$$\int_0^1 dv \exp(-isD_\lambda) = \exp(-ism^2) - \tfrac{1}{2}is\lambda^2\mathbf{p}^2 \int_0^1 dv \, v^2 \exp(-isD_\lambda). \tag{4-15.93}$$

The first term on the right side produces, in (4–15.91), a linear function of λ^2 that does not contribute to the λ difference we need. In effect, then,

$$e^2\phi^* I(\lambda^2, u = 1)_{A^2}\phi \rightarrow i \, 2\pi Z^2\alpha^3|\varphi|^2 \int_0^1 dv \, v^2 \int \frac{(d\mathbf{p})}{(2\pi)^3} \frac{1}{\mathbf{p}^2} (\lambda^2)^2 \exp(-isD_\lambda) \tag{4-15.94}$$

and

$$i \int_0^\infty ds \, e^2\phi^* I(\lambda^2, u = 1)_{A^2}\phi = i \, 2\pi Z^2\alpha^3|\varphi|^2 \int_0^1 dv \, v^2 \int \frac{(d\mathbf{p})}{(2\pi)^3} \frac{1}{\mathbf{p}^2} \frac{1}{D_\lambda} (\lambda^2)^2. \tag{4-15.95}$$

The function $(\lambda^2)^2$, with its first λ^2 derivative, vanishes at $\lambda = 0$. Hence, the energy displacement contribution derived here is

$$\delta_5 E = 2\pi Z^2\alpha^3|\varphi|^2 \int_0^1 dv\, v^2 \int \frac{(d\mathbf{p})}{(2\pi)^3} \frac{1}{\mathbf{p}^2} \frac{1}{D_1}$$

$$= Z^2\alpha^3|\varphi|^2 \frac{1}{m} \int_0^1 dv\, v^2(1-v^2)^{-1/2}, \qquad (4\text{-}15.96)$$

or

$$\delta_5 E = \tfrac{1}{4}\pi Z^2\alpha^3|\varphi|^2 \frac{1}{m}. \qquad (4\text{-}15.97)$$

The sum of these five pieces [Eqs. (4–15.40, 70, 74, 90, 97)] is

$$\delta E = \left(8 - \frac{9}{16}\right)\pi Z^2\alpha^3|\varphi|^2 \frac{1}{m}. \qquad (4\text{-}15.98)$$

When the vacuum polarization effect [Eqs. (4–15.3, 4)] is included, this reads

$$\delta E = \frac{8}{3} Z\alpha^2|\varphi|^2 \frac{1}{m}\left[3\pi Z\alpha\left(1 - \frac{9}{128} + \frac{1}{3}\frac{1}{128}\right)\right]. \qquad (4\text{-}15.99)$$

So written, the factor in brackets represents the $Z\alpha$ modification to the additive constants $1/12 - 1/40$. Concerning the magnitude of this effect, we shall only remark here that the large numerical factor $\sim 3\pi$ roughly matches the dominant logarithm of the principal contribution so that the fractional measure of this additional upward displacement of s levels is, for $Z = 1$, fairly well given by $\alpha = 7.3 \times 10^{-3}$. Precise numbers will be reserved for the analogous discussion of the more experimentally accessible spin $\frac{1}{2}$ system.

4–16 H-PARTICLE ENERGY DISPLACEMENTS. SPIN $\frac{1}{2}$ RELATIVISTIC THEORY I

The spin $\frac{1}{2}$ counterpart of the vacuum amplitude (4–14.2) is

$$e^2 \int \frac{(dk)}{(2\pi)^4} \psi_1\gamma^0\gamma^\mu \frac{1}{k^2} \frac{m - \gamma(\Pi - k)}{(\Pi - k)^2 - eq\sigma F + m^2} \gamma_\mu\psi_2. \qquad (4\text{-}16.1)$$

We employ the analogue of (4–14.10),

$$\frac{1}{k^2} \frac{1}{(\Pi - k)^2 - eq\sigma F + m^2} = -\int_0^\infty ds\, s \int_0^1 du \exp[-is\chi(u)], \qquad (4\text{-}16.2)$$

with

$$\chi(u) = u[(\Pi - k)^2 - eq\sigma F + m^2] + (1 - u)k^2$$

$$= (k - u\Pi)^2 + u(1 - u)(\Pi^2 - eq\sigma F + m^2) + u^2(m^2 - eq\sigma F). \quad (4\text{-}16.3)$$

A useful rearrangement is given by

$$\gamma^\mu(m - \gamma(\Pi - k)) \exp(- is\chi) \gamma_\mu$$
$$= \gamma^\mu(m - (1 - u)\gamma\Pi + \gamma(k - u\Pi)) \exp(- is\chi) \gamma_\mu$$
$$= \gamma^\mu(m - (1 - u)\gamma\Pi)\gamma_\mu \exp(- is\chi) + \gamma^\mu(m - (1 - u)\gamma\Pi)[\exp(- is\chi), \gamma_\mu]$$
$$+ \gamma^\mu\gamma(k - u\Pi) \exp(- is\chi) \gamma_\mu, \quad (4\text{-}16.4)$$

where we shall write

$$\gamma^\mu(m - (1 - u)\gamma\Pi) = (m + (1 - u)\gamma\Pi)\gamma^\mu + 2(1 - u)\Pi^\mu. \quad (4\text{-}16.5)$$

Then, symmetrization between left and right produces

$$\gamma^\mu(m - \gamma(\Pi - k)) \exp(- is\chi) \gamma_\mu$$
$$= (- 4m - 2(1 - u)\gamma\Pi).\exp(- is\chi) + (1 - u)[\Pi^\mu, [\exp(- is\chi), \gamma_\mu]]$$
$$+ \tfrac{1}{2}(m + (1 - u)\gamma\Pi)\gamma^\mu[\exp(- is\chi), \gamma_\mu] - \tfrac{1}{2}[\exp(- is\chi), \gamma_\mu]\gamma^\mu(m + (1 - u)\gamma\Pi)$$
$$+ \gamma^\mu\gamma(k - u\Pi).\exp(- is\chi) \gamma_\mu. \quad (4\text{-}16.6)$$

We shall also discard all terms involving $\gamma\Pi + m$ appearing on the left or right, which are not involved in applications to energy displacements. This results in

$$\gamma^\mu(m - \gamma(\Pi - k)) \exp(- is\chi) \gamma_\mu$$
$$\rightarrow - 2m(1 + u) \exp(- is\chi) + (1 - u)[\Pi^\mu, [\exp(- is\chi), \gamma_\mu]]$$
$$+ \tfrac{1}{2}mu[\gamma^\mu, [\exp(- is\chi), \gamma_\mu]] + \gamma^\mu\gamma(k - u\Pi).\exp(- is\chi) \gamma_\mu, \quad (4\text{-}16.7)$$

where it is helpful to note that

$$\gamma(k - u\Pi).\exp(- is\chi) = \gamma.((k - u\Pi).\exp(- is\chi)) + \tfrac{1}{4}[u\Pi, [\exp(- is\chi), \gamma]]. \quad (4\text{-}16.8)$$

In the absence of an electromagnetic field only the first term on the right side of (4-16.7) survives the formation of commutators and the performance of the k integration. Under these conditions,

$$e^2 \int \frac{(dk)}{(2\pi)^4} \exp[- is\chi(u)] = - i\frac{\alpha}{4\pi} \frac{1}{s^2} \exp(- ism^2u^2) \exp(- is\mathscr{H}), \quad (4\text{-}16.9)$$

where

$$\mathscr{H} = u(1 - u)(m^2 - (\gamma\Pi)^2). \quad (4\text{-}16.10)$$

Then, we have

$$- \int ds\, s\, du\, e^2 \int \frac{(dk)}{(2\pi)^4} \gamma^\mu (m - \gamma(\Pi - k)) \exp(-is\chi)\, \gamma_\mu$$

$$= -i\frac{\alpha}{2\pi} m \int_0^1 du(1+u) \int \frac{ds}{s} \exp[-is(m^2 u^2 + \mathscr{H})], \qquad (4\text{-}16.11)$$

or, after a partial integration with respect to u that discards local terms (which vanish for nonoverlapping ψ_1 and ψ_2):

$$\frac{\alpha}{2\pi} m \int_0^1 du(u + \tfrac{1}{2}u^2)[2m^2 u + (1 - 2u)(m^2 - (\gamma\Pi)^2)] \int_0^\infty ds \exp[-is(m^2 u^2 + \mathscr{H})]$$

$$\to -i\frac{\alpha}{2\pi} m \int_0^1 du(u + \tfrac{1}{2}u^2) \left[\frac{2m^2 u + (1 - 2u)(m^2 - (\gamma\Pi)^2)}{m^2 u^2 + u(1 - u)(m^2 - (\gamma\Pi)^2)} - \frac{2}{u} + \frac{2}{mu^2}(\gamma\Pi + m) \right].$$

$$(4\text{-}16.12)$$

The last version includes the contact terms necessary to satisfy the normalization condition; they serve to introduce the factor $(\gamma\Pi + m)^2$. In any application to particle fields obeying $(\gamma\Pi + m)\psi = 0$, this structure vanishes. We are interested only in explicitly field-dependent terms.

As a first example, consider a weak, homogeneous electromagnetic field. In this situation there is no effect of the noncommutativity of Π in the k-integration of $\exp(-is\chi)$, since it can only involve $[\Pi_\mu, \Pi_\nu]F^{\mu\nu}$, which is quadratic in the field. It is merely necessary to include the spin term in modifying (4–16.9),

$$e^2 \int \frac{(dk)}{(2\pi)^4} \exp(-is\chi) = -i\frac{\alpha}{4\pi} \frac{1}{s^2} \exp(-ism^2 u^2) \exp(-is\mathscr{H}) (1 + isu^2 eq\sigma F),$$

$$(4\text{-}16.13)$$

where just the first two terms in the expansion of that exponential have been retained. The vector obtained through k-integration of $(k - u\Pi)^\mu. \exp(-is\chi)$ can only be a multiple of $F^{\mu\nu}\Pi_\nu$ (these elementary observations are also confirmed by the results of explicit integration, of course). But

$$\gamma_\mu F^{\mu\nu}\Pi_\nu = \tfrac{1}{2}i[\sigma F, \gamma\Pi + m], \qquad (4\text{-}16.14)$$

which does not contribute to energy displacements. Also, the terms in (4–16.7, 8) involving commutators with Π disappear in the homogeneous field situation. When we note that

$$\tfrac{1}{2}[\gamma^\mu, [\sigma F, \gamma_\mu]] = \gamma^\mu \sigma F \gamma_\mu + 4\sigma F = 4\sigma F, \qquad (4\text{-}16.15)$$

which is to be applied to the full spin term of

$$\mathscr{H} - u^2 eq\sigma F = u(1-u)(\Pi^2 + m^2) - ueq\sigma F, \qquad (4\text{–}16.16)$$

we obtain the following two field-dependent contributions to the momentum integral of (4–16.7), as simplified by $\mathscr{H} \to 0$:

$$e^2 \int \frac{(dk)}{(2\pi)^4} \gamma^\mu (m - \gamma(\Pi - k)) \exp(-is\chi)\,\gamma_\mu$$

$$\to -\frac{\alpha}{2\pi}\frac{m}{s} u^2(1+u)\exp(-ism^2u^2)\,eq\sigma F + \frac{\alpha}{\pi}\frac{m}{s} u^2 \exp(-ism^2u^2)\,eq\sigma F$$

$$= \frac{\alpha}{2\pi}\frac{m}{s} u^2(1-u)\exp(-ism^2 u^2)\,eq\sigma F. \qquad (4\text{–}16.17)$$

The parametric integrals indicated in (4–16.11) then give

$$i\,\frac{\alpha}{2\pi}\frac{1}{m}\int_0^1 du(1-u)eq\sigma F = i\,\frac{\alpha}{2\pi}\frac{eq}{2m}\sigma F, \qquad (4\text{–}16.18)$$

and the additional action term obtained by space-time extrapolation,

$$\int (dx)\tfrac12 \psi(x)\gamma^0 \frac{\alpha}{2\pi}\frac{eq}{2m}\sigma F\psi(x), \qquad (4\text{–}16.19)$$

states the familiar $\alpha/2\pi$ supplement to the magnetic moment. This derivation is surely as short as any.

The discussion of energy displacements in a Coulomb field will follow the pattern laid down for spin 0. Thus, we now consider

$$\chi_\lambda(u) = (k - \lambda u\Pi)^2 + u(1-u)(m^2 - (\gamma\Pi)^2) + u^2(m^2 - \lambda^2 eq\sigma F) \qquad (4\text{–}16.20)$$

and the integral

$$I(\lambda^2) = \int \frac{(dk)}{(2\pi)^4} \exp[-is\chi_\lambda(u)]. \qquad (4\text{–}16.21)$$

The latter is such that

$$e^2 I(0) = -i\,\frac{\alpha}{4\pi}\frac{1}{s^2}\exp(-ism^2u^2)\exp(-is\mathscr{H}) \qquad (4\text{–}16.22)$$

reproduces the null electromagnetic field structure of Eq. (4–16.9). When applied to the first term on the right side of (4–16.7), with the contact additions stated in (4–16.12), no contribution to energy displacements emerges. The computation of the first λ^2 derivative differs from the spin 0 discussion only in the appearance of the spin term with coefficient λ^2. Thus, based on Eq. (4–15.17), we have

$$\frac{dI(\lambda^2)}{d\lambda^2} = isu^2 \int \frac{(dk)}{(2\pi)^4}\left[2\int \tfrac12 dv\,dw\,w\,\exp\{-is\chi_\lambda[(1+v)/2]w\}\,\Pi \exp[-is\chi_\lambda(1-w)]\,\Pi\right.$$

$$\times \exp\{- is\chi_\lambda[(1 - v)/2]w\}$$

$$- \int \tfrac{1}{2}dv \exp\{- is\chi_\lambda[(1 + v)/2]\} (\Pi^2 - eq\sigma F) \exp\{- is\chi_\lambda[(1 - v)/2]\}\Bigg],$$

(4–16.23)

and

$$e^2 \frac{dI(\lambda^2)}{d\lambda^2}\bigg|_{\lambda=0} = \frac{\alpha}{4\pi} \frac{u^2}{s} \exp(- ism^2u^2) \Bigg[2\int \tfrac{1}{2}dv\,dw\, w \exp\{- is\mathscr{H}[(1 + v)/2]w\} \Pi$$

$$\times \exp[- is\mathscr{H}(1 - w)] \Pi \exp\{- is\mathscr{H}[(1 - v)/2]w\}$$

$$- \int \tfrac{1}{2}dv \exp\{- is\mathscr{H}[(1 + v)/2]\} (\Pi^2 - eq\sigma F)$$

$$\times \exp\{- is\mathscr{H}[(1 - v)/2]\}\Bigg].$$

(4–16.24)

When fields obeying $(\gamma\Pi + m)\psi = 0$ directly multiply this structure, it reduces to

$$\frac{\alpha}{4\pi} \frac{u^2}{s} \exp(- ism^2u^2) \left[eq\sigma F + 2\int_0^1 dw\, w\Pi\{\exp[- is\mathscr{H}(1 - w)] - 1\}\Pi \right]$$ (4–16.25)

and we get

$$- \int ds\, s\, du(- 2m)(1 + u)e^2 \frac{dI}{d\lambda^2}\bigg|_{\lambda=0}$$

$$= - i\frac{\alpha}{2\pi} \frac{1}{m} \int_0^1 du(1 + u) \left[eq\sigma F - 2\int_{-1}^1 \frac{1}{2} dv \frac{1 - v^2}{4} \Pi \frac{\mathscr{H}}{m^2u^2 + \frac{1 + v}{2}\mathscr{H}} \Pi \right],$$

(4–16.26)

in which we have replaced the parameter w:

$$w = \tfrac{1}{2}(1 - v).$$ (4–16.27)

Next, we turn to the commutator of $\exp(- is\chi)$ with γ,

$$[\exp(- is\chi), \gamma]$$

$$= - is \int \tfrac{1}{2} dv \exp\{- is\chi[(1 + v)/2]\} [- ueq\sigma F, \gamma] \exp\{- is\chi[(1 - v)/2]\}$$

$$= - 2us \int \tfrac{1}{2} dv \exp\{- is\chi[(1 + v)/2]\}eqF\gamma \exp\{- is\chi[(1 - v)/2]\},$$ (4–16.28)

where

$$(F\gamma)^\mu = F^{\mu\nu}\gamma_\nu. \tag{4-16.29}$$

A related quantity is produced by inserting χ_λ in place of χ on the right side of (4-16.28). The evaluation at $\lambda = 0$ gives

$$[\exp(-is\chi), \gamma]_0 = -2us\exp(-isk^2)\exp(-ism^2u^2)\int \tfrac{1}{2}dv\exp\{-is\mathscr{H}[(1+v)/2]\}$$

$$\times eqF\gamma \exp\{-is\mathscr{H}[(1-v)/2]\} \tag{4-16.30}$$

and

$$e^2\int \frac{(dk)}{(2\pi)^4}[\exp(-is\chi), \gamma]_0$$

$$= i\frac{\alpha}{2\pi}\frac{u}{s}\exp(-ism^2u^2)\int \tfrac{1}{2}dv\exp\{-is\mathscr{H}[(1+v)/2]\}\,eqF\gamma\exp\{-is\mathscr{H}[(1-v)/2]\}. \tag{4-16.31}$$

From this we derive the reduced $[\mathscr{H}\psi = 0]$ form of the double commutator expression involving Π:

$$-\int ds\, s\, du(1-u)e^2\int \frac{(dk)}{(2\pi)^4}[\Pi, [\exp(-is\chi), \gamma]_0]$$

$$= -i\frac{\alpha}{2\pi}\int \tfrac{1}{2}dv\, du\, u(1-u)$$

$$\times \frac{1}{i}\left[\Pi\frac{1}{m^2u^2 + \dfrac{1+v}{2}\mathscr{H}}eqF\gamma - eqF\gamma\frac{1}{m^2u^2 + \dfrac{1+v}{2}\mathscr{H}}\Pi\right]. \tag{4-16.32}$$

Let us note here that

$$\frac{1}{i}eqF\gamma = [\gamma\Pi + m, \Pi], \tag{4-16.33}$$

which lets us replace (4-16.32) with the equivalent form

$$-i\frac{\alpha}{\pi}\int \tfrac{1}{2}dv\, du\, u(1-u)\Pi\frac{\gamma\Pi + m}{m^2u^2 + \dfrac{1+v}{2}\mathscr{H}}\Pi \tag{4-16.34}$$

The denominator terms appearing in (4-16.26) and (4-16.34) differ through the occurrence, in the numerator, of the second order and first order forms of the Dirac differential operator, respectively. The connection between them, which follows from

$$m^2 - (\gamma\Pi)^2 = 2m(\gamma\Pi + m) - (\gamma\Pi + m)^2, \tag{4-16.35}$$

is given by

$$\frac{1}{2m} \Pi \frac{\mathcal{H}}{m^2 u^2 + \dfrac{1+v}{2}\mathcal{H}} \Pi$$

$$= u(1 - u)\left[\Pi \frac{\gamma\Pi + m}{m^2 u^2 + \dfrac{1+v}{2}\mathcal{H}} \Pi - \frac{e^2}{2m} F\gamma \frac{1}{m^2 u^2 + \dfrac{1+v}{2}\mathcal{H}} F\gamma\right]. \tag{4-16.36}$$

For the first two terms of (4–16.7) we have now indicated structures that, according to the spin 0 experience, should correctly display the low energy characteristics of the bound system. The remaining terms are not sensitive to low energy phenomena. We show this by an evaluation for weak, slowly varying fields. That has already been done for the third term of (4–16.7), containing the double commutator with γ, and we restate the result:

$$-\int ds\, s\, du\, \tfrac{1}{2}mue^2 \int \frac{(dk)}{(2\pi)^4} [\gamma, [\exp(-is\chi), \gamma]] \cong i \frac{\alpha}{\pi} \frac{eq}{m} \sigma F. \tag{4-16.37}$$

Notice, by the way, that the two contributions to the magnetic moment are again in evidence [Eqs. (4–16.26) and (4–16.37)]. We now consider the last term of (4–16.7), and its decomposition as given by (4–16.8). Beginning with the second part of (4–16.8), we use (4–16.31), with its reduced weak field consequence

$$e^2 \int \frac{(dk)}{(2\pi)^4} [\Pi, [\exp(-is\chi), \gamma]_0] \rightarrow i \frac{\alpha}{2\pi} \frac{u}{s} \exp(-ism^2 u^2)\,(\Pi eqF\gamma - eqF\gamma\Pi), \tag{4-16.38}$$

from which follows

$$-\int ds\, s\, du\, \tfrac{1}{4}ue^2 \int \frac{(dk)}{(2\pi)^4} \gamma^\mu[\Pi, [\exp(-is\chi), \gamma]_0]\gamma_\mu = -i \frac{\alpha}{4\pi} \frac{1}{m^2} \frac{1}{i} [\Pi, eqF\gamma]. \tag{4-16.39}$$

In order to deal with $(k - u\Pi).\exp(-is\chi)$, we start with the identity

$$0 = \int \frac{(dk)}{(2\pi)^4} \frac{\partial}{\partial k} \exp(-is\chi)$$

$$= -2is \int \frac{(dk)}{(2\pi)^4} \int_{-1}^{1} \tfrac{1}{2}\,dv \exp\{-is\chi[(1+v)/2]\}\,(k - u\Pi)\exp\{-is\chi[(1-v)/2]\}. \tag{4-16.40}$$

Partial integration with respect to v then yields

$$\int_{-1}^{1} \tfrac{1}{2}\, dv \exp\{- is\chi[(1 + v)/2]\}\, (k - u\Pi)\, \exp\{- is\chi[(1 - v)/2]\}$$

$$= (k - u\Pi).\exp(- is\chi) - isu \int_{-1}^{1} \tfrac{1}{2}\, dv\, \tfrac{1}{2} v \exp\{- is\chi[(1 + v)/2]\}\, [\chi, \Pi]$$

$$\times \exp\{- is\chi[(1 - v)/2]\}, \tag{4-16.41}$$

which means that, in the context of the k integration, one can make the substitution

$$(k - u\Pi).\exp(- is\chi)$$

$$\rightarrow isu \int \tfrac{1}{2}\, dv\, \tfrac{1}{2} v \exp\{- is\chi[(1 + v)/2]\}\, [\chi, \Pi]\, \exp\{- is\chi[(1 - v)/2]\}. \tag{4-16.42}$$

Since the commutator $[\chi, \Pi]$ is already linear in the field, the weak field simplification permits the evaluation

$$\gamma.((k - u\Pi).\exp(- is\chi))$$

$$\cong isu \int \tfrac{1}{2}\, dv\, \tfrac{1}{2} v \exp\{- is\chi[(1 + v)/2]\}\, \gamma.[\chi, \Pi]\, \exp\{- is\chi[(1 - v)/2]\}, \tag{4-16.43}$$

where

$$\gamma.[\chi, \Pi] = [\chi, \gamma\Pi] - [\chi, \gamma].\Pi$$

$$= [\chi, \gamma].(k - \Pi) = - 2iu(k - \Pi).eqF\gamma. \tag{4-16.44}$$

The next-to-last step exploits the fact that χ, being constructed from $\gamma(\Pi - k)$, commutes with the latter operator. We note here that

$$- 2i\Pi.eqF\gamma = [\gamma\Pi + m, \Pi^2] = [\gamma\Pi + m, eq\sigma F], \tag{4-16.45}$$

and this term is therefore dropped. The k integral of what remains can be evaluated with the aid of our ξ transformation device. It involves the integral

$$\int \frac{(dk)}{(2\pi)^4} \exp\{- isk^2[(1 + v)/2]\}\, k.\exp(- iup\xi)\, \exp\{- isk^2[(1 - v)/2]\}$$

$$\rightarrow \frac{1}{(4\pi)^2} \frac{1}{is^2} \tfrac{1}{2}uvp \exp\{- isu^2[(1 - v^2)/4]p^2\}, \tag{4-16.46}$$

which, for slowly varying fields, produces the evaluation

$$e^2 \int \frac{(dk)}{(2\pi)^4} \gamma^\mu \gamma.((k - u\Pi).\exp(- is\chi))\gamma_\mu$$

$$= \frac{\alpha}{4\pi} \frac{u^3}{s} \exp(- ism^2u^2) \int_{-1}^{1} \frac{1}{2}\, dv\, v^2 \frac{1}{i} [\Pi, eqF\gamma] \tag{4-16.47}$$

and then

$$- \int ds \, s \, du \, e^2 \int \frac{(dk)}{(2\pi)^4} \gamma^\mu \gamma \cdot ((k - u\Pi) \cdot \exp(- is\chi)) \gamma_\mu$$

$$= i \frac{\alpha}{4\pi} \frac{1}{m^2} \int_{-1}^{1} \tfrac{1}{2} dv \, v^2 \int_0^1 du \, u \, \frac{1}{i} \, [\Pi, eqF\gamma]$$

$$= i \frac{\alpha}{24\pi} \frac{1}{m^2} \frac{1}{i} \, [\Pi, eqF\gamma]. \tag{4-16.48}$$

The various pieces are put together into an operator standing between $- i\psi_1\gamma^0$ and ψ_2, which, as a supplement to $\gamma\Pi + m$ may be called the mass operator modification. This operator is

$$- \frac{\alpha}{2\pi} \frac{eq}{2m} \sigma F + \frac{\alpha}{2\pi} \int \tfrac{1}{2} dv \, du \, u(1 - u)[1 + v^2 - u(1 - v^2)]\Pi \, \frac{\gamma\Pi + m}{m^2u^2 + \dfrac{1+v}{2} \mathscr{H}} \Pi$$

$$+ \frac{5}{12} \frac{\alpha}{\pi} \frac{1}{m^2} \Pi(\gamma\Pi + m)\Pi$$

$$+ \frac{\alpha}{4\pi} \frac{1}{m} \int \tfrac{1}{2} dv(1 - v^2) \, du \, u(1 - u^2)e^2F\gamma \, \frac{1}{m^2u^2 + \dfrac{1+v}{2} \mathscr{H}} F\gamma.$$

$$\tag{4-16.49}$$

It is convenient, however, to rearrange part of the second term, as indicated by

$$u^2\Pi \, \frac{\gamma\Pi + m}{m^2u^2 + \dfrac{1+v}{2} \mathscr{H}} \Pi$$

$$= \frac{1}{m^2} \Pi(\gamma\Pi + m)\Pi - \frac{1}{m^2} \frac{1+v}{2} \Pi \, \frac{(\gamma\Pi + m)\mathscr{H}}{m^2u^2 + \dfrac{1+v}{2} \mathscr{H}} \Pi$$

$$= \frac{1}{m^2} \Pi(\gamma\Pi + m)\Pi - \frac{1}{m^2} u(1 - u) \frac{1+v}{2} e^2F\gamma \, \frac{m - \gamma\Pi}{m^2u^2 + \dfrac{1+v}{2} \mathscr{H}} F\gamma. \tag{4-16.50}$$

That replaces (4-16.49) with

$$- \frac{\alpha}{2\pi} \frac{eq}{2m} \sigma F + \frac{\alpha}{2\pi} \int \tfrac{1}{2} dv(1 + v^2) \, du \, u(1 - u)\Pi \, \frac{\gamma\Pi + m}{m^2u^2 + \dfrac{1+v}{2} \mathscr{H}} \Pi$$

$$+ \frac{\alpha}{4\pi} \frac{1}{m^2} \Pi(\gamma\Pi + m)\Pi$$

$$+ \frac{\alpha}{4\pi} \frac{1}{m} \int \tfrac{1}{2} dv (1 - v^2) \, du \, u(1 - u^2) e^2 F\gamma \, \frac{1}{m^2 u^2 + \dfrac{1+v}{2} \mathcal{H}} F\gamma$$

$$+ \frac{\alpha}{4\pi} \frac{1}{m^2} \int \tfrac{1}{2} dv (1 + v)(1 - v^2) \, du \, u(1 - u)^2 e^2 F\gamma \, \frac{m - \gamma\Pi}{m^2 u^2 + \dfrac{1+v}{2} \mathcal{H}} F\gamma.$$

$$(4\text{-}16.51)$$

The last two terms are explicitly quadratic in the field. If they are discarded, and the u integral of the second term approximated in the manner of (4–14.99), we get

$$- \frac{\alpha}{2\pi} \frac{eq}{2m} \sigma F + \frac{2\alpha}{3\pi} \frac{1}{m^2} \Pi(\gamma\Pi + m)\left(\log \frac{m^2}{m^2 - (\gamma\Pi)^2} + \frac{11}{24}\right)\Pi, \quad (4\text{-}16.52)$$

where the change of the added constant from its spin 0 value of $\frac{1}{12}$ comes entirely from the third term of (4–16.51),

$$\frac{1}{12} + \frac{3}{2} \frac{1}{4} = \frac{11}{24}. \qquad (4\text{-}16.53)$$

To examine the nonrelativistic limit of (4–16.52) it may be simpler to reintroduce the commutator (4–16.33), so that

$$\Pi(\gamma\Pi + m)\left(\log \frac{m^2}{m^2 - (\gamma\Pi)^2} + \frac{11}{24}\right)\Pi$$

$$= \Pi\left(\log \frac{m^2}{m^2 - (\gamma\Pi)^2} + \frac{11}{24}\right)\frac{1}{i} \, eqF\gamma$$

$$\cong \mathbf{p}\cdot\left(\log \frac{\frac{1}{2}m}{H - E} + \frac{11}{24}\right)[V, \mathbf{p}]\gamma^0 - V\left(\log \frac{\frac{1}{2}m}{H - E} + \frac{11}{24}\right)[V, \boldsymbol{\gamma}\cdot\mathbf{p}]$$

$$\cong \mathbf{p}\cdot(H - E)\left(\log \frac{\frac{1}{2}m}{H - E} + \frac{11}{24}\right)\mathbf{p}. \qquad (4\text{-}16.54)$$

The last steps involve the nonrelativistic reduction to a state for which $\gamma^{0\ast} = +1$. Note that, in contrast with the spin 0 situation, the term explicitly quadratic in the potential V disappears in the nonrelativistic limit to the extent that it will not contribute a $Z\alpha$ correction, since $\boldsymbol{\gamma}$ only connects states with opposite values of γ^0. With the nonrelativistic version stated in Eqs. (4–16.52) and (4–16.54), we have arrived, in a unified way, at the dominant part of the energy displacement where the effective upper limit of photon energy is given, as in Eq. (4–11.52), by

$$\log K = \log \frac{1}{2} m + \frac{11}{24}. \tag{4-16.55}$$

Again the explicit effect of vacuum polarization is to alter the additive constant,

$$\frac{11}{24} \rightarrow \frac{11}{24} - \frac{1}{5}. \tag{4-16.56}$$

Now we must produce the $Z\alpha$ modification to this known result. Let us first assess the errors introduced in the reduction from (4-16.51) to (4-16.52). The path of the spin 0 discussion is followed in assigning the following correction to the second term of (4-16.51):

$$\delta_1 E = \psi^* \gamma^0 \frac{\alpha}{2\pi} \int_{-1}^{1} \tfrac{1}{2} dv (1 + v^2) \frac{1+v}{2} e^2 F\gamma \left(-m^2 \frac{d}{dm^2} \right) \frac{1}{m^2} \int_0^1 du$$

$$\times \frac{m - \gamma \Pi}{m^2 u + \dfrac{1+v}{2}(1-u)(\Pi^2 + m^2)} F\gamma\psi, \tag{4-16.57}$$

where it suffices to use the field at the origin, $\psi(0)$, with the property $\gamma^{0\prime} = +1$. Accordingly,

$$F\gamma(m - \gamma\Pi)F\gamma \rightarrow F_{k0}\gamma^0 m(1 + \gamma^0)F_{k0}\gamma^0, \tag{4-16.58}$$

since neither $\boldsymbol{\gamma} \cdot \mathbf{p}$ nor the time component of the product contribute:

$$F^{0k}\gamma_k m(1 + \gamma^0)F_{0l}\gamma^l \rightarrow 0. \tag{4-16.59}$$

This gives

$$\delta_1 E = 16\pi Z^2 \alpha^3 |\psi(0)|^2 m \int_{-1}^{1} \tfrac{1}{2} dv (1 + v^2) \frac{1+v}{2} \left(-m^2 \frac{d}{dm^2} \right) \frac{1}{m^2} \int_0^1 du \int \frac{(d\mathbf{p})}{(2\pi)^3} \frac{1}{\mathbf{p}^2}$$

$$\times \frac{1}{m^2 u + \dfrac{1+v}{2}(1-u)\mathbf{p}^2}$$

$$= 6 Z^2 \alpha^3 |\psi(0)|^2 \frac{1}{m^2} \int_{-1}^{1} \tfrac{1}{2} dv (1 + v^2) \left(\frac{1+v}{2} \right)^{1/2} \int_0^1 du \, u^{-1/2}(1-u)^{-1/2}$$

$$= \left(5 + \frac{9}{35} \right) \pi Z^2 \alpha^3 |\psi(0)|^2 \frac{1}{m^2}. \tag{4-16.60}$$

The last term of (4-16.51) also involves the combination (4-16.58) and gives the energy displacement

$$\delta_2 E = 16\pi Z^2 \alpha^3 |\psi(0)|^2 \frac{1}{m} \int_{-1}^{1} \tfrac{1}{2} dv (1 - v^2) \frac{1+v}{2} \int_0^1 du(1-u)^2 \int \frac{(d\mathbf{p})}{(2\pi)^3} \frac{1}{\mathbf{p}^2}$$

$$\times \ \frac{1}{m^2 u + \dfrac{1+v}{2}(1-u)\mathbf{p}^2}$$

$$= 4Z^2\alpha^3|\psi(0)|^2 \frac{1}{m^2} \int_{-1}^{1} \tfrac{1}{2}dv(1-v^2)\left(\frac{1+v}{2}\right)^{1/2}\int_0^1 du\, u^{-1/2}(1-u)^{3/2}$$

$$= \frac{24}{35}\,\pi Z^2\alpha^3|\psi(0)|^2\frac{1}{m^2}. \tag{4–16.61}$$

The next-to-last term of (4–16.51), on the other hand, contains the product (omitting an inert factor)

$$F\gamma F\gamma = F_{k0}\gamma^0 F_{k0}\gamma^0 - F_{0k}\gamma^k F_{0l}\gamma^l \tag{4–16.62}$$

where, since

$$-\gamma_k\cdot\gamma_l = \delta_{kl}, \tag{4–16.63}$$

the second term effectively doubles the first one. This yields

$$\delta_3 E = 8\pi Z^2\alpha^3|\psi(0)|^2\frac{1}{m}\int_{-1}^{1}\tfrac{1}{2}dv(1-v^2)\int_0^1 du(1-u^2)\int\frac{(d\mathbf{p})}{(2\pi)^3}\frac{1}{\mathbf{p}^2}$$

$$\times \ \frac{1}{m^2 u + \dfrac{1+v}{2}(1-u)\mathbf{p}^2}$$

$$= 2Z^2\alpha^3|\psi(0)|^2\frac{1}{m^2}\int_{-1}^{1}\tfrac{1}{2}dv(1-v^2)\left(\frac{1+v}{2}\right)^{-1/2}\int_0^1 du\, u^{-1/2}(1-u)^{1/2}(1+u)$$

$$= \tfrac{4}{3}\pi Z^2\alpha^3|\psi(0)|^2\frac{1}{m^2}. \tag{4–16.64}$$

The sum of these three effects is

$$\delta_a E = (\delta_1 + \delta_2 + \delta_3)E = \left(7 + \frac{29}{105}\right)\pi Z^2\alpha^3|\psi(0)|^2\frac{1}{m^2}. \tag{4–16.65}$$

The next set of corrections refer to the errors made in such replacements as setting $I(0) + dI/d\lambda^2|_{\lambda=0}$ in place of $I(1)$. As in the spin 0 discussion, we carry out an expansion in powers of the vector potential. We therefore write $\chi_\lambda(u)$ [Eq. (4–16.20)] as

$$\chi_\lambda(u) = \zeta_\lambda(u) + 2\lambda u(k - \lambda up).eqA - 2u(1-u)p.eqA - (u(1-u) + \lambda^2 u^2)eq\sigma F$$

$$+ (u(1-u) + \lambda^2 u^2)e^2 A^2, \tag{4–16.66}$$

where $\zeta_\lambda(u)$ retains the meaning given in (4–15.52). The linear term in the expansion of $I(\lambda^2)$ is [cf. Eq. (4–15.54)]

$$I(\lambda^2)_A = -is \int \frac{(dk)}{(2\pi)^4} \frac{dv}{2} \exp\{-is[(1+v)/2]\zeta_\lambda\} [2\lambda u(k-\lambda up).eqA$$

$$- 2u(1-u)p.eqA$$

$$- (u(1-u) + \lambda^2 u^2)eq\sigma F] \exp\{-is[(1-v)/2]\zeta_\lambda\}. \qquad (4\text{–}16.67)$$

Since this differs in a transparent way from the spin 0 structure, we can proceed immediately to the analogue of the matrix element (4–15.59), except that the application to the first term of (4–16.7) requires the additional factor of $1 + u$:

$$- \int ds\, s\, du(1+u)e^2 I(\lambda^2)_A$$

$$= i\frac{\alpha}{4\pi} \int \tfrac{1}{2}dv\, du(1+u)\,\frac{u(1-u)(p'+p'')eqA + (u(1-u) + \lambda^2 u^2)eq\sigma F}{D_\lambda}.$$

$$(4\text{–}16.68)$$

Were we to evaluate this expression in the state $\psi(0)$, the outcome would be a linear function of λ^2, which would be annulled by the λ difference of interest here. As with spin 0, we must consider an iterated field.

Using the second order form of the Dirac equation,

$$(\Pi^2 - eq\sigma F + m^2)\psi = 0, \qquad (4\text{–}16.69)$$

we find, as the analogues of (4–15.43),

$$\psi = \psi(0) + \frac{1}{p^2+m^2}(2eqp.A + eq\sigma F)\psi(0) \qquad (4\text{–}16.70)$$

and

$$\psi^*\gamma^0 = \psi(0)^* + \psi(0)^*(2eqp.A + eq\sigma F)\frac{1}{p^2+m^2}, \qquad (4\text{–}16.71)$$

where we recall that $\psi(0)$ represents a state of null spatial momentum and definite parity, $\gamma^{0'} = +1$. In analogy with (4–15.61), this gives

$$- \int ds\, s\, du(1+u)e^2\psi^*\gamma^0 I(\lambda^2)_A\psi$$

$$= i\, 8\pi Z^2\alpha^3|\psi(0)|^2 \int \tfrac{1}{2}\, dv\, du(1+u)$$

$$\times \int \frac{(dp)}{(2\pi)^3}\left(\frac{1}{\mathbf{p}^2}\right)^3 \frac{u(1-u)4m^2 - (u(1-u) + \lambda^2 u^2)\mathbf{p}^2}{D_\lambda}, \qquad (4\text{–}16.72)$$

since

$$\sigma^{0k}F_{0k}\sigma^{0l}F_{0l} \rightarrow - F_{0k}F_{0k} \tag{4-16.73}$$

in the state $\psi(0)$. Introducing the required λ difference produces the energy displacement contribution

$$\delta_4 E = (\delta'_4 + \delta''_4 + \delta'''_4)E, \tag{4-16.74}$$

where

$$\delta'_4 E = 64\pi Z^2\alpha^3|\psi(0)|^2 m^3 \int \tfrac{1}{2}\, dv\, du(1+u)u(1-u)\left(u^2\frac{1-v^2}{4}\right)^2 \int \frac{(dp)}{(2\pi)^3}\frac{1}{p^2}\frac{1}{D_0^2 D_1},$$

$$\delta''_4 E = 16\pi Z^2\alpha^3|\psi(0)|^2 m \int \tfrac{1}{2}\, dv\, du(1+u)u^3\frac{1-v^2}{4}\int \frac{(dp)}{(2\pi)^3}\frac{1}{p^2}\frac{1}{D_0 D_1}, \tag{4-16.75}$$

and

$$\delta'''_4 E = - 16\pi Z^2\alpha^3|\psi(0)|^2 m \int \tfrac{1}{2}\, dv\, du(1+u)u(1-u)u^2\frac{1-v^2}{4}\int \frac{(dp)}{(2\pi)^3}\frac{1}{p^2}\frac{1}{D_0^2}; \tag{4-16.76}$$

the relation

$$u^2\frac{1-v^2}{4}p^2 = D_1 - D_0 \tag{4-16.77}$$

has been used in this separation. The part called $\delta'_4 E$ differs from the analogous spin 0 structure [Eq. (4-15.65)] only in the additional factor of $1 + u$. In particular, the multiplicative factors are identical, since

$$|\psi(0)|^2 = 2m|\varphi|^2. \tag{4-16.78}$$

Accordingly, the modification of Eq. (4-15.68) gives

$$\delta'_4 E = 16 Z^2\alpha^3|\psi(0)|^2\frac{1}{m^2}\frac{3}{35}\int_0^1 dw\, w^{7/2}\int_0^1 du\, u^{-1/2}(1-u^2)(1-uw)^{-1/2}$$

$$= \left(\frac{3}{32}+\frac{3}{35}\right)\pi Z^2\alpha^3|\psi(0)|^2\frac{1}{m^2}. \tag{4-16.79}$$

The momentum integral of $\delta''_4 E$ has also been met before, in Eq. (4-15.72). Suitably changing the numerical factor and u-dependence of (4-15.73) supplies

$$\delta''_4 E = 4 Z^2\alpha^3|\psi(0)|^2\frac{1}{m^2}\frac{1}{3}\int_0^1 dw\, w^{3/2}\int_0^1 du\, u^{-1/2}(1+u)(1-uw)^{-1/2}$$

$$= \frac{7}{12}\pi Z^2\alpha^3|\psi(0)|^2\frac{1}{m^2}. \tag{4-16.80}$$

Performing the momentum integration of the third part produces

$$\delta'''_4 E = -2Z^2\alpha^3|\psi(0)|^2 \frac{1}{m^2} \int_{-1}^{1} \tfrac{1}{2}dv \frac{1-v}{2}\left(\frac{1+v}{2}\right)^{1/2} \int_{0}^{1} du(1+u)u^{-1/2}(1-u)^{1/2}$$

$$= -\tfrac{1}{3}\pi Z^2\alpha^3|\psi(0)|^2 \frac{1}{m^2}. \tag{4-16.81}$$

Adding the three pieces, we get

$$\delta_4 E = \left(\frac{11}{32} + \frac{3}{35}\right)\pi Z^2\alpha^3|\psi(0)|^2 \frac{1}{m^2}. \tag{4-16.82}$$

Let us complete the discussion of the first term in (4–16.7) by considering the part of $I(\lambda^2)$ that is explicitly quadratic in the potential. For that, we return to the spin 0 result of Eq. (4–15.80) and introduce the additional spin term:

$$e^2 I(\lambda^2)_{A^2} = i\frac{\alpha}{4\pi} \int dw\, w \exp(-ism^2u^2) \left\{4\lambda^2u^2\left(-\frac{i}{2s}\right)eqA\right.$$

$$\times \exp[-is(1-w)u(1-u)(p''^2+m^2)]\, eqA$$

$$+\, [2u(1-u)p'eqA + (u(1-u)+\lambda^2u^2)eq\sigma F]$$

$$\times \exp[-is(1-w)u(1-u)(p''^2+m^2)]$$

$$\times \left.[2u(1-u)p'eqA + (u(1-u)+\lambda^2u^2)eq\sigma F]\right\}$$

$$\times \exp[-is\lambda^2u^2w(1-w)(p'-p'')^2]. \tag{4-16.83}$$

Then, as the analogue of (4–15.81), we get

$$e^2\psi^*\gamma^0 I(\lambda^2)_{A^2}\psi = i\,8\pi Z^2\alpha^3|\psi(0)|^2 \int_{-1}^{1} \tfrac{1}{2}dv \frac{1-v}{2} \int \frac{(dp)}{(2\pi)^3}\left(\frac{1}{p^2}\right)^2$$

$$\times \left[i\lambda^2\frac{u^2}{s} + 2m^2u^2(1-u)^2 - \tfrac{1}{2}(u(1-u)+\lambda^2u^2)^2 p^2\right]\exp(-isD_\lambda). \tag{4-16.84}$$

As in (4–15.83), but with the additional factor of $1+u$, performing the s-integration now produces

$$-\int ds\, s\, du(1+u)e^2\psi^*\gamma^0 I(\lambda^2)_{A^2}\psi$$

$$= i\,8\pi Z^2\alpha^3|\psi(0)|^2 \int \tfrac{1}{2}dv \frac{1-v}{2} du(1+u)$$

$$\times \int \frac{(d\mathbf{p})}{(2\pi)^3} \left(\frac{1}{\mathbf{p}^2}\right)^2 \left[-\lambda^2 u^2 \frac{1}{D_\lambda} + 2(1-u)^2 \left(-m^2 \frac{d}{dm^2}\right)\frac{1}{D_\lambda}\right.$$

$$\left. - \tfrac{1}{2}(1 - u + \lambda^2 u)^2 \frac{\mathbf{p}^2}{m^2}\left(-m^2 \frac{d}{dm^2}\right)\frac{1}{D_\lambda}\right]. \tag{4–16.85}$$

The introduction of the λ difference employed for this term has the following effect on the various contributions in (4–16.85):

$$\frac{\lambda^2}{D_\lambda} \rightarrow \frac{1}{D_1} - \frac{1}{D_0} = -\frac{u^2 \dfrac{1 - v^2}{4}\, \mathbf{p}^2}{D_0 D_1},$$

$$\frac{1}{D_\lambda} \rightarrow \frac{1}{D_1} - \frac{1}{D_0} + \frac{u^2 \dfrac{1 - v^2}{4}\, \mathbf{p}^2}{D_0^2} = \frac{\left(u^2 \dfrac{1 - v^2}{4}\, \mathbf{p}^2\right)^2}{D_0^2 D_1}, \tag{4–16.86}$$

as in (4–15.84), and

$$\frac{(1 - u + \lambda^2 u)^2}{D_\lambda} \rightarrow \frac{1}{D_1} - \frac{1 - u^2}{D_0} + \frac{(1 - u)^2 u^2 \dfrac{1 - v^2}{4}\, \mathbf{p}^2}{D_0^2}. \tag{4–16.87}$$

We shall divide the associated energy shift $\delta_5 E$ into two parts that are simply related to the spin 0 calculation,

$$\delta'_5 E = 16\pi Z^2 \alpha^3 |\psi(0)|^2 m \int \tfrac{1}{2} dv \, \frac{1 - v}{2}\, du(1 + u) u^4 \frac{1 - v^2}{4} \int \frac{(d\mathbf{p})}{(2\pi)^3} \frac{1}{\mathbf{p}^2} \frac{1}{D_0 D_1},$$

$$\delta''_5 E = 32\pi Z^2 \alpha^3 |\psi(0)|^2 m \int \tfrac{1}{2} dv \, \frac{1 - v}{2}\, du(1 + u)(1 - u)^2 \left(u^2 \frac{1 - v^2}{4}\right)^2$$

$$\times \left(-m^2 \frac{d}{dm^2}\right)\int \frac{(d\mathbf{p})}{(2\pi)^3} \frac{1}{D_0^2 D_1}, \tag{4–16.88}$$

and a third part that is of direct spin origin:

$$\delta'''_5 E = -8\pi Z^2 \alpha^3 |\psi(0)|^2 \frac{1}{m} \int \tfrac{1}{2} dv \, \frac{1 - v}{2}\, du(1 + u)\left(-m^2 \frac{d}{dm^2}\right)\int \frac{(d\mathbf{p})}{(2\pi)^3} \frac{1}{\mathbf{p}^2}$$

$$\times \left[\frac{1}{D_1} - \frac{1 - u^2}{D_0} + \mathbf{p}^2 \frac{(1 - u)^2 u^2 \dfrac{1 - v^2}{4}}{D_0^2}\right]. \tag{4–16.89}$$

We adapt (4–15.88) to give

$$\delta'_5 E = 4 Z^2 \alpha^3 |\psi(0)|^2 \frac{1}{m^2} \int_0^1 dw(\tfrac{1}{3} w^{3/2} - \tfrac{1}{5} w^{5/2}) \int_0^1 du \, u^{1/2}(1 + u)(1 - uw)^{-1/2} \tag{4–16.90}$$

while (4–15.89) supplies

$$\delta''{}_5 E = 12 Z^2 \alpha^3 |\psi(0)|^2 \frac{1}{m^2} \int dw \left(\frac{1}{15} w^{5/2} - \frac{1}{35} w^{7/2} \right)$$

$$\times \int du \, u^{-1/2}(1 + u)(1 - u)^2 (1 - uw)^{-3/2}. \qquad (4\text{–}16.91)$$

In these integrals we meet the special situation of Eq. (4–15.69), with $a = b$:

$$\int_0^1 du \int_0^1 dw \, u^{a-1} w^{a-1} (1 - uw)^{c-1} = \frac{\Gamma(a)\Gamma(c)}{\Gamma(a+c)} [\psi(a+c) - \psi(a)]. \quad (4\text{–}16.92)$$

Here,

$$\psi(z) = \frac{d}{dz} \log \Gamma(z) \qquad (4\text{–}16.93)$$

will be determined, for our purposes, by the two properties

$$\psi(z) = \psi(z - 1) + \frac{1}{z - 1}, \qquad (4\text{–}16.94)$$

following from

$$\Gamma(z) = (z - 1)\Gamma(z - 1), \qquad (4\text{–}16.95)$$

and

$$\psi(1) - \psi(\tfrac{1}{2}) = 2 \log 2. \qquad (4\text{–}16.96)$$

The latter is a particular example ($n = 2$, $z = \tfrac{1}{2}$) of the formula

$$n\psi(nz) - \sum_{k=0}^{n-1} \psi\left(z + \frac{k}{n} \right) = n \log n \qquad (4\text{–}16.97)$$

that is implied by the Γ-function multiplication property

$$\Gamma(nz) = (2\pi)^{-(n-1)/2} n^{nz-1/2} \prod_{k=0}^{n-1} \Gamma\left(z + \frac{k}{n} \right). \qquad (4\text{–}16.98)$$

Thus,

$$\delta'{}_5 E = \left(\log 2 - \frac{13}{24} \right) \pi Z^2 \alpha^3 |\psi(0)|^2 \frac{1}{m^2} \qquad (4\text{–}16.99)$$

and

$$\delta''{}_5 E = \left(2 + \frac{3}{16} - \frac{1}{35} - 3 \log 2 \right) \pi Z^2 \alpha^3 |\psi(0)|^2 \frac{1}{m^2}. \qquad (4\text{–}16.100)$$

Turning to the spin term δ'''_5E, we perform the momentum integrations to get

$$\delta'''_5E = Z^2\alpha^3|\psi(0)|^2\frac{1}{m^2}\left[\int_0^1 dw(w^{1/2} - \tfrac{1}{3}w^{3/2})\int_0^1 du\, u^{-1/2}(1+u)(1-uw)^{-3/2}\right.$$

$$- \int_0^1 dw\, w^{-1/2}(1-w)\int_0^1 du\, u^{1/2}(1-u)^{-1/2}(1+u)$$

$$\left.- \tfrac{1}{2}\int_0^1 dw\, w^{-1/2}(1-w)^2\int_0^1 du\, u^{-1/2}(1-u)^{1/2}(1+u)\right], \qquad (4\text{-}16.101)$$

where a partial integration with respect to w has already been effected in the first of the three integrals. The result of these integrations is

$$\delta'''_5E = (1 + \tfrac{1}{12} - 2\log 2)\pi Z^2\alpha^3|\psi(0)|^2\frac{1}{m^2}, \qquad (4\text{-}16.102)$$

and the sum of the three parts appears as

$$\delta_5E = \left(\frac{8}{3} + \frac{1}{16} - \frac{1}{35} - 4\log 2\right)\pi Z^2\alpha^3|\psi(0)|^2\frac{1}{m^2}. \qquad (4\text{-}16.103)$$

The total effect associated with the first term of (4-16.7) is displayed in

$$(\delta_4 + \delta_5)E = \left(\frac{8}{3} + \frac{13}{32} + \frac{2}{35} - 4\log 2\right)\pi Z^2\alpha^3|\psi(0)|^2\frac{1}{m^2}. \qquad (4\text{-}16.104)$$

For the second term of (4-16.7), we begin with

$$[\exp(-is\chi), \gamma]_\lambda$$

$$= -2us\int \tfrac{1}{2}dv\exp\{-is\chi_\lambda[(1+v)/2]\}\,eqF\gamma\exp\{-is\chi_\lambda[(1-v)/2]\} \qquad (4\text{-}16.105)$$

and remark, as in the spin 0 discussion, that terms linear in F suffice for the evaluation of the additional commutator with p in $\Pi = p - eqA$. As for the additional commutator with A, the sole component, A^0, selects $F_{0\lambda}\gamma^k$ which has a vanishing matrix element in the $\gamma^{0'} = +1$ state $\psi(0)$. Effectively, then,

$$i[\Pi, [\exp(-is\chi), \gamma]_\lambda] = 2us\int \tfrac{1}{2}dv\exp\{-is\zeta_\lambda[(1+v)/2]\}\,eq\gamma J\exp\{-is\zeta_\lambda[(1-v)/2]\}, \qquad (4\text{-}16.106)$$

and, stated as a matrix element, we have

$$e^2 \int \frac{(dk)}{(2\pi)^4} i[\Pi, [\exp(-is\chi), \gamma]_\lambda] = - i\frac{\alpha}{2\pi} \frac{u}{s} \int_{-1}^{1} \tfrac{1}{2}dv\, eq\gamma J \exp(-isD_\lambda),$$

$$(4\text{--}16.107)$$

which is followed by

$$- e^2 \int ds\, s\, du(1-u) \int \frac{(dk)}{(2\pi)^4} \psi^* \gamma^0 i[\Pi, [\exp(-is\chi), \gamma]_\lambda]\psi$$

$$= \frac{\alpha}{2\pi} \int \tfrac{1}{2}dv\, du\, u(1-u)\psi^* \gamma^0 \frac{eq\gamma J}{D_\lambda} \psi.$$

$$(4\text{--}16.108)$$

The energy shift of interest is produced by a single λ difference:

$$\delta_6 E = \frac{\alpha}{2\pi} \int \tfrac{1}{2}dv\, du\, u(1-u)\psi^* \gamma^0 eq\gamma J \left(\frac{1}{D_1} - \frac{1}{D_0}\right)\psi$$

$$= - 32\pi Z^2\alpha^3 |\psi(0)|^2 m \int \tfrac{1}{2}\, dv\, du\, u(1-u)u^2 \frac{1-v^2}{4} \int \frac{(dp)}{(2\pi)^3} \frac{1}{\mathbf{p}^2} \frac{1}{D_0 D_1}.$$

$$(4\text{--}16.109)$$

Comparison with the analogous Eq. (4–15.72) shows that, with the usual equivalence,

$$2m|\varphi|^2 = |\psi(0)|^2,$$

$$(4\text{--}16.110)$$

these are identical, and therefore

$$\delta_6 E = - \tfrac{1}{2}\pi Z^2\alpha^3 |\psi(0)|^2 \frac{1}{m^2}.$$

$$(4\text{--}16.111)$$

Now for the third term of (4–16.7). Both here and in the fourth term we employ a direct power series, in A, expansion, from which the related first approximations, Eq. (4–16.37) and Eqs. (4–16.39, 48), respectively, are to be removed. The linear term of (4–16.68), with $\lambda = 1$, is adapted to produce

$$- \int ds\, s\, du\, \tfrac{1}{2}mue^2 I_A = i\frac{\alpha}{8\pi} m \int \tfrac{1}{2}\, dv\, du\, u\, \frac{u(1-u)(p' + p'')eqA + ueq\sigma F}{D_1}.$$

$$(4\text{--}16.112)$$

The commutator with γ selects only the spin term, and [Eq. (4–16.15)]

$$[\gamma, [\sigma F, \gamma]] = 8\sigma F.$$

$$(4\text{--}16.113)$$

Then the associated energy displacement is

$$\delta_7 E = - \frac{\alpha}{\pi} m \int \tfrac{1}{2}\, dv\, du\, u^2\psi^* \gamma^0 eq\sigma F \left(\frac{1}{D_1} - \frac{1}{m^2 u^2}\right)\psi,$$

$$(4\text{--}16.114)$$

which makes explicit the subtraction of Eq. (4-16.37) (multiplied by i to extract the energy shift). The evaluation of the matrix element gives

$$\delta_7 E = -32\pi Z^2\alpha^3|\psi(0)|^2\frac{1}{m}\int dw\,du\,uw(1-uw)\int\frac{(d\mathbf{p})}{(2\pi)^3}\frac{1}{\mathbf{p}^2}\frac{1}{D_1}$$

$$= -8Z^2\alpha^3|\psi(0)|^2\frac{1}{m^2}\int dw\,du\,w^{1/2}u^{-1/2}(1-uw)^{1/2}, \qquad (4\text{-}16.115)$$

and

$$\delta_7 E = -3\pi Z^2\alpha^3|\psi(0)|^2\frac{1}{m^2}. \qquad (4\text{-}16.116)$$

Concerning the term quadratic in A, we note that the γ commutator restricts attention in I_{A^2} to terms of the form σF and $\sigma F\cdots\sigma F$. For the first possibility, as modified by (4-16.113), the use of the state $\psi(0)$, with $\gamma^{0\prime}=+1$, excludes a finite matrix element for $\sigma^{0k}=i\gamma^0\gamma^k$. It is the rotational invariance of $\psi(0)$, on the other hand, that reduces $\sigma^{0k}F_{0k}\cdots\sigma^{0l}F_{0l}$ to $-F_{0k}\cdots F_{0k}$, which, as a multiple of the unit matrix, commutes with γ. Thus, there is no contribution from I_{A^2}, and the full implication of the third term in (4-16.7) is conveyed by $\delta_7 E$.

The structure of the double commutator with Π and γ that occurs in (4-16.8) is already contained in the equations beginning with (4-16.106). The appropriate modification of Eq. (4-16.108), with $\lambda=1$, reads

$$-e^2\int ds\,s\,du\,\tfrac{1}{4}u\int\frac{(dk)}{(2\pi)^4}\psi^*\gamma^0\gamma^\mu i[\Pi,[\exp(-is\chi),\gamma]]\gamma_\mu\psi$$

$$=\frac{\alpha}{4\pi}\int\tfrac{1}{2}dv\,du\,u^2\psi^*\gamma^0\frac{eq\gamma J}{D_1}\psi, \qquad (4\text{-}16.117)$$

and the implied energy shift [removing that described in Eq. (4-16.39)] is

$$\delta_8 E = \frac{\alpha}{4\pi}\int\tfrac{1}{2}dv\,du\,u^2\psi^*\gamma^0 eq\gamma J\left(\frac{1}{D_1}-\frac{1}{m^2u^2}\right)\psi$$

$$= -16\pi Z^2\alpha^3|\psi(0)|^2\frac{1}{m}\int dw\,du\,uw(1-uw)\int\frac{(d\mathbf{p})}{(2\pi)^3}\frac{1}{\mathbf{p}^2}\frac{1}{D_1}. \qquad (4\text{-}16.118)$$

Since this is just $\tfrac{1}{2}\delta_7 E$, we have

$$\delta_8 E = -\tfrac{3}{2}\pi Z^2\alpha^3|\psi(0)|^2\frac{1}{m^2}. \qquad (4\text{-}16.119)$$

Turning into the home stretch, we use the substitution (4-16.42) to produce the linear term

$$\gamma^{\mu}\gamma.((k - u\Pi).\exp(- is\chi))_A \gamma_{\mu}$$

$$= su \int \tfrac{1}{2} \, dv \, v \exp\{- is\zeta[(1 + v)/2]\} \, [- 2ueq\gamma F.(k - up) + 2u(1 - u)eq\gamma F.p]$$

$$\times \exp\{- is\zeta[(1 - v)/2]\}. \tag{4-16.120}$$

Something else occurs here that has been omitted. It is a multiple of

$$\gamma^{\lambda}.\sigma^{\mu\nu} \, \partial_{\lambda} F_{\mu\nu} = i\gamma_5 \gamma_{\kappa} \varepsilon^{\kappa\lambda\mu\nu} \partial_{\lambda} F_{\mu\nu}, \tag{4-16.121}$$

which vanishes in the absence of magnetic charge. The momentum integral (4–15.56) is utilized in deriving

$$e^2 \int \frac{(dk)}{(2\pi)^4} \gamma^{\mu}\gamma.((k - u\Pi).\exp(- is\chi))_A \gamma_{\mu}$$

$$= - i \frac{\alpha}{4\pi} \frac{u}{s} \int \tfrac{1}{2} \, dv \, v \exp(- isD_1) \, [- u^2 veq\gamma F(p' - p'') + u(1 - u)eq\gamma F(p' + p'')], \tag{4-16.122}$$

and then

$$- \int ds \, s \, du \, e^2 \int \frac{(dk)}{(2\pi)^4} \gamma^{\mu}\gamma.((k - u\Pi).\exp(- is\chi))_A \gamma_{\mu}$$

$$= \frac{\alpha}{4\pi} \int \tfrac{1}{2} \, dv \, v \, du \, u \frac{1}{D_1} [u^2 vieq\gamma J + u(1 - u)eq\gamma F(p' + p'')]. \tag{4-16.123}$$

The implied energy displacement, with the necessary subtraction, is

$$\delta_9 E = \delta'_9 E + \delta''_9 E, \tag{4-16.124}$$

where

$$\delta'_9 E = - \frac{\alpha}{4\pi} \int \tfrac{1}{2} \, dv \, v^2 \int du \, u^3 \psi^* \gamma^0 eq\gamma J \left(\frac{1}{D_1} - \frac{1}{m^2 u^2} \right) \psi \tag{4-16.125}$$

and

$$\delta''_9 E = \frac{\alpha}{4\pi} \int \tfrac{1}{2} \, dv \, v \int du \, u^2 (1 - u) i\psi^* \gamma^0 \frac{eq\gamma F(p' + p'')}{D_1} \psi. \tag{4-16.126}$$

The first of these parts is evaluated as $(1 + v = 2w)$

$$\delta'_9 E = 16\pi Z^2 \alpha^3 |\psi(0)|^2 \frac{1}{m} \int_0^1 dw(2w - 1)^2 \int_0^1 du \, u^2 w(1 - uw) \int \frac{(dp)}{(2\pi)^3} \frac{1}{p^2} \frac{1}{D_1}$$

$$= 4Z^2 \alpha^3 |\psi(0)|^2 \frac{1}{m^2} \int_0^1 dw \int_{0}^{1} du(2w - 1)^2 u^{1/2} w^{1/2} (1 - uw)^{1/2}, \tag{4-16.127}$$

and, since we shall see that $\delta''_9E = 0$, this gives

$$\delta_9E = \left(\log 2 - \frac{9}{16}\right)\pi Z^2\alpha^3|\psi(0)|^2\frac{1}{m^2}. \tag{4-16.128}$$

The argument is completed by observing that

$$i\psi^*\gamma^0\gamma F(p' + p'')\psi = \psi^*\gamma^0[\gamma p + m, \sigma F]\psi$$
$$= \psi^*(0)\gamma^0[eq\gamma A, \sigma F]\psi(0), \tag{4-16.129}$$

for

$$[\gamma^0, \sigma^{0k}] = 2i\gamma^k \tag{4-16.130}$$

has a vanishing matrix element in the state $\psi(0)$.

Last of all is the explicitly quadratic term related to (4-16.123). Let us note first that the vector potentials which supplement the momenta in (4-16.120) do not enter into this calculation since

$$\gamma FA = \gamma_k F^{k0}A_0 \tag{4-16.131}$$

vanishes in the $\psi(0)$ state. It is only the linear terms in the expansion of the exponentials, $\exp[-is\chi\frac{1}{2}(1 \pm v)]$, that contribute. The factors that appear in this way are

$$-2ueqF.(k - up) + 2u(1 - u)eqF.p + ueq\partial\sigma F \tag{4-16.132}$$

and

$$2u(k - up).eqA - 2u(1 - u)p.eqA - ueq\sigma F. \tag{4-16.133}$$

We remark first that the spin terms appearing here can be discarded. It has already been seen that

$$\gamma.(\partial\sigma F) = 0, \tag{4-16.134}$$

which is an identity. When the two spin terms are multiplied together we encounter, for example,

$$\gamma^\mu.\sigma F\partial_\mu\sigma F \rightarrow \gamma^k.(\sigma^{0l}\sigma^{0m})F_{0l}\partial_k F_{0m}, \tag{4-16.135}$$

where we have only to note that the matrix appearing here anticommutes with γ^0 and therefore effectively vanishes in the state $\psi(0)$. The same conclusion extends to $\sigma F = \sigma^{0k}F_{0k}$, occurring singly, since $\gamma^\mu.\sigma^{0k}$ either vanishes, for $\mu = 0$, or, for $\mu = l$, involves the three-dimensional vector matrix $[\gamma_k, \gamma_l]$ whose coefficient must vanish in the rotationally invariant state $\psi(0)$. Further simplifications accompany the k-integration. Recalling the discussion of (4-15.77), with its dependence on the characteristic property of a Lorentz gauge, we note again that the redefinition of the integration variable produces the effective reduction

$$\left(k - u\frac{p'+p''}{2}\right)A\gamma F\left(k - u\frac{p'+p''}{2}\right) \to kA\gamma Fk, \qquad (4\text{-}16.136)$$

which involves the vanishing, under integration, of a term containing a single factor of k. The outcome of the integration for (4-16.136), proportional to $\gamma FA = \gamma^k F_{k0}A^0$, yields no contribution in the state $\psi(0)$. Terms involving the single k factor of $(k - up).eqA$ also vanish on integration. The two product terms that survive are written out in

$$\gamma^\mu\gamma.((k - u\Pi).\exp(- is\chi))_{A^2}\gamma_\mu$$

$$= - i\, 2s^2u \int \tfrac{1}{2}\, dv\, dw\, w\left(\frac{1}{2} - \frac{1-v}{2}\,w\right)$$

$$\times \exp\{- is\zeta w[(1 + v)/2]\}\,[- 2u(1 - u)p.eqA]$$

$$\times \exp[- is\zeta(1 - w)]\,[- 2ueq\gamma F.(k - up) + 2u(1 - u)eq\gamma F.p]$$

$$\times \exp\{- i\zeta w[(1 - v)/2]\} - i\, 2s^2u \int \tfrac{1}{2}\, dv\, dw\, w\left(w\frac{1+v}{2} - \frac{1}{2}\right)$$

$$\times \exp\{- is\zeta w[(1 + v)/2]\}$$

$$\times [- 2ueq\gamma F.(k - up) + 2u(1 - u)eq\gamma F.p]$$

$$\times \exp[- is\zeta(1 - w)]\,[- 2u(1 - u)p.eqA]\exp\{- is\zeta w[(1 - v)/2]\}. \qquad (4\text{-}16.137)$$

Note that, as in Eqs. (4-15.13-16), we have introduced the more symmetrical set of parameters to express the result of expanding the exponentials. The k-integration that is anticipated in (4-16.137) gives

$$e^2 \int \frac{(dk)}{(2\pi)^4}\gamma^\mu\gamma.((k - u\Pi).\exp(- is\chi))_{A^2}\gamma_\mu$$

$$= - \frac{\alpha}{2\pi}\,u \int dw\, w\,\frac{1-w}{2}\,[- u(1 - u)(p' + p'')eqA]\exp(- isD_1)$$

$$\times [- 2u^2(\tfrac{1}{2} - w)ieq\gamma J + u(1 - u)eq\gamma F(p' + p'')] - \frac{\alpha}{2\pi}\,u \int dw\, w\,\frac{w-1}{2}$$

$$\times [2u^2(\tfrac{1}{2} - w)ieq\gamma J + u(1 - u)eq\gamma F(p' + p'')]$$

$$\times \exp(- isD_1)\,[- u(1 - u)(p' + p'')\, eqA], \qquad (4\text{-}16.138)$$

where, as effectively happens in the discussion leading up to Eq. (4-15.87), we have made the substitution

$$w \to 1 - w. \qquad (4\text{-}16.139)$$

In the term

$$\gamma F(p' + p'') = \gamma^0 F_{0k}(p' + p'')^k + \gamma^k F_{k0}(p' + p'')^0, \qquad (4\text{-}16.140)$$

only the matrix γ^0 can contribute to the expectation value. Its coefficient takes alternative forms depending upon which of the two terms in (4-16.138) is being considered. For the first one (recall that $\mathbf{p}' = 0$)

$$F_{0k}(p' + p'')^k \rightarrow F_{0k}(p'' - p')^k = iJ^0, \qquad (4\text{-}16.141)$$

while the other becomes

$$F_{0k}(p' + p'')^k \rightarrow - F_{0k}(p' - p'')^k = - iJ^0. \qquad (4\text{-}16.142)$$

The two terms of (4-16.138) are then seen to be identical, which we present as

$$- i \int ds\, s\, du\, e^2 \int \frac{(dk)}{(2\pi)^4} \psi^* \gamma^0 \gamma^\mu \gamma.((k - u\Pi).\exp(- is\chi))_{A^2} \gamma_\mu \psi$$

$$= \frac{\alpha}{\pi} m \int dw\, w(1 - w)\, du\, u^2(1 - u)(u - 2u^2 w) \psi^* \gamma^0 eq J^0 \frac{1}{D_1^{\,2}} eq A^0 \psi. \qquad (4\text{-}16.143)$$

This is the last contribution to the energy shift:

$$\delta_{10}E = 16\pi Z^2 \alpha^3 |\psi(0)|^2 \frac{1}{m} \int_0^1 dw\, w(1 - w) \int_0^1 du\, u(1 - u)(1 - 2uw)$$

$$\times \left(- m^2 \frac{d}{dm^2}\right) \int \frac{(dp)}{(2\pi)^3} \frac{1}{p^2} \frac{1}{D_1}$$

$$= 2Z^2 \alpha^3 |\psi(0)|^2 \frac{1}{m^2} \int_0^1 dw \int_0^1 du\, w^{1/2}(1 - w) u^{-1/2}(1 - u)(1 - 2uw)(1 - uw)^{-1/2}.$$

$$(4\text{-}16.144)$$

Now that we have seen all of the contributions, it is worthy of comment that every one of these double parametric integrals is reducible to (4-15.69), or its specialization (4-16.92). In this example, we get

$$\delta_{10}E = (\log 2 - \tfrac{1}{2})\pi Z^2 \alpha^3 |\psi(0)|^2 \frac{1}{m^2}. \qquad (4\text{-}16.145)$$

We add the various pieces, $\delta_4 E, \ldots, \delta_{10}E$, as given in Eqs. (4-16.104, 111, 116, 119, 128, 145), to get

$$\delta_b E = (\delta_4 + \cdots + \delta_{10})E = \left(- 3 + \frac{11}{32} - \frac{29}{105} - 2\log 2\right) \pi Z^2 \alpha^3 |\psi(0)|^2 \frac{1}{m^2},$$

$$(4\text{-}16.146)$$

and then, recalling $\delta_a E$ [Eq. (4–16.65)], we find that the complete effect, without reference to vacuum polarization, is

$$(\delta_a + \delta_b)E = \left(4 + \frac{11}{32} - 2\log 2\right)\pi Z^2\alpha^3|\psi(0)|^2\,\frac{1}{m^2}.\qquad(4\text{–}16.147)$$

This is about 20 percent less than the spin 0 result of (4–15.98). To finish the calculation, we introduce the appropriate spin $\frac{1}{2}$ structure into the vacuum polarization calculation of Eqs. (4–15.3–5), so that

$$\int(dx)m|\mathbf{x}|\delta\mathcal{D}(\mathbf{x}) = \frac{2\alpha}{3\pi}\,m\int_{(2m)^2}^{\infty}\frac{dM^2}{(M^2)^{5/2}}\left(1 + \frac{2m^2}{M^2}\right)\left(1 - \frac{4m^2}{M^2}\right)^{1/2}$$

$$= \frac{\alpha}{\pi}\,\frac{1}{(2m)^2}\int_0^1 dv(1 - v^2)^{1/2}v^2(1 - \tfrac{1}{3}v^2) = \frac{5\alpha}{96}\,\frac{1}{(2m)^2},\qquad(4\text{–}16.148)$$

which implies the energy shift

$$\frac{5}{48}\,\pi Z^2\alpha^3|\psi(0)|^2\,\frac{1}{m^2}.\qquad(4\text{–}16.149)$$

The final result, then, is

$$\delta E = \tfrac{4}{3}Z\alpha^2|\psi(0)|^2\,\frac{1}{m^2}\left[3\pi Z\alpha\left(1 + \frac{11}{128} - \tfrac{1}{2}\log 2 + \frac{5}{192}\right)\right],\qquad(4\text{–}16.150)$$

which has again been so written that the quantity in brackets represents the effective addition to the list of additive constants that supplement the dominant logarithm.

Concerning the precise magnitude of the additional upward displacement of s-levels, we recall that, for $n = 2$,

$$\tfrac{4}{3}Z\alpha^2|\psi(0)|^2\,\frac{1}{m^2} = \frac{1}{3\pi}\,Z^4\alpha^3\left(\frac{M}{M+m}\right)^3\text{Ry}$$

$$= 135.644Z^4\left(\frac{M}{M+m}\right)^3\text{MHz},\qquad(4\text{–}16.151)$$

which exhibits the reduced mass effect in $\psi(0)$. For hydrogen, this gives:

$$\text{H:}\quad \delta E_{2s1/2} = 7.13\text{ MHz}.\qquad(4\text{–}16.152)$$

Added to the earlier theoretical value of Eq. (4–11.114), we now have the following predicted value

$$\text{H:}\quad E_{2s1/2} - E_{2p1/2} = 1057.68\text{ MHz}.\qquad(4\text{–}16.153)$$

This time, the agreement with the experimental value of 1057.90 ± 0.10 MHz is even better than might have been anticipated since there still remain various secondary effects to be considered, most notably the modification of relative order α, as contrasted with the $Z\alpha$ effects considered here.

Incidentally, the result obtained in this section reproduces that derived quite some time ago by two independent sets of workers (Harvard and Cornell). The first announcements were: R. Karplus, A. Klein, J. Schwinger, *Phys. Rev.* **84**, 597 (1951); M. Baranger, *Phys. Rev.* **84**, 866 (1951). The complete descriptions of the methods are found in: Karplus, Klein, and Schwinger, *Phys. Rev.* **86**, 288 (1952); Baranger, Bethe, and Feynman, *Phys. Rev.* **92**, 482 (1953). There is a genetic relationship between the present work and the earlier one of the Harvard group, but the latter calculation is much more complicated in its details.

4–17 H-PARTICLE ENERGY DISPLACEMENTS. SPIN ½ RELATIVISTIC THEORY II

Atomic nuclei carry electromagnetic properties other than charge. Magnetic dipole moments, electric quadrupole moments, and so forth can, and do exist, as limited only by the nuclear spin. Correspondingly, the energy spectrum exhibits an additional, hyperfine structure. For the s-levels of a hydrogenic atom, with their restriction to spin angular momentum, only the nuclear magnetic dipole moment can be effective. It produces a doublet structure in the $1s_{1/2}$ level, which splitting has been measured with great accuracy in hydrogen, deuterium, and tritium. Our intent in this section is to review the elementary theory of this effect and discuss its electrodynamic modifications to relative order α and $Z\alpha^2$.

The nuclear magnetic moment is expressed in terms of the spin S and the g-factor g_S as

$$\mu = \frac{e}{2M_p} g_S S, \tag{4–17.1}$$

where M_p is the proton mass. The divergenceless spatial current, from which the moment is computed as [cf. Eq. (3–10.59)]

$$\mu = \tfrac{1}{2} \int (dr)\, r \times J, \tag{4–17.2}$$

s given by

$$J(r) = \nabla \times \mu\, \delta(r). \tag{4–17.3}$$

he associated vector potential and magnetic field are

$$A(r) = \nabla \times \frac{\mu}{4\pi r} = \frac{\mu \times r}{4\pi r^3},$$

$$H(r) = \nabla \times \left(\nabla \times \frac{\mu}{4\pi r} \right) = \nabla \nabla \cdot \left(\frac{\mu}{4\pi r} \right) + \mu \, \delta(r). \qquad (4\text{-}17.4)$$

We shall usually be concerned only with the rotational average of the coordinate dependence in the magnetic field, which is

$$\langle H(r) \rangle = \tfrac{1}{3} \nabla^2 \frac{\mu}{4\pi r} + \mu \, \delta(r) = \tfrac{2}{3} \mu \, \delta(r). \qquad (4\text{-}17.5)$$

Let us begin with the energy displacement computed from the primitive interaction, in a nonrelativistic approximation. Referring to the action expression for a spin $\tfrac{1}{2}$ particle in an electromagnetic field [Eq. (3–10.63) with $g_{\text{prim.}} = 2$, as is appropriate to an electron], we infer the energy displacement associated with the weak vector potential $A(r)$ to be

$$\delta E = - \int (dr) \psi(r)^* \gamma^0 eq\gamma \cdot A(r)\psi(r)$$

$$= e \int (dr)\psi(r)^* \gamma^0 \gamma \cdot A(r)\psi(r), \qquad (4\text{-}17.6)$$

where the latter form incorporates the charge assignment of an electron,

$$(eq)' = - e. \qquad (4\text{-}17.7)$$

The field $\psi(r)$ of the bound state omits the time dependent factor $\exp[- ip^0 x^0]$. We exploit the Dirac equations

$$(\gamma\Pi + m)\psi = 0, \qquad \psi^* \gamma^0 (\gamma\Pi + m) = 0, \qquad (4\text{-}17.8)$$

to rewrite (4–17.6) as

$$\delta E = - \frac{e}{2m} \int (dr)\psi^* \gamma^0 (\gamma\Pi\gamma \cdot A + \gamma \cdot A\gamma\Pi)\psi$$

$$= \frac{e}{2m} \int (dr)\psi^* \gamma^0 (2p \cdot A + \sigma \cdot H)\psi, \qquad (4\text{-}17.9)$$

which uses the fact that $\Pi = p$ when the Coulomb field is represented by the scalar potential A^0, and that symmetrization of the $p \cdot A$ product is unnecessary since

$$\nabla \cdot A = 0. \qquad (4\text{-}17.10)$$

In the nonrelativistic approximation, $\psi(r)$ describes a state of zero orbital angular momentum ($s_{1/2}$), and definite intrinsic parity, $\gamma^{0\prime} = + 1$. This implies the vanishing of the orbital term

$$\mathbf{p} \cdot \mathbf{A} = \frac{1}{4\pi r^3} \mathbf{\mu} \cdot \mathbf{L}, \qquad \mathbf{L} = \mathbf{r} \times \mathbf{p}, \tag{4–17.11}$$

and permits the magnetic field to be replaced by the average of (4–17.5). The immediate result is

$$\delta E = \frac{e}{2m} \frac{2}{3} \mathbf{\sigma} \cdot \mathbf{\mu} |\psi(0)|^2, \tag{4–17.12}$$

where

$$\mathbf{\sigma} \cdot \mathbf{\mu} = \frac{e}{2M_p} g_s \mathbf{\sigma} \cdot \mathbf{S} \tag{4–17.13}$$

is assigned one of the eigenvalues associated with the total angular momentum

$$\mathbf{F} = \mathbf{S} + \tfrac{1}{2}\mathbf{\sigma}. \tag{4–17.14}$$

They are given by

$$(\mathbf{\sigma} \cdot \mathbf{S})' = \mathbf{F}^2 - \mathbf{S}^2 - \tfrac{3}{4} = \begin{cases} F = S + \tfrac{1}{2}: & S \\ F = S - \tfrac{1}{2}: & -S - 1 \end{cases}, \tag{4–17.15}$$

and the splitting of the s state is measured by

$$\Delta E = E_{S+(1/2)} - E_{S-(1/2)} = \frac{2}{3} \frac{e}{2m} \frac{e}{2M_p} (2S + 1) g_s |\psi(0)|^2$$

$$= \frac{4}{3} \frac{m}{M_p} Z^3 \alpha^2 (2S + 1) g_s \left(\frac{M}{M + m} \right)^3 \text{Ry.} \tag{4–17.16}$$

The latter form refers specifically to the 1s state, and includes the reduced mass effect for a nucleus of mass M.

The measured value of the proton magnetic moment, expressed in the nuclear magneton $e/2M_p$, is

$$\mu_p = \tfrac{1}{2}g_p = 2.79278 \pm 0.00002. \tag{4–17.17}$$

The magnitude of the s state splitting computed from (4–17.16) is, then,

$$\text{H:} \quad \Delta E_{\text{nonrel.}} = 1418.83 \text{ MHz}, \tag{4–17.18}$$

which is to be compared with the measured value (many more significant figures are available):

$$\text{H:} \quad \Delta E_{\text{meas.}} = 1420.406 \text{ MHz}. \tag{4–17.19}$$

The major part of the discrepancy of 1.58 MHz is removed by invoking the $\alpha/2\pi$ modification of the electron moment. This gives the modified theoretical value

$$\text{H:} \quad \Delta E_{\text{nonrel.}+\alpha} = 1420.48 \text{ MHz.} \tag{4-17.20}$$

In connection with the residual discrepancy of 0.07 MHz, we shall discuss in this section the theoretical modifications of relative orders $(Z\alpha)^2$ and $Z\alpha^2$.

The first of these, $\sim (Z\alpha)^2$, is the purely relativistic correction to the non-relativistic formula. Perhaps the simplest way to compute it is to use the relativistic hydrogenic wave functions in evaluating (4-17.6), rather than trying to estimate the corrections in (4-17.9). The solution of the Dirac equation,

$$\left[\boldsymbol{\gamma} \cdot \frac{1}{i} \boldsymbol{\nabla} - \gamma^0 \left(p^0 + \frac{Z\alpha}{r} \right) + m \right] \psi(\mathbf{r}) = 0, \tag{4-17.21}$$

that applies to the ground state, is a mixture of $s_{1/2}$ and $p_{1/2}$ waves of opposite intrinsic parity,

$$\psi = \psi_\text{s} + \psi_\text{p}, \tag{4-17.22}$$

which are coupled together by the gradient term in (4-17.21). This is described by the pair of equations $[\gamma^0 \boldsymbol{\gamma} = i\gamma_5 \boldsymbol{\sigma}]$

$$\left(p^0 - m + \frac{Z\alpha}{r} \right) \psi_\text{s} = \gamma_5 \boldsymbol{\sigma} \cdot \boldsymbol{\nabla} \psi_\text{p},$$

$$\left(p^0 + m + \frac{Z\alpha}{r} \right) \psi_\text{p} = \gamma_5 \boldsymbol{\sigma} \cdot \boldsymbol{\nabla} \psi_\text{s}. \tag{4-17.23}$$

We exhibit the spin-angle dependence of these two components in

$$\psi_\text{s}(\mathbf{r}) = f(r) v, \qquad \psi_\text{p}(\mathbf{r}) = - \gamma_5 \boldsymbol{\sigma} \cdot \mathbf{n} g(r) v, \tag{4-17.24}$$

where $\mathbf{n}$ is the unit radial vector

$$\mathbf{n} = \mathbf{r}/r, \tag{4-17.25}$$

and v is an arbitrary unit spinor with $\gamma^{0\prime} = +1$. This is verified by extracting the purely radial equations

$$\frac{df}{dr} + \left(p^0 + m + \frac{Z\alpha}{r} \right) g = 0,$$

$$\left(\frac{d}{dr} + \frac{2}{r} \right) g - \left(p^0 - m + \frac{Z\alpha}{r} \right) f = 0; \tag{4-17.26}$$

the latter one has involved the following operator algebra $[\mathbf{r} \times \boldsymbol{\nabla} g = 0]$

$$\boldsymbol{\sigma} \cdot \boldsymbol{\nabla} \boldsymbol{\sigma} \cdot \mathbf{n} g = \boldsymbol{\nabla} \cdot \mathbf{n} g = \left(\mathbf{n} \cdot \boldsymbol{\nabla} + \frac{2}{r} \right) g. \tag{4-17.27}$$

The pair of similar equations is effectively diagonalized by writing

$$g(r) = \gamma f(r), \tag{4-17.28}$$

and identifying analogous coefficients in the two equations:

$$\gamma(p^0 + m) = -\frac{1}{\gamma}(p^0 - m), \qquad \gamma Z\alpha = 2 - \frac{1}{\gamma} Z\alpha. \tag{4-17.29}$$

This gives (only one root of the quadratic γ equation is physically acceptable)

$$\gamma = \frac{1}{Z\alpha}[1 - (1 - (Z\alpha)^2)^{1/2}] \cong \tfrac{1}{2}Z\alpha, \qquad Z\alpha \ll 1, \tag{4-17.30}$$

and

$$p^0 = m\frac{1 - \gamma^2}{1 + \gamma^2} = m(1 - Z\alpha\gamma) = m(1 - (Z\alpha)^2)^{1/2}$$

$$\cong m - \tfrac{1}{2}(Z\alpha)^2 m. \tag{4-17.31}$$

The radial dependence now obtained by solving

$$\frac{df}{dr} + \left(Z\alpha m + \frac{Z\alpha\gamma}{r}\right)f = 0 \tag{4-17.32}$$

is

$$f(r) = Nr^{-Z\alpha\gamma}\exp(-Z\alpha mr). \tag{4-17.33}$$

The coefficient N is determined (apart from a phase factor) by the normalization condition [cf. Eq. (3–15.33)]

$$1 = \int (d\mathbf{r})\psi(\mathbf{r})^*\psi(\mathbf{r}) = \int_0^\infty 4\pi r^2\, dr\, (f^2 + g^2) = (1 + \gamma^2)4\pi N^2\int_0^\infty dr\, r^{2-2Z\alpha\gamma}\exp(-2Z\alpha mr)$$

$$= (1 + \gamma^2)4\pi N^2\left(\frac{1}{2Z\alpha m}\right)^{3-2Z\alpha\gamma}\Gamma(3 - 2Z\alpha\gamma). \tag{4-17.34}$$

When $2Z\alpha\gamma \cong 4\gamma^2$ is neglected, we regain the familiar nonrelativistic normalization constant $|\psi(0)|^2$,

$$N^2 \cong |\psi(0)|^2 = \frac{1}{\pi}(Z\alpha m)^3. \tag{4-17.35}$$

The energy shift computed from (4–17.6), or

$$\delta E = e\int (d\mathbf{r})\psi(\mathbf{r})^*i\gamma_5\boldsymbol{\sigma}\cdot\mathbf{A}(\mathbf{r})\psi(\mathbf{r}), \tag{4-17.36}$$

is given by

$$\delta E = e \int (d\mathbf{r}) f(r) g(r) v^* i [\boldsymbol{\sigma} \cdot \mathbf{A}(\mathbf{r}), \boldsymbol{\sigma} \cdot \mathbf{n}] v$$

$$= 2e \int (d\mathbf{r}) f(r) g(r) v^* \boldsymbol{\sigma} \cdot \mathbf{n} \times \mathbf{A}(\mathbf{r}) v, \qquad (4\text{--}17.37)$$

where

$$\mathbf{n} \times \mathbf{A}(\mathbf{r}) = \frac{1}{4\pi r^2} (\boldsymbol{\mu} - \mathbf{n}\mathbf{n} \cdot \boldsymbol{\mu}) \to \frac{2}{3} \frac{\boldsymbol{\mu}}{4\pi r^2}. \qquad (4\text{--}17.38)$$

The last step above has introduced the rotational average. Accordingly, in an eigenstate of $\boldsymbol{\sigma} \cdot \mathbf{S}$, we have

$$\delta E = \tfrac{1}{3} e \boldsymbol{\sigma} \cdot \boldsymbol{\mu} \int_0^\infty dr \, f(r) g(r). \qquad (4\text{--}17.39)$$

It is simplest to divide this radial integral by the unit normalization integral of (4–17.34),

$$\int_0^\infty dr \, fg = \frac{1}{4\pi} \frac{\gamma}{1 + \gamma^2} \frac{\int_0^\infty dr \, f^2}{\int_0^\infty dr \, r^2 f^2} = \frac{1}{4\pi} \frac{\gamma}{1 + \gamma^2} (2Z\alpha m)^2 \frac{\Gamma(1 - 2Z\alpha\gamma)}{\Gamma(3 - 2Z\alpha\gamma)}$$

$$= \frac{1}{4m} |\psi(0)|^2 \frac{1}{(1 - Z\alpha\gamma)(1 - 2Z\alpha\gamma)}, \qquad (4\text{--}17.40)$$

and then

$$\delta E = \frac{e}{2m} \frac{2}{3} \boldsymbol{\sigma} \cdot \boldsymbol{\mu} |\psi(0)|^2 \frac{1}{(1 - Z\alpha\gamma)(1 - 2Z\alpha\gamma)}. \qquad (4\text{--}17.41)$$

For small $Z\alpha$, the relativistic correction factor is

$$\frac{1}{(1 - Z\alpha\gamma)(1 - 2Z\alpha\gamma)} = 1 + 3Z\alpha\gamma \simeq 1 + \tfrac{3}{2}(Z\alpha)^2. \qquad (4\text{--}17.42)$$

Considered by itself, this $(Z\alpha)^2$ effect acts in the wrong direction to resolve the discrepancy between (4–17.19) and (4–17.20). It would increase the difference to 0.19 MHz.

We therefore turn to the discussion of the modification of order $Z\alpha^2$, that is, of order $Z\alpha$ relative to the calculation of Eq. (4–17.20). With all the accumulated expertise of the last section to draw on, this is a comparatively simple task. The first topic is vacuum polarization, which is much more important in this situation than with the Coulomb field, owing to the more concentrated nature of the nuclear magnetic field. The alteration of the vector potential A^μ by δA^μ changes energy values by the amount [it is the more general form of Eq. (4–17.9)]

$$\delta E = -\frac{1}{2m} \int (d\mathbf{r}) \psi^* \gamma^0 (2eq\Pi.\delta A + eq\sigma\delta F)\psi, \tag{4-17.43}$$

where

$$\delta A(\mathbf{r}) = \int (d\mathbf{r}') \, \delta\mathscr{D}(\mathbf{r} - \mathbf{r}') J(\mathbf{r}'), \tag{4-17.44}$$

or

$$\delta A(\mathbf{p}) = \mathbf{p}^2 \, \delta\mathscr{D}(\mathbf{p}) A(\mathbf{p}), \tag{4-17.45}$$

specifies the nature of the alteration. Perhaps one should observe first that, in contrast with the Coulomb field situation, no vacuum polarization modification of the magnetic field coupling appears when the momentum dependence of the wave function ψ is neglected [this is the use of $\psi(0)$ only]. That is because

$$\delta H(\mathbf{p}) = \mathbf{p}^2 \, \delta\mathscr{D}(\mathbf{p}) \, H(\mathbf{p}) \rightarrow \mathbf{p}^2 \, \delta\mathscr{D}(\mathbf{p})\tfrac{2}{3}\mu \tag{4-17.46}$$

vanishes at $\mathbf{p} = 0$. We are therefore interested in the iterated field

$$\psi = \psi(0) + \frac{1}{p^2 + m^2} (2eqp.A + eq\sigma F)\psi(0). \tag{4-17.47}$$

It implies the energy shift

$$\delta E = -\frac{1}{m} \psi(0)^* (2eqp\delta A + eq\sigma\delta F) \frac{1}{p^2 + m^2} (2eqpA + eq\sigma F)\psi(0)$$

$$= -\frac{e^2}{m} \psi(0)^* \left(2p\delta A \frac{1}{p^2 + m^2} \sigma F + \sigma\delta F \frac{1}{p^2 + m^2} 2pA \right)\psi(0), \tag{4-17.48}$$

where we have recognized that the two cross-product terms contribute equally, and extracted the linear spin dependence. Throughout this discussion, terms quadratic in the spin can be discarded since

$$\boldsymbol{\sigma} \cdot \mathbf{H} i \gamma_5 \boldsymbol{\sigma} \cdot \mathbf{E} \tag{4-17.49}$$

disappears in the $\gamma^{0\prime} = +1$ state $\psi(0)$. Contained here is the effect of the magnetic field on Coulomb field vacuum polarization, and the effect of the Coulomb field on magnetic field vacuum polarization. The two effects are equal, for both produce the term

$$-2mA^0\mathbf{p}^2 \, \delta\mathscr{D} \frac{1}{p^2 + m^2} \boldsymbol{\sigma} \cdot \mathbf{H}. \tag{4-17.50}$$

Thus,

$$\delta E_{\text{vac.pol.}} = 16\pi\alpha \, \tfrac{2}{3}\boldsymbol{\sigma} \cdot \boldsymbol{\mu}|\psi(0)|^2 Ze \int \frac{(d\mathbf{p})}{(2\pi)^3} \frac{1}{\mathbf{p}^2} \delta\mathscr{D}(\mathbf{p}), \qquad (4\text{-}17.51)$$

where [cf. Eq. (4-3.28)]

$$\delta\mathscr{D}(\mathbf{p}) = \frac{\alpha}{4\pi} \int_0^1 dv \, \frac{v^2(1 - \tfrac{1}{3}v^2)}{m^2 + \mathbf{p}^2 \dfrac{1 - v^2}{4}}. \qquad (4\text{-}17.52)$$

We shall write all these hyperfine structure energy shifts in units of the non-relativistic value (4-17.12), which is designated as

$$F = \frac{e}{2m} \frac{2}{3} \, \boldsymbol{\sigma} \cdot \boldsymbol{\mu}|\psi(0)|^2. \qquad (4\text{-}17.53)$$

Accordingly,

$$\delta E_{\text{vac.pol.}}/F = 32\pi Z\alpha m \int \frac{(d\mathbf{p})}{(2\pi)^3} \frac{1}{\mathbf{p}^2} \delta\mathscr{D}(\mathbf{p})$$

$$= Z\alpha^2 \frac{4}{\pi} \int_0^1 dv \, v^2(1 - \tfrac{1}{3}v^2)(1 - v^2)^{-1/2} = \tfrac{3}{4}Z\alpha^2. \qquad (4\text{-}17.54)$$

The discussion to follow parallels that of Section 4-16, but is simpler since there is no infrared sensitivity in virtue of the short range nature of the magnetic field. In consequence, we embark directly on the power series expansion, and employ the λ device only in the first term of (4-16.7) in order to extract (4-16.12), the nonexplicitly field dependent part. The latter does not contribute to energy shifts. There is just one point to keep in mind. Since this calculation specifically seeks corrections to the description of the electron in which the $\alpha/2\pi$ addition to the magnetic moment is already taken into account, the two terms that produce the $\alpha/2\pi$ effect must be removed.

We begin with Eq. (4-16.68), which is repeated here, apart from an additional factor:

$$-\int ds \, s \, du(1 + u)(- 2m)e^2 I(\lambda^2)_A$$

$$= -i\frac{\alpha}{2\pi} m \int \tfrac{1}{2}dv \, du(1 + u)$$

$$\times \frac{u(1 - u)(p' + p'')eqA + (u(1 - u) + \lambda^2 u^2)eq\sigma F}{D_\lambda}. \qquad (4\text{-}17.55)$$

Taking the first λ difference gives the energy shift contribution

$$\delta_1 E = \delta'_1 E + \delta''_1 E, \qquad (4\text{-}17.56)$$

with

$$\delta'_1 E = \frac{\alpha}{2\pi} m \int \tfrac{1}{2} dv\, du (1+u)\psi^* \gamma^0 u (1-u)((p'+p'')eqA + eq\sigma F)\left(\frac{1}{D_1} - \frac{1}{D_0}\right)\psi$$

(4-17.57)

and

$$\delta''_1 E = \frac{\alpha}{2\pi} m \int \tfrac{1}{2} dv\, du (1+u)\psi^* \gamma^0 u^2 eq\sigma F\left(\frac{1}{D_1} - \frac{1}{m^2 u^2}\right)\psi.$$

(4-17.58)

In the latter, we have removed the term that contributes to the magnetic moment in the situation of slowly varying fields [it is equivalent to the first term on the right side of (4-16.17)]. Introducing the iterated particle field into (4-17.57) and extracting the linear spin term, we get

$$\delta'_1 E = \frac{2\alpha}{\pi} m \int \tfrac{1}{2} dv\, du (1+u)u(1-u)\psi(0)^*(p'+p'')eqA \frac{1}{p^2+m^2}\left(\frac{1}{D_1}-\frac{1}{D_0}\right)eq\sigma F\psi(0)$$

$$= F\, 32Z\alpha^2 m^3 \int \tfrac{1}{2} dv\, du (1+u)u(1-u)u^2 \frac{1-v^2}{4}\int \frac{(dp)}{(2\pi)^3}\frac{1}{\mathbf{p}^2}\frac{1}{D_0 D_1},$$

(4-17.59)

or, with the appropriate modification of the structure in Eqs. (4-15.72, 73),

$$\frac{\delta'_1 E}{F} = Z\alpha^2 \frac{8}{\pi}\frac{1}{3}\int_0^1 dw\, w^{3/2}\int_0^1 du\, u^{-1/2}(1-u^2)(1-uw)^{-1/2}.$$

(4-17.60)

The double parametric integrals appearing here have been met before, which is the general experience in the present calculation, and give

$$\frac{\delta'_1 E}{F} = (2 - 2\log 2)Z\alpha^2.$$

(4-17.61)

Turning to (4-17.58), we find that

$$\delta''_1 E = \frac{\alpha}{\pi} m \int \tfrac{1}{2} dv\, du (1+u)u^2\psi(0)^*(p'+p'')eqA \frac{1}{p^2+m^2} eq\sigma F\left(\frac{1}{D_1} - \frac{1}{m^2 u^2}\right)\psi(0)$$

$$= F\, 16Z\alpha^2 m \int dw\, du (1+u)uw(1-uw)\int \frac{(dp)}{(2\pi)^3}\frac{1}{\mathbf{p}^2}\frac{1}{D_1},$$

(4-17.62)

and

$$\frac{\delta''_1 E}{F} = Z\alpha^2 \frac{4}{\pi}\int dw\, du (1+u)u^{-1/2}w^{1/2}(1-uw)^{1/2}$$

$$= \left(\frac{5}{4} + \log 2\right)Z\alpha^2.$$

(4-17.63)

The sum of the two pieces is

$$\frac{\delta_1 E}{F} = \left(\frac{13}{4} - \log 2\right) Z\alpha^2,$$ (4–17.64)

which, as it happens, is the negative of the final result (without vacuum polarization).

The linear spin term in Eq. (4–16.83) leads, as the counterpart of (4–16.85), to

$$- \int ds\, s\, du(1 + u)\psi^* \gamma^0 e^2 I(\lambda^2)_{A^2}\psi = -i\, 8Z\alpha^2 F \int \tfrac{1}{2}\, dv\, \frac{1 - v}{2}\, du(1 - u^2)(1 - u + \lambda^2 u)$$

$$\times \left(- m^2 \frac{d}{dm^2}\right) \int \frac{(d\mathbf{p})}{(2\pi)^3} \frac{1}{\mathbf{p}^2} \frac{1}{D_\lambda},$$ (4–17.65)

and then to the energy shift contribution:

$$\frac{\delta_2 E}{F} = - 16Z\alpha^2 m \int dw(1 - w)\, du(1 - u^2)\left(- m^2 \frac{d}{dm^2}\right) \int \frac{(d\mathbf{p})}{(2\pi)^3} \frac{1}{\mathbf{p}^2}\left(\frac{1}{D_1} - \frac{1 - u}{D_0}\right).$$ (4–17.66)

We divide this into two parts,

$$\frac{\delta'_2 E}{F} = - 16Z\alpha^2 m \int dw(1 - w)\, du(1 - u^2)\left(- m^2 \frac{d}{dm^2}\right) \int \frac{(d\mathbf{p})}{(2\pi)^3} \frac{1}{\mathbf{p}^2}\left(\frac{1}{D_1} - \frac{1}{D_0}\right)$$

$$= - Z\alpha^2 \frac{2}{\pi} \int dw(1 - w)w^{-1/2}\, du\, u^{-3/2}(1 - u^2)\left(\frac{1}{(1 - uw)^{1/2}} - \frac{1}{(1 - u)^{1/2}}\right)$$

$$= Z\alpha^2 \frac{2}{\pi} \int_0^1 dw \int_0^1 du(w^{1/2} - \tfrac{1}{3}w^{3/2})u^{-1/2}(1 - u^2)(1 - uw)^{-3/2}$$

$$= \left(\frac{1}{2} + \frac{5}{3} - 2\log 2\right) Z\alpha^2,$$ (4–17.67)

and

$$\frac{\delta''_2 E}{F} = - 16Z\alpha^2 m \int dw(1 - w)\, du\, u(1 - u^2)\left(- m^2 \frac{d}{dm^2}\right) \int \frac{(d\mathbf{p})}{(2\pi)^3} \frac{1}{\mathbf{p}^2} \frac{1}{D_0}$$

$$= - Z\alpha^2 \frac{2}{\pi} \int_0^1 dw\, w^{-1/2}(1 - w) \int_0^1 du\, u^{-1/2}(1 - u)^{-1/2}(1 - u^2)$$

$$= - \frac{5}{3} Z\alpha^2.$$ (4–17.68)

The sum is

$$\frac{\delta_2 E}{F} = (\tfrac{1}{2} - 2\log 2)Z\alpha^2, \tag{4–17.69}$$

which adds to $\delta_1 E$ in giving the complete contribution of the first term (I) in (4–16.7):

$$\frac{\delta_1 E}{F} = \left(\frac{15}{4} - 3\log 2\right)Z\alpha^2. \tag{4–17.70}$$

The discussion of the second term in the list begins with Eq. (4–16.108), evaluated for $\lambda = 1$,

$$-e^2 \int ds\, s\, du(1-u) \int \frac{(dk)}{(2\pi)^4}\, \psi^* \gamma^0 i[\Pi, [\exp(-is\chi), \gamma]]\psi$$
$$= \frac{\alpha}{2\pi} \int \tfrac{1}{2}\, dv\, du\, u(1-u)\psi^* \gamma^0 \frac{eq\gamma J}{D_1}\, \psi, \tag{4–17.71}$$

although one must also remark that the vector potential in Π does not contribute a spin term. The introduction of the iterated field produces the energy shift

$$\delta_{\text{II}} E = \frac{\alpha}{2\pi} \int \tfrac{1}{2}\, dv\, du\, u(1-u)\psi(0)^* \frac{1}{D_1}\left(eq\gamma J \frac{1}{p^2+m^2} eq\sigma F + eq\sigma F \frac{1}{p^2+m^2} eq\gamma J\right)\psi(0), \tag{4–17.72}$$

where one must be careful to note that there are two contributions in each product, as indicated by

$$-\gamma^0 J^0(\ )\sigma\cdot\mathbf{H} - \gamma\cdot\mathbf{J}(\ )i\gamma^0\gamma\cdot\mathbf{E}, \tag{4–17.73}$$

or $[\gamma^0\psi(0) = \psi(0)]$

$$-J^0(\ )\sigma\cdot\mathbf{H} + \sigma\times(\nabla\times\mathbf{H})\cdot(\ )\mathbf{E}. \tag{4–17.74}$$

The vector identity

$$\sigma\times(\nabla\times\mathbf{H}) = \nabla(\sigma\cdot\mathbf{H}) - \sigma(\nabla\cdot\mathbf{H}) + \nabla\times(\sigma\times\mathbf{H}), \tag{4–17.75}$$

in combination with the properties

$$\nabla\cdot\mathbf{E} = J^0, \qquad \nabla\cdot\mathbf{H} = 0, \qquad \nabla\times\mathbf{E} = 0, \tag{4–17.76}$$

which are applied after partial integration, shows the equivalence of the two terms in (4–17.73), since the multiplication order is not significant for this matrix element. That gives

$$\delta_{\text{II}} E = -\frac{2\alpha}{\pi} \int \tfrac{1}{2}\, dv\, du\, u(1-u)\psi(0)^* \frac{1}{D_1} eq J^0 \frac{1}{p^2+m^2} eq\sigma\cdot\mathbf{H}\psi(0)$$

$$= - 16Z\alpha^2 mF \int dw \, du \, u(1 - u) \int \frac{(d\mathbf{p})}{(2\pi)^3} \frac{1}{\mathbf{p}^2} \frac{1}{D_1},$$ (4–17.77)

or

$$\frac{\delta_{\mathrm{II}}E}{F} = - Z\alpha^2 \frac{4}{\pi} \int_0^1 dw \int_0^1 du \, w^{-1/2} u^{-1/2}(1 - u)(1 - uw)^{-1/2}$$

$$= (2 - 8\log 2)Z\alpha^2.$$ (4–17.78)

For the third term of (4–16.7), we follow the discussion beginning with Eq. (4–16.112) and leading to (4–16.114), where the second contribution to the $\alpha/2\pi$ magnetic moment has already been excised. This yields

$$\delta_4 E = - \frac{\alpha}{\pi} m \int \tfrac{1}{2} \, dv \, du \, u^2 \psi^* \gamma^0 eq\sigma F \left(\frac{1}{D_1} - \frac{1}{m^2 u^2} \right) \psi$$

$$= - 32Z\alpha^2 mF \int dw \, du \, uw(1 - uw) \int \frac{(d\mathbf{p})}{(2\pi)^3} \frac{1}{\mathbf{p}^2} \frac{1}{D_1},$$ (4–17.79)

or

$$\frac{\delta_4 E}{F} = - Z\alpha^2 \frac{8}{\pi} \int_0^1 dw \int_0^1 du \, w^{1/2} u^{-1/2}(1 - uw)^{1/2},$$ (4–17.80)

which will be recognized as the integral of (4–16.115). Accordingly, we have

$$\delta_4 E/F = - 3Z\alpha^2.$$ (4–17.81)

As for the term explicitly quadratic in the field, and linear in the spin, we recall that the double commutator of $\tfrac{1}{2}mu[\gamma, [\exp(- is\chi), \gamma]]$ multiplies the spin term by a factor of 8 [Eq. (4–16.113)], so that this structure can be obtained from the corresponding one in the first term of (4–16.7) by the substitution

$$1 + u \to - 2u,$$ (4–17.82)

applied to the $\lambda = 1$ form. Picking out that term from (4–17.67) then gives

$$\frac{\delta_5 E}{F} = Z\alpha^2 \frac{4}{\pi} \int_0^1 dw \int_0^1 du \, w^{-1/2}(1 - w)u^{-1/2}(1 - u)(1 - uw)^{-1/2}$$

$$= (- 6 + 12\log 2)Z\alpha^2,$$ (4–17.83)

and, on adding (4–17.81), we get

$$\delta_{\mathrm{III}}E/F = (- 9 + 12\log 2)Z\alpha^2.$$ (4–17.84)

The first three terms [Eqs. (4–17.70, 78, 84)] give

$$(\delta_{\mathrm{I}} + \delta_{\mathrm{II}} + \delta_{\mathrm{III}})E/F = (- (13/4) + \log 2)Z\alpha^2,$$ (4–17.85)

which is the negative of the particular contribution exhibited in Eq. (4–17.64). One may infer from the remark made there that this is the complete expression. And so it is, since it turns out that the fourth term of (4–16.7) has a vanishing contribution. It would be very pleasant to prove this in one masterful stroke, but we shall have to use more prosaic means. The double commutator term in (4–16.8) has the consequence already stated in Eq. (4–16.117),

$$\delta_6 E = \frac{\alpha}{4\pi} \int \tfrac{1}{2}\, dv\, du\, u^2 \psi^* \gamma^0 \frac{eq\gamma J}{D_1} \psi$$

$$= - 8Z\alpha^2 mF \int \tfrac{1}{2}\, dv\, du\, u^2 \int \frac{(d\mathbf{p})}{(2\pi)^3} \frac{1}{\mathbf{p}^2} \frac{1}{D_1}, \tag{4–17.86}$$

and

$$\frac{\delta_6 E}{F} = - Z\alpha^2 \frac{2}{\pi} \int_0^1 dw \int_0^1 du\, w^{-1/2} u^{1/2} (1 - uw)^{-1/2}$$

$$= - Z\alpha^2. \tag{4–17.87}$$

Next we consider the linear field term of Eq. (4–16.123) which, for convenience, is repeated here in the form of an energy shift:

$$- i \int ds\, s\, du\, e^2 \int \frac{(dk)}{(2\pi)^4} \psi^* \gamma^0 \gamma^\mu \gamma .((k - u\Pi).\exp(- is\chi))_A \gamma_\mu$$

$$= \frac{\alpha}{4\pi} \int \tfrac{1}{2}\, dv\, v\, du\, u\, \psi^* \gamma^0 [- u^2 v eq\gamma J + u(1 - u)ieq\gamma F(p' + p'')] \frac{1}{D_1} \psi. \tag{4–17.88}$$

The first of the two parts,

$$\delta'_7 E = - \frac{\alpha}{4\pi} \int \tfrac{1}{2}\, dv\, v^2\, du\, u^3 \psi^* \gamma^0 \frac{eq\gamma J}{D_1} \psi, \tag{4–17.89}$$

can be inferred from the analogous evaluation of (4–17.71), leading to (4–17.77). Making the appropriate substitutions, we get

$$\delta'_7 E = 8Z\alpha^2 mF \int dw\, du(2w - 1)^2 u^3 \int \frac{(d\mathbf{p})}{(2\pi)^3} \frac{1}{\mathbf{p}^2} \frac{1}{D_1}, \tag{4–17.90}$$

which we prefer to combine with $\delta''_7 E$ before further evaluation.

There is some danger, in discussing the second term of (4–17.88), of being misled by the notation. First, recall that p' and p'' are row and column indices, respectively, in a typical matrix element. If we then wish to use p' for representing the momentum in $\psi(0)$, a transposition is needed in one of the two terms produced by the insertion of the iterated field. Attention must be paid also to the convention adopted in reducing the D_λ of (4–15.60) to the D_λ of (4–15.62). Thus far, we have escaped all this tight-rope walking because, apart from D_λ, the integrands have been even functions of the variable v; the second term of (4–17.88) is not in that category. It is now seen that

$$\psi^*\gamma^0 \frac{v}{D_1} ieq\gamma F(p' + p'')\psi = \psi(0)^* ieq\gamma F(p' + p'') \frac{v}{D_1(-v)} \frac{1}{p''^2 + m^2} eq\sigma F\psi(0)$$

$$+ \psi(0)^* eq\sigma F \frac{1}{p''^2 + m^2} \frac{v}{D_1(v)} ieq\gamma F(p' + p'')\psi(0),$$

$$(4\text{–}17.91)$$

where we have temporarily written $D_1(v)$ to indicate the D_1 function of Eq. (4–15.64), in which

$$\mathbf{p}^2 = (p' - p'')^2 = p''^2 + m^2. \tag{4–17.92}$$

We shall change the integration variable v to $-v$, in the first term of (4–17.91), and employ the reductions illustrated in Eqs. (4–16.141, 142). After recognizing that the two terms of (4–17.91) are effectively equal, we get

$$\psi^*\gamma^0 \frac{v}{D_1} ieq\gamma F(p' + p'')\psi \rightarrow -2v\psi(0)^* \left[eq\sigma \cdot \mathbf{H} \frac{1}{p''^2 + m^2} \frac{1}{D_1} eqJ^0 \right.$$

$$\left. + ieq\gamma \cdot \mathbf{E} \frac{1}{p''^2 + m^2} \frac{1}{D_1} eq\gamma \cdot \mathbf{J} \right]\psi(0)$$

$$\rightarrow -4v\psi(0)^* eq\sigma \cdot \mathbf{H} \frac{1}{p''^2 + m^2} \frac{1}{D_1} eqJ^0\psi(0), \quad (4\text{–}17.93)$$

using the equivalence noted in connection with (4–17.73). Accordingly,

$$\delta''_7 E = -8Z\alpha^2 mF \int dw(2w - 1)\, du\, u^2(1 - u) \int \frac{(d\mathbf{p})}{(2\pi)^3} \frac{1}{\mathbf{p}^2} \frac{1}{D_1}, \quad (4\text{–}17.94)$$

which, added to (4–17.90), gives

$$\frac{\delta_7 E}{F} = -8Z\alpha^2 m \int dw\, du(2w - 1)u^2(1 - 2uw) \int \frac{(d\mathbf{p})}{(2\pi)^3} \frac{1}{\mathbf{p}^2} \frac{1}{D_1}$$

$$= -Z\alpha^2 \frac{2}{\pi} \int_0^1 dw \int_0^1 du\, w^{-1/2}(2w - 1)u^{1/2}(1 - 2uw)(1 - uw)^{-1/2}$$

$$= (-1 + 2\log 2)Z\alpha^2. \tag{4–17.95}$$

Finally, there is the explicitly quadratic term constructed from the linear factors of (4–16.132) and (4–16.133). Since we want the linear spin dependence, the terms of interest are obtained by multiplying $-ueq\sigma F$ into the two nonspin terms of (4–16.132), and by multiplying the spin terms. The latter is an exception to the rule, because multiplication with γ^μ is also involved. We first note that

$$\gamma^\nu \gamma^\mu . \sigma F \gamma_\nu = -2g^{\mu\nu}\sigma F . \gamma_\nu = -2\gamma^\mu . \sigma F, \qquad (4\text{–}17.96)$$

since

$$\gamma^\nu \sigma F \gamma_\nu = 0. \qquad (4\text{–}17.97)$$

For the spin product we have, omitting inessential factors and introducing permissible partial integrations,

$$\gamma^\mu . (\sigma F \, \partial_\mu \, \sigma F) = \gamma^k . (\boldsymbol{\sigma} \cdot \mathbf{H} \, \partial_k (-i)\gamma^0 \boldsymbol{\gamma} \cdot \mathbf{E} + (-i)\gamma^0 \boldsymbol{\gamma} \cdot \mathbf{E} \, \partial_k \, \boldsymbol{\sigma} \cdot \mathbf{H})$$

$$= \tfrac{1}{2} i\gamma^0 [\gamma^k \boldsymbol{\sigma} \cdot \mathbf{H} \gamma^l \partial_k E_l + \partial_k E_l \gamma^l \boldsymbol{\sigma} \cdot \mathbf{H} \gamma^k$$

$$- \boldsymbol{\sigma} \cdot \mathbf{H} \, \partial_k E_l \gamma^l \gamma^k - \gamma^k \gamma^l \partial_k E_l \boldsymbol{\sigma} \cdot \mathbf{H}]. \qquad (4\text{–}17.98)$$

The next step makes use of the null value of $\boldsymbol{\nabla} \times \mathbf{E}$, which is the symmetry

$$\partial_k E_l = \partial_l E_k. \qquad (4\text{–}17.99)$$

It enables one to make the replacement

$$\gamma^k \gamma^l \to \tfrac{1}{2}\{\gamma^k, \gamma^l\} = -\delta^{kl}, \qquad (4\text{–}17.100)$$

and to employ the relation $[\boldsymbol{\gamma} = i\gamma^0\gamma_5\boldsymbol{\sigma}]$

$$-\tfrac{1}{2}(\gamma^k \sigma_m \gamma^l + \gamma^l \sigma_m \gamma^k) = \tfrac{1}{2}(\sigma_k \sigma_m \sigma_l + \sigma_l \sigma_m \sigma_k)$$

$$= \delta_{km}\sigma_l + \delta_{lm}\sigma_k - \delta_{kl}\sigma_m, \qquad (4\text{–}17.101)$$

which follows directly from the anticommutation property

$$\tfrac{1}{2}\{\sigma_k, \sigma_l\} = \delta_{kl}. \qquad (4\text{–}17.102)$$

The additional fact that $\boldsymbol{\nabla} \cdot \mathbf{H} = 0$ leads to the effective evaluation

$$\gamma^\mu . (\sigma F \, \partial_\mu \, \sigma F) \to 2i\gamma^0 \boldsymbol{\sigma} \cdot \mathbf{H} J^0, \qquad (4\text{–}17.103)$$

and then

$$\gamma^\nu \gamma^\mu . (\sigma F \, \partial_\mu \, \sigma F)\gamma_\nu \to -4i\boldsymbol{\sigma} \cdot \mathbf{H} J^0. \qquad (4\text{–}17.104)$$

As the counterpart of Eq. (4–16.138), we now have

$$e^2 \int \frac{(dk)}{(2\pi)^4} \gamma^\mu \gamma . ((k - u\Pi).\exp(-is\chi))_{A^2}\gamma_\mu$$

$$= -\frac{\alpha}{4\pi} u \int dw \, w \frac{1-w}{2} \{2ueq\sigma F \exp(-isD_1)$$

$$\times \left[- 2u^2(\tfrac{1}{2} - w)ieq\gamma J + u(1 - u)eq\gamma F(p' + p'')\right]$$

$$+ u^2 4ieq\boldsymbol{\sigma} \cdot \mathbf{H} \exp(- isD_1) eqJ^0\} - \frac{\alpha}{4\pi} u \int dw \, w\frac{w - 1}{2}\{[2u^2(\tfrac{1}{2} - w)ieq\gamma J$$

$$+ u(1 - u)eq\gamma F(p' + p'')] \exp(- isD_1) 2ueq\boldsymbol{\sigma} F$$

$$- u^2 4ieq\boldsymbol{\sigma} \cdot \mathbf{H} \exp(- isD_1) eqJ^0\}, \tag{4–17.105}$$

and the same reductions used before produce the energy shift expression

$$- i \int ds \, s \, du \, e^2 \int \frac{(dk)}{(2\pi)^4} \, \psi^* \gamma^0 \gamma^\mu \gamma.((k - u\Pi).\exp(- is\chi))_{A^2}\gamma_\mu\psi$$

$$= \frac{\alpha}{2\pi} \int dw \, w \, \frac{1 - w}{2} \, du \, u\psi(0)^* \frac{1}{D_1{}^2} [4u^2(1 - 2uw)eq\boldsymbol{\sigma} \cdot \mathbf{H}eqJ^0 + 4u^2 eq\boldsymbol{\sigma} \cdot \mathbf{H}eqJ^0]\psi(0),$$

$$\tag{4–17.106}$$

or

$$\frac{\delta_8 E}{F} = 16Z\alpha^2 m \int dw \, du \, w(1 - w)u^3(1 - uw) \int \frac{(d\mathbf{p})}{(2\pi)^3} \frac{1}{D_1{}^2}$$

$$= Z\alpha^2 \frac{2}{\pi} \int_0^1 dw \int_0^1 du \, w^{-1/2}(1 - w)u^{1/2}(1 - uw)^{-1/2}$$

$$= (2 - 2\log 2)Z\alpha^2. \tag{4–17.107}$$

And, indeed, the sum of the pieces in Eqs. (4–17.87, 95, 107) does vanish,

$$\delta_{\mathrm{IV}}E = 0. \tag{4–17.108}$$

The final outcome, obtained by adding to (4–17.85) the energy shift of vacuum polarization origin, Eq. (4–17.54), is

$$\delta E/F = - (\tfrac{5}{2} - \log 2)Z\alpha^2. \tag{4–17.109}$$

Taken by itself, it decreases the theoretical value of the hydrogen hyperfine splitting by 0.137 MHz. But the combination of the relativistic effect in (4–17.42) and the electrodynamic effect just computed results in a comparatively small decrease, measured by

$$\mathrm{H:} \quad - (1 - \log 2)\alpha^2 = - 1.64 \times 10^{-5}. \tag{4–17.110}$$

This represents a decrease of 0.023 MHz, altering the theoretical value of (4–17.20) to

$$\mathrm{H:} \quad \Delta E_{\mathrm{rel.} + \alpha + Z\alpha^2} = 1420.46 \text{ MHz.} \tag{4–17.111}$$

What shall we say about the residual discrepancy of 0.05 MHz? Quite simply that we have now moved outside the realm of pure electrodynamics into the domain of the strong interactions that govern the properties of the proton.

It is known from high energy electron scattering experiments that, with regard to its electric and magnetic properties, the proton acts as an object completely distributed over a certain spatial volume. In other words, there are electric charge and magnetic moment form factors that must be determined almost entirely by the nonelectromagnetic interactions associated with the subnuclear particles to which the proton is coupled. It is clear qualitatively that the abandonment of the point charge, point dipole description must decrease the magnitude of the interaction responsible for the hyperfine structure splitting. That is in the right direction to remove the remaining discrepancy. We shall estimate the magnitude of this effect.

In the nonrelativistic theory, the distribution of nucleon magnetism, $\rho_m(\mathbf{r})$, thus far taken to be a delta function, is integrated over the square of the electron wave function. Accordingly, we now have the replacement

$$|\psi(0)|^2 \rightarrow \int (d\mathbf{r})\rho_m(\mathbf{r})|\psi(\mathbf{r})|^2. \tag{4-17.112}$$

The short range behavior of the wave function is determined by the nucleon electric charge. If this is distributed in accordance with the function $\rho_e(\mathbf{r})$, we must also take that into account:

$$1 - \frac{r}{a_0} \rightarrow 1 - \frac{1}{a_0}\int (d\mathbf{r}')|\mathbf{r} - \mathbf{r}'|\rho_e(\mathbf{r}'); \tag{4-17.113}$$

note that there is no change once one is sufficiently outside the charge distribution. The implication of both effects is

$$|\psi(0)|^2 \rightarrow (1 - 2Z\alpha mR)|\psi(0)|^2, \tag{4-17.114}$$

where

$$R = \int (d\mathbf{r})(d\mathbf{r}')\rho_m(\mathbf{r})|\mathbf{r} - \mathbf{r}'|\rho_e(\mathbf{r}') \tag{4-17.115}$$

is an average nucleon radius. We give it another form by using the representation of Eq. (4-15.48),

$$|\mathbf{r} - \mathbf{r}'| = -8\pi \int \frac{(d\mathbf{p})}{(2\pi)^3} \exp[i\mathbf{p} \cdot (\mathbf{r} - \mathbf{r}')]\frac{1}{\mathbf{p}^2 - \varepsilon^2}\frac{1}{\mathbf{p}^2}, \tag{4-17.116}$$

whence

$$R = - 8\pi \int \frac{(dp)}{(2\pi)^3} \frac{1}{p^2 - \varepsilon^2} \frac{1}{p^2} \rho_m(- \mathbf{p})\rho_e(\mathbf{p})$$

$$= 8\pi \int \frac{(dp)}{(2\pi)^3} \left(\frac{1}{p^2}\right)^2 (1 - \rho_m(- \mathbf{p})\rho_e(\mathbf{p})), \qquad (4\text{--}17.117)$$

or

$$R = \frac{4}{\pi} \int_0^\infty dp \frac{1}{p^2} (1 - \rho_m(p)\rho_e(p)). \qquad (4\text{--}17.118)$$

We have utilized the principal value integral

$$\int_0^\infty dp \frac{1}{p^2 - \varepsilon^2} = 0, \qquad (4\text{--}17.119)$$

and the spherical symmetry of the distribution functions.

The empirical data on the proton are represented approximately by

$$\rho_m(p) \cong \rho_e(p) \cong \frac{1}{[1 + (p^2/M_0^2)]^2}, \qquad (4\text{--}17.120)$$

where

$$M_0 \cong 0.90 \, M_{\mathrm{p}}. \qquad (4\text{--}17.121)$$

Evaluating the integral in (4–17.118) gives

$$R = \frac{35}{8} \frac{1}{M_0} = 1.0 \times 10^{-13} \, \mathrm{cm}, \qquad (4\text{--}17.122)$$

which implies the following fractional decrease in the hyperfine splitting:

$$\mathrm{H}: \quad \frac{35}{4} \alpha \frac{m}{M_0} = 3.8 \times 10^{-5}. \qquad (4\text{--}17.123)$$

This represents a decrease by 0.05 MHz, which, to the accuracy that has been retained, completely reconciles theory and experiment.

Having broached the subject of finite nuclear size, let us also estimate its effect in the relative displacement of s and p levels. The change in the Coulomb interaction energy is

$$\delta V(\mathbf{r}') = - Z\alpha \int (d\mathbf{r}) \left[\frac{1}{|\mathbf{r}' - \mathbf{r}|} - \frac{1}{r'}\right] \rho_e(\mathbf{r}), \qquad (4\text{--}17.124)$$

and this induces the energy shift

$$\delta E = \int (d\mathbf{r}') \, \delta V(\mathbf{r}') \, |\psi(\mathbf{r}')|^2 \cong - Z\alpha |\psi(0)|^2 \int (d\mathbf{r})(d\mathbf{r}') \left[\frac{1}{|\mathbf{r} - \mathbf{r}'|} - \frac{1}{r'}\right] \rho_e(\mathbf{r}).$$

$$(4\text{--}17.125)$$

If we combine the property

$$- \nabla^2 \int (d\mathbf{r}') \left[\frac{1}{|\mathbf{r} - \mathbf{r}'|} - \frac{1}{r'} \right] = 4\pi \qquad (4\text{--}17.126)$$

with the vanishing of the integral at $\mathbf{r} = 0$, it is inferred that

$$\int (d\mathbf{r}') \left[\frac{1}{|\mathbf{r} - \mathbf{r}'|} - \frac{1}{r'} \right] = - \frac{2\pi}{3} r^2. \qquad (4\text{--}17.127)$$

This gives

$$\delta E = \tfrac{2}{3} \pi Z \alpha |\psi(0)|^2 \langle r^2 \rangle, \qquad (4\text{--}17.128)$$

where

$$\langle r^2 \rangle = \int (d\mathbf{r}) r^2 \rho_e(\mathbf{r}) = - \nabla_p^2 \rho_e(p)|_{p=0}$$

$$= \frac{12}{M_0^2} = (0.81 \times 10^{-13}\,\mathrm{cm})^2. \qquad (4\text{--}17.129)$$

The energy shift is only significant in s states, and the upward displacement of the ns level is

$$\delta E_{ns} = \frac{16}{n^3} Z^4 \alpha^2 \frac{m^2}{M_0^2}\,\mathrm{Ry}. \qquad (4\text{--}17.130)$$

For the 2s level of hydrogen, this amounts to

$$\delta E_{2s1/2} = 0.13\,\mathrm{MHz}, \qquad (4\text{--}17.131)$$

which further narrows the gap between the theoretical value, now adjusted to

$$\mathrm{H}: \quad E_{2s1/2} - E_{2p1/2} = 1057.81\,\mathrm{MHz}, \qquad (4\text{--}17.132)$$

and the experimental value of 1057.90 ± 0.10 MHz. But we must again warn that potentially bigger effects than this one have not yet been considered (although it is a fair inference that they are largely counterbalancing).

Index

9 780738 2005